Theories and Practices in Tourism, Marketing and Management

İrfan Yazıcıoğlu / Alper Işın / Özgür Yayla /
Yener Oğan / Eren Yalçın (eds.)

Theories and Practices in Tourism, Marketing and Management

PETER LANG

Berlin - Bruxelles - Chennai - Lausanne - New York - Oxford

Library of Congress Cataloging-in-Publication
A CIP catalog record for this book has been applied for at the Library of Congress.

Bibliographic information published by the Deutsche Nationalbibliothek
The Deutsche Nationalbibliothek lists this publication in the Deutsche Nationalbibliografie; detailed bibliographic data is available on the Internet at http://dnb.d-nb.de.

ISBN 978-3-631-92510-2 (Print)
E-ISBN 978-3-631-92322-1 (E-PDF)
E-ISBN 978-3-631-92832-5 (E-PUB)
DOI.10.3726/b22456

Published by Peter Lang GmbH, Berlin, Germany

info@peterlang.com - www.peterlang.com

Contents

Seyit Ahmet Solmaz*

Chapter 1 An Examination of Contemporary Supply Chain Structures in the Hospitality Industry

Introduction

Today, hotel businesses are no longer simple structures that only meet the overnight stay needs of customers but have become complex businesses that respond to the needs of customers in many different areas such as food and beverage, entertainment, sports, meeting-congress, banqueting and spa-wellness. In this respect, hotels are businesses that must adapt to changing technology, market dynamics, travel motives and changing economic structure (Rutherford & O'Fallon, 2007).

Hotel businesses, which must adapt rapidly to the changing structure of the tourism industry, stand out as businesses that offer physical products that will appeal to the expectations of post-modern tourists and services to support them in the same place with their dynamic management structure. In this respect, hotels are both an industrial business and a service business. Hotels, which offer a wide range of products and services to their customers due to their different service units, need to cooperate with many suppliers and develop close relationships based on trust (Tektaş & Kavak, 2010; Solmaz, 2019). These relationships positively affect both the financial performance of hotels and their satisfaction as an industrial buyer (Kim & Cha, 2002; Solmaz, 2012; Solmaz, 2016). However, the distribution channel in hotel businesses is short, and there is a short time between the procurement and preparation of products and their presentation to the guest. In addition, the obligation of hotels to provide similar quality, consistent products, and services to their guests always makes suppliers more important for hotel businesses renders suppliers of greater importance to hotel businesses (Akbaba & Çavuşoğlu, 2017: 48–49).

Hotels are businesses that provide uninterrupted service on a 24/7 basis throughout the year. In such a situation, coordination and mutual information exchange between hotels and their suppliers becomes extremely critical (Solmaz,

* Assoc. Prof. Dr., Harran University, Faculty of Tourism, Department of Tourism Guidance, seyitahmetsolmaz@harran.edu.tr

2019). It is obvious that hotels, which purchase approximately 3000 to 10,000 different products and services with orders ranging from 50,000 to 100,000 (Uyar, 2015), need a highly efficient supply chain structure. At this point, SCM, which is an approach to manage all the goods, services, information, and financial flows in the distribution channels from suppliers to final consumers (Cooper & Ellram, 1993), emerges as a modern management technique that should be used for hotel businesses aiming to provide customer satisfaction at a high level.

In this study, the purchasing characteristics of hotel businesses, purchasing structures, supply chain structures developed for the hospitality industry and the roles of the actors in this structure have been tried to be discussed in general.

The Critical Role of Purchasing in Hotel Operations

The concept of purchasing generally means procuring the things we need in daily life. For hotel professionals, this definition is quite limited. For hotels, purchasing refers to the agreements made to procure the goods and services from suppliers in line with the profitability targets of the business (Quayle, 2006: 3). In these agreements, it is clear which characteristics each product group should have. In line with these characteristics, the procurement process carried out at the right place and time, at the right price, at the most favorable cost, in accordance with the quality expectations of the business can be defined as professional hotel purchasing (Monczka et al., 2009: 39).

Purchasing is a strategically important function in supply chain management in terms of its contribution to the effectiveness and efficiency of the business and its positive impact on profitability (Fredendall, 2001: 213). It should be stated that purchasing for a hotel is much more than receiving offers from suppliers and making decisions (Evans, 2007). While Porter (1980) emphasizes the importance of strategic purchasing in terms of cost leadership, Koçoğlu and Avcı (2014) state that purchasing will contribute positively to the capital turnover rate with the cost savings it will provide. Reducing inventory costs, creating value for consumers, and ensuring competitiveness are among the other critical effects of purchasing for businesses (Akturan, 2009).

Purchasing Structure and Process in Hotel Businesses

Purchasing the materials needed in the operational departments of a hotel (front office, housekeeping, food, and beverage) and all the products needed by the support departments depends primarily on effective teamwork and effective supply chain management. Despite this, the organization of purchasing departments in hotel businesses is quite simple and the products needed by the entire hotel are purchased with a small number of personnel (Uyar, 2015).

The organization in international chain hotels is slightly more complex than in independent hotels (Feinstein & Stefanelli, 2008). In an independent five-star hotel, purchasing activities are usually organized by a purchasing manager reporting to the general manager and a purchasing chief, purchasing specialists and storekeepers, whereas in chain hotels, due to centralized purchasing, there is a vice president of purchasing and a corporate purchasing director reporting to the head office (Warfel & Morrison, 2005).

As mentioned above, hotels, which are service-producing factories, purchase from a wide range of areas. The products purchased for a hotel can be listed as general categories as follows. These also draw a general framework of hotel suppliers. As seen in Table 1, hotels supply dozens of different goods and services in 10 different main categories.

It is not easy to produce a final product from all these different inputs and to ensure customer satisfaction. Therefore, it is imperative for hotels to establish effective communication with their suppliers. In fact, at this point, it should be noted that all this purchasing organization and supplier communication is now carried out online via the internet (Kothari, Hu, & Roehl, 2005). In addition, there are many automation programs for hotels and restaurants to make effective purchasing and inventory tracking. These automation programs are also adapted to the central automation programs used by the front office and can be configured according to the service and organizational structure and other characteristics of each hotel.

Table 1. Hotel Purchasing Categories and Main Supplier Groups*

Main Purchasing and Supplier Groups	Sub-Categories
Food Group Suppliers	*a) Delicatessen Ingredients* *b) Milk and Dairy Products* *c) Oils* *d) Vegetables and Fruits* *e) Meats* *f) Frozen Products* *g) Dry Foods* *h) Other Direct Food*
Beverage Group Suppliers	*a) Alcoholic Beverages* *b) Soft Drinks*

(*continued*)

Table 1. Continued

Main Purchasing and Supplier Groups	Sub-Categories
Hotel Fixture and Equipment Suppliers	*a) Industrial Kitchen, Bar and Service Equipment* *b) Indoor and Outdoor, Garden Furniture* *c) Hotel Textile, Fabric, Curtain and Furnishing* *d) Mattress, Base and Baby Beds* *e) Building Ground-Maintenance Materials and Equipment* *f) Laundry Machinery and Equipment* *g) Floor Coating Systems (Pvc, Parquet, Carpet etc.)*
Cleaning and Boucle Group Suppliers	*a) General Cleaning Materials* *b) Housekeeping, Laundry, and Stewart Cleaning* *c) Bouquet Materials and Products with Cartridge Apparatus* *d) Promotion Products* *e) Armbands, Hotel Room Cards, and Towel Cards*
Stationery and Printed Document Suppliers	*a) Kitchen Packaging Materials* *b) Stationery Supplies* *c) Service Ornaments and Menus* *d) Printed Documents and Publicity Materials*
Animation Equipment Suppliers	*a) Animation Consumables* *b) Sound, Light, Led and Video Systems* *c) Fitness, Conditioning and Sports Equipment* *d) Party Organization Materials*
Pool Chemicals and Fuel Materials Suppliers	*a) Orientation and Information* *b) Pool Chemicals* *c) Fuels*
Landscape Equipment and Theming Suppliers	*a) Pesticides* *b) Landscape Contracting and Plant Design-Decoration* *c) Seasonal Indoor and Outdoor Plants* *d) Garden Equipment - Flower / Grass Seeds*
Staff Equipment Suppliers	*a) Staff Uniform* *b) Personnel Shoes and Work Shoes* *c) Personnel Attendance Control Systems*
Other Suppliers of Goods and Services	*Consulting Services, Healthcare, Education Support, Insurance, Technology, Telecommunications, etc.*

**This table was created by the author*

In the classical operation, the purchasing process for a hotel business commences with the demands of different departments within the hotel. These requests, which are transmitted to the purchasing department by department managers, are collated in the purchasing department and the order for the relevant products or services is dispatched to the supplier company. Thereafter, the order is monitored, received, stored, and transferred to the relevant departments. In this context, the critical stock amount for each product is determined by department managers and the purchasing team. Purchase requests are considered when this stock falls below this amount. In most cases, the automation programs used provide a direct warning when the critical stock falls below the critical stock.

Figure 1. Purchasing Cycle in Hotel Business

In addition, in larger organizations (in cases of hotel-specific production), on platforms called B2B panels, the buyer and supplier can mutually track stock on a common network. There are also some upper limits on purchasing for many product groups. At the same time, each department of the hotel determines its own product specifications for the materials it needs. These specifications are the main criteria for the purchasing department during the supplier research and selection phase.

Supply Chain Concept and Structures

The supply chain concept emerged at a time when businesses began to focus on their core competencies and realized that collaborating with other firms provided them with many advantages. The resource-based view (Wernerfelt, 1984;

Rumelt, 1984; Srivastava, 2005), which argues that competitive advantage is possible only if the firm has unique knowledge, skills, resources and competencies that are difficult to imitate, has led to the development of the understanding of working with the most competent companies in its own business field by focusing on its own competencies and outsourcing in areas other than this. It can be stated that the concept of supply chain management emerged because of this change in philosophy.

In the current highly competitive business environment, companies are striving to reduce costs, enhance profitability, and ensure coordination and control in their operational processes. To ensure the smooth flow of products and services from the initial supplier to the final consumer, it is crucial to cultivate close and effective relationships with the businesses involved in the supply chain. Indeed, these factors have contributed to the proliferation of supply chains (Mentzer et al., 2001; Kaya & Azaltun, 2012).

The concept of the supply chain was first formally introduced in the field of logistics literature (Cooper and Ellram, 1993; Lumnus and Vokurka, 1999). It can be defined as a network that facilitates the flow of products and services to the consumer in a way that creates value, encompassing all businesses involved in the process from the initial supplier to the final distributor (Christopher, 1992). This network encompasses a diverse range of entities, including raw material suppliers, enterprises engaged in the transformation of raw materials and semi-finished products into processed products, producer/manufacturer enterprises, and distributors, wholesalers, and retailers responsible for the distribution of final products to the end consumer (Paksoy, 2005; Sarıoğlan, 2011).

In essence, a supply chain is defined as a network of two or more legally distinct entities that are connected through material, informational, and financial flows. These entities may include firms engaged in the production of parts, components, and finished products, as well as logistics service providers. Additionally, the final customer may also be considered a part of the supply chain (Stadtler, 2005).

The configuration of supply chains varies according to the size of the chain and the number of links connected to it, the scale of the enterprise, the field of activity, and the goods and services it produces. However, in the literature, supply chains are generally analysed in two ways: as "*single-stage supply chains*" and "*multi-stage supply chains*" (Solmaz, 2016).

a) Single-Stage Supply Chains

A single-stage supply chain is a method used by companies that are usually disconnected from the production stage. In this method, the raw materials and materials received become products after going through a manufacturing or transformation process. This type of supply chain does not have many information processing and decision-making functions. Single-stage supply chains are also the focal point of supply chains. It is a supply chain used in small businesses as it is not overly complex (Metz, 1998).

b) Multi-Stage Supply Chains
A multi-stage supply chain typically comprises a multitude of processes. It is a common occurrence in businesses where outsourcing is prevalent, particularly in those where the supply chain is of significant complexity. Indeed, multi-stage supply chains are a combination of single-stage supply chains. The multi-stage supply chain model is analogous to the model proposed by Mentzer et al. (2001) in the final supply chain structure. This model encompasses a significantly larger number of entities than the basic supply chain, including suppliers' suppliers and third-party service providers, among others.

Supply Chain Structures in Hospitality Industry

The purchasing structures of different hotels may appear similar, but in fact they are quite distinct. The size and service features of the hotel are key determinants of the purchasing structure. For instance, the service concept of a city hotel and a coastal hotel is not the same. Consequently, the type and quantity of goods and services demanded by these two types of hotels also vary. This results in the supply chain structure being established in a different manner. The size of the hotel is the most important variable that determines the supply chain structure. The size of the hotel affects its bargaining power with suppliers. When suppliers in the hospitality industry are evaluated, it is evident that the size of the business is a crucial criterion for them to accept the hotel into their customer portfolio.

As illustrated in Table 1, hotels must procure goods and services from 10 distinct product groups to maintain their operations. This is a mandatory requirement for the continued functioning of the hotel. However, when the sub-suppliers in each main supplier group are considered, a hotel must enter into numerous agreements with a multitude of suppliers for a vast array of products and must also oversee the fulfilment of orders. Despite the potential for online ordering, inventory tracking and purchasing processes to be conducted with the aid of automation programs, the reality is that the process is often fraught with difficulties. This has given rise to a new category of suppliers in the tourism industry, which we will refer to as "Hotel Supermarkets" in this article.

A hotel supply chain is comprised of three actors: suppliers, hotels, and customers. While this chain is relatively straightforward, it is not a common occurrence in real life. The primary reason for this is that the tourism industry does not have a direct distribution system. The nature of the touristic product necessitates that the final consumer, the tourist, be transported to the location where the product and service are available. This service is provided by tour operators and travel agencies, which are significant actors in the tourism distribution channel (Kozak, 2016).

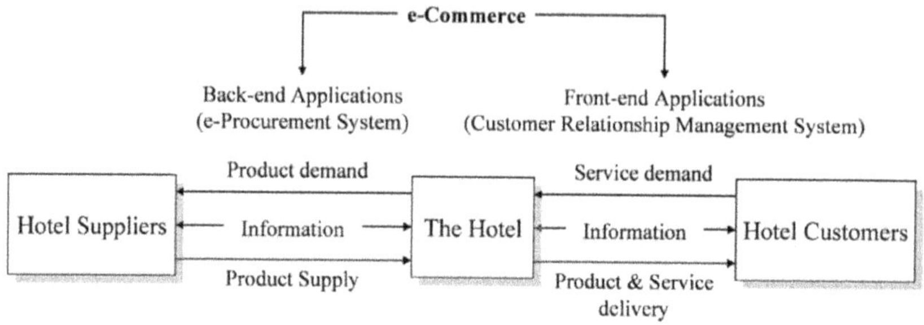

Figure 3. Simple Supply Chain for Hotels

Source: Kothari, T., Hu, C. Roehl, W. S. (2005), "e-Procurement: an emerging tool for the hotel supply chain management", *International Journal of Hospitality Management* (24) 3: 369–389.

Figure 4, which models the basic supply chain structure for hotels, represents a more realistic approach. In this model, which is mostly used by small hotels, the hotel makes direct agreements with all suppliers. This method is useful for small businesses as the quantities of products to be supplied are relatively small. On the other hand, hotels below a certain scale cannot establish a relationship with hotel supermarkets anyway.

The second actor in this model is the hotel company that produces the service. It produces an integrated accommodation service by combining the product and service groups obtained from all suppliers. The service produced by the hotel is delivered to tourists, who are the final consumers, by tour operators and agencies. At this point, there is a flow of goods and services from the first supplier to the final consumer. In the opposite direction, there is an information flow and a financial flow. There is an exception in this flow for the tourism industry. The second model, designated the extended supply chain model (Figure 5), will elaborate on this exception.

Figure 5 represents a model developed for large-scale hotels. As can be seen, there are hotel supermarkets between the first-level suppliers and the hotel in the extended chain. These suppliers act as intermediaries, procuring several product groups listed in Table 1 from the first-level suppliers and selling them to the hotels.

The current supply structure in the hospitality industry is as follows. Hotel supermarkets facilitate the supply chain management of hotel businesses. The larger the supply management company (hotel supermarket), the wider the supplier network it works with.

"Hilton Supply Management" can be given as an example of these businesses. The company within the Hilton hotel group manages a global supply chain with more than 3000 suppliers worldwide. In addition to its own hotel operations, Hilton Supply Management oversees the supply of products and services to over 17,000 businesses worldwide. These suppliers collectively generate over USD 13 billion in purchasing volume (mysupplymanagement.com).

Although not as extensive as Hilton SM, there are also supply management companies that provide the entire product range for only the food and beverage group or only the fixture and equipment group. Furthermore, for smaller-scale purchases, Metro Markets Group offers products from nearly 5,000 suppliers and serves them to its customers with its "HORECA" (Hotel, Restaurant, Catering) concept (metro-tr.com).

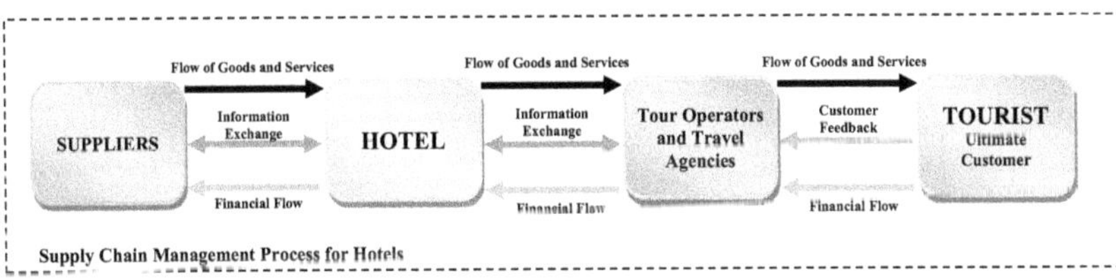

Figure 4. Basic Supply Chain Structure for Hotels*

*Figures 4 and 5 were developed by the author based on interviews with managers and a review of the relevant literature.

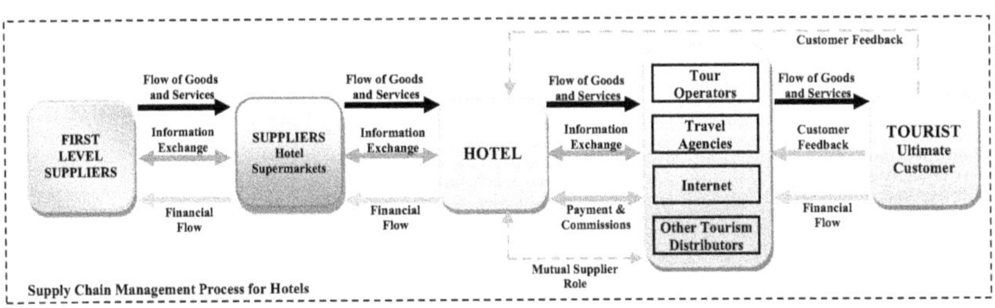

Figure 5. Extended Supply Chain Structure for Hotels*

*Figures 4 and 5 were developed by the author based on interviews with managers and a review of the relevant literature.

In addition to the sources, there are numerous websites that facilitate the connection between hotels and suppliers (e.g., otel-ihtiyac.com; hotelasistan.com). These websites offer a valuable service to businesses seeking to procure products directly from suppliers.

Although not as extensive as Hilton SM, there are also supply management companies that provide the entire product range for only the food and beverage group or only the fixture and equipment group. Furthermore, for smaller-scale purchases, Metro Markets Group offers products from nearly 5,000 suppliers and serves them to its customers with its "HORECA" (Hotel, Restaurant, Catering) concept (metro-tr.com). In addition to these, there are also numerous websites that facilitate the connection between hotels and suppliers (e.g. otel-ihtiyac.com; hotelasistan.com). These websites provide a valuable service for businesses that wish to procure products directly from suppliers.

In the context of tourism, it is crucial to reverse the distribution channel, whereby consumers are taken to the place where the product is produced rather than the product being delivered to the consumer. Consequently, another significant business group that occurs between the final consumer and the hotel in hotel supply chains and fulfils the distribution activity is that of travel agencies, tour operators and other tourism intermediaries. The exception previously mentioned does not apply to these businesses. There is a mutual buyer-supplier relationship between tour operators and hotels. This is essentially a strategic partnership for both parties (Türkay et al. 2011; Solmaz and Türkay, 2017: 166). While the hotel provides rooms for the package tours prepared by tour operators, the operator provides customers for the hotel and ensures that its service reaches the final consumer. Consequently, the financial flow between these two businesses is in the form of mutual fee and commission payments.

Conclusion

This section briefly describes the supply chain structures for hotel businesses, highlighting the ways in which the unique structure of the hotel industry differs from other industrial enterprises and sectors. The current supply chain structure in the hotel industry has been re-modeled through interviews with managers in the field and interpretation of the relevant literature. This study explains the most up-to-date supply chain structure and functioning for hotels. In this respect, it is thought to make a meaningful contribution to the existing literature. Furthermore, this study has also revealed the purchasing process, structure, and general purchasing characteristics of hotels in general. The strategic importance of purchasing for hotels to gain competitive advantage has been analyzed. As a result, supply chain management and purchasing are at the forefront of critical business activities that need to be handled meticulously by hotel businesses trying to ensure service continuity and customer satisfaction. The effective and efficient operation of this process is contingent upon the establishment and management of effective supply chain structures.

Acknowledgements

I would like to extend my gratitude to my esteemed colleagues Chef Yunus Rahmi KARA and Erdal KIZILYAR, Purchasing Manager of 39 Kalamış Marina, for their invaluable contributions to the content of this chapter. Their valuable insights have contributed significantly to the shaping of this chapter.

References

Akbaba, A. & Çavuşoğlu, F. (2017). Konaklama İşletmeciliği. in *Turizm İşletmeciliği* (Eds. Akbaba, A., Öter, Z., Güler, M. E. & Altıntaş, V.). Detay Yayıncılık, Ankara.

Akturan, U. (2009). Tedarik Zinciri Yönetiminde Satın Alma Fonksiyonunun Değişen Karakteri: Stratejik Satın Alma. *Öneri Dergisi*, Cilt 8, Sayı 31, ss. 103–111.

Christopher, M. G. (1992). *Logistics and Supply Chain Management*. Pitman. UK.

Cooper, M. C. & Ellram, L. M. (1993). Characteristic of Supply Chain Management and Implication for Purchasing and Logistic Strategy. *The International Journal of Logistics Management*, Vol. 4, No. 2, pp. 13–24.

Evans, C. L. (2007). The Hotel Purchasing Function. in *Hotel Management and Operations*. (Eds. Rutherford, D. G. & O'Fallon, M. C.). John Wiley & Sons, Inc.

Feinstein, A. H. & Stefanelli, J. M. (2008). *Purchasing: Selection and Procurement for the Hospitality Industry*. John Wiley & Sons, Inc.

Fredendall, L. D. (2001). Basics of Supply Chain Management. *APICS: The Educational Society for Resource Management*, UK.

Kaya, E. & Azaltun, M. (2012). Role of information systems in supply chain management and its application on five-star hotels in Istanbul. *Journal of Hospitality and Tourism Technology*. Vol 3, No 2, pp. 138–146. https://doi.org/10.1108/17579881211248808

Kim, W. G. & Cha, Y. (2002). Antecedents and Consequences of Relationship Quality in Hotel Industry. *Hospitality Management*, Vol. 21, pp. 321–338.

Koçoğlu, C. M. & Avcı, M. (2014). Satın Alma Yönetimi: Teorik Bir Çalışma. *Kastamonu Üniversitesi İktisadi ve İdari Bilimler Fakültesi Dergisi*, Sayı 3., ss. 33–47.

Kothari, T., Hu, C. & Roehl, W. S. (2005). e-Procurement: and Emerging Tool for the Hotel Supply Chain Management. *Hospitality Management*, Vol. 24, pp. 369–389.

Kozak, N. (2016). *Turizm Pazarlaması*. Detay Yayıncılık, Ankara.

Lumnus, R. R. & Vokurka, R. J. (1999). Defining Supply Chain Management: A Historical Perspective and Practical Guidelines. *Industrial Management and Data Systems*, Vol. 99, No. 1, pp. 11–17.

Mentzer, J. T. et al. (2001). "Defining Supply Chain Management", *Journal of Business Logistics*, Vol. 22, No. 2. Pp. 1–26.

Metz, P. J. (1998). Demystifying Supply Chain Management. *Supply Chain Management Review*. 2(4): 1–10.

Monczka, R. B., Handfield, B., Giunipero, L. C. & Patterson J. L. (2009). *Purchasing and Supply Chain Management*. 4. Edition, South-Western Cengage Learning.

Paksoy, T. (2005). Tedarik Zinciri Yönetiminde Dağıtım Ağlarının Tasarımı ve Optimizasyonu: Malzeme İhtiyaç Kısıtı Altında Stratejik Bir Üretim-Dağıtım Modeli. *Selçuk Üniversitesi, Sosyal Bilimler Enstitüsü Dergisi*, Sayı 14, s. 435–454.

Porter, M. E. (1980). *Competitive Strategy: Techniques for Analyzing Industries and Competitors*. New York: Free Press.

Quayle, M. (2006). Purchasing and Supply Chain Management: Strategies and Realities. IRM Press.

Rumelt, R. P. (1984). Towards a strategic theory of the firm. In R. Lamb (ed.) *Competitive Strategic Management*. Prentice-Hall, pp. 556–570.

Rutherford, D. G. & O'Fallon, M. C. (2007). *Hotel Management and Operations* (4. Edition). John Wiley & Sons, Inc.

Sarıoğlan, M. (2011). Konaklama İşletmelerinde Tedarik Zinciri Yönetimi Kapsamında Tedarikçi Seçim Kriterleri Üzerine Görgül Bir Araştırma. *Balıkesir Üniversitesi Sosyal Bilimler Enstitüsü Dergisi*, 14 (25): 239–253.

Solmaz, S. A. & Türkay, O. (2017). Uluslararası Seyahat İşletmelerinde Tedarik Zinciri Yönetimi ve Lojistik. içinde *Uluslararası Seyahat İşletmeciliği*. (Eds. Zengin, B., Sarıışık, M. & Avcıkurt, C.). Detay Yayıncılık. Ankara.

Solmaz, S. A. (2012). Pazar Yönlülük ve Tedarikçi İlişkilerinin İşletme Performansına Etkileri: Otel İşletmeleri Örneği (Master Thesis), Sakarya University, Institute of Social Science.

Solmaz, S. A. (2016). Alıcı-Tedarikçi İlişki Kalitesi ve Müşteri Değerinin Endüstriyel Tüketici Tatmini Üzerindeki Etkileri: Konaklama İşletmeleri Üzerine Bir Araştırma (Phd Thesis). Sakarya University, Institute of Social Science.

Solmaz, S. A. (2019). Effects of Cooperation and Communication in Buyer-Supplier Relations on Trust and Commitment to Suppliers: A Research on 4- and 5-Star Hotels in Istanbul. *Journal of Tourism and Gastronomy Studies.* 7(1), 428–449. https://doi.org/10.21325/jotags.2019.371

Srivastava, S. C. (2005). Managing Core Competence of the Organization. *Vikalpa.* Vol 30, No 4, pp. 49–63. https://doi.org/10.1177/0256090920050405.

Stadtler, H. (2005). Supply Chain Management. In *Supply Chain Management and Advanced Planning.* 3. Edition (Eds. Stadtler, H. & Kilger, C.). Springer.

Tektaş, Ö. Ö. ve Kavak, B. (2010). Endüstriyel Ürünlerin Satın Alınması Sürecinde Tedarikçi ile Olan İlişki Kalitesinin Algılanan Değer Üzerindeki Etkisi: Beş Yıldızlı Otellerde Bir Araştırma. *Anatolia: Turizm Araştırmaları Dergisi*, Cilt 21, Sayı 1, s. 51–63.

Türkay, O., Solmaz, S. A. & Şengül, S. (2011). Strategic Analysis of the External Environment and the Importance of the Information: Research on Perceptions of Hotel Managers. *Procedia Social and Behavioral Science.* Vol 24, pp. 1060–1069. https://doi.org/10.1016/j.sbspro.2011.09.103

Uyar, H. (2015). Otel İşletmelerinin Yiyecek-İçecek Satın Alma Davranışlarının Belirlenmesi: Beş Yıldızlı Otel İşletmeleri Üzerine Bir Araştırma (Phd Thesis). Sakarya University, Institute of Social Science.

Warfel, M. C. & Morrison, L. C. (2005). Purchasing for Food Service Managers. 5. Edition. Berkeley, CA: McCutchan.

Wernerfelt, B. (1984). Resource-Based View of the Firm. *Strategic Management Journal.* Vol 5, No 2, pp. 171–180.

Internet References

https://www.mysupplymanagement.com *Date Accessed: 14.05.2024.*

https://www.metro-tr.com *Date Accessed: 15.05.2024.*

https://otel-ihtiyac.com/#/oi/suppliers *Date Accessed: 10.05.2024.*

https://hotelasistan.com *Date Accessed: 10.05.2024.*

Gözdegül Başer*

Chapter 2 Opportunities and Threats of Generative Artificial Intelligence for Customer Experience: A Review of the Customer Experience Journey[1]

Introduction

In today's world, technology and the internet have fundamentally altered consumer behavior, making customers more digital, mobile, and socially connected while engaging with new customer experience channels and feedback mechanisms (SAS; Kumar & Anjaly, 2017). This shift has established a new digital frontier, creating a novel form of customer experience (Bansal, 2023). The swift evolution of technology, driven by Artificial Intelligence (AI), has significantly changed the landscape, blurring the boundaries between reality and science fiction, as well as between utopia and dystopia (Naude, 2021).

Artificial Intelligence (AI) relies on big data processing and algorithm management (Margas, 2023). It is defined as human-like intelligence capable of performing tasks, activities, and mimicking human problem-solving abilities (Russell & Norvig, 2016; Lai & Hung, 2018). The three foundational AI concepts are: (1) Machine Learning (ML), (2) Neural Networks (NN), and (3) Deep Learning (DL) (Groumpos, 2019). AI is increasingly contributing to the service industry among others. For instance, AI-powered robots can "understand" their environment, solve problems, and take actions to achieve goals (Iliadi, 2023). AI learns from external datasets and applies this knowledge to achieve specific objectives through adaptive processes (Kılıçhan & Yilmaz, 2020). By identifying patterns and relationships in data, AI can make predictions or decisions

* PhD., Antalya Bilim University, School of Tourism, Department of Tourism Management, gozdegul.baser@antalya.edu.tr; Visiting Scholar, Washington State University, Carson College of Business, School of Hospitality Business Management, gozdegul.baser@wsu.edu

1 This study is a part of the project "The use of Artificial Intelligence (AI) in Hospitality and Tourism Service Management", 2219 - International Postdoctoral Research Fellowship Program for Turkish Citizens, sponsored by Tübitak (The Scientific and Technological Research Council of Türkiye).

(Claveria & Torra, 2014). AI significantly impacts marketing by personalizing services and product recommendations based on customers' past purchases and decisions (Ameen et al., 2021). Generative Artificial Intelligence (GAI), a subdiscipline of AI, involves techniques that generate new content like text, images, or audio from training data. GAI can create text resembling writers' work or images mimicking illustrators, aiding humans as intelligent question-answering systems (Feuerriegel et al., 2024).

Customer focus has always been a fundamental aspect of hospitality and tourism. Delivering a strong customer experience has been a primary management goal and a crucial driver of commercial success and competitive advantage, central to marketing theory and practice (Lemon & Verhoef, 2016; Bolton et al., 2018; McColl-Kennedy et al. 2019). Today's consumers face more complex choices than ever and have a wider range of channels to explore them (Meyer & Schwager, 2007).

This chapter aims to examine and discuss the integration of Generative Artificial Intelligence (GAI) in creating an AI and GAI enabled customer experience, and how the customer journey can be redesigned in the hospitality and tourism industry. A conceptual model for the customer journey is explained. The opportunities and threats posed by GAI are explored, along with potential future trends and challenges.

Customer Experience

The concept of Customer Experience (CX) began gaining attention in the mid-1980s and became more prominent in the 1990s following Pine and Gilmore's book "The Experience Economy" (1999). In this book, the authors introduced "experiences" as a new type of economic offering (Gentile, Spiller & Noci, 2007). CX is defined as a comprehensive, multi-faceted concept that encompasses a customer's cognitive, emotional, behavioral, sensorial, and social reactions to a company's offerings throughout the entire purchase journey (Lemon & Verhoef, 2016). According to McColl-Kennedy (2019), CX includes elements of value creation, distinct emotions and cognitive responses at various touchpoints.

Cognitive aspects of CX involve factors like the functionality, speed, and availability of a service. Emotional aspects can range from positive to negative feelings, such as delight, regret, anger, outrage, joy, or surprise (Keiningham et al., 2017). Social aspects of the customer experience pertain to the influence of other people, including family, friends, and a customer's broader social network (Verhoef et al., 2009). Sensorial aspects are shaped by both offline and online experiences. Offline experiences include elements like artifacts, lighting, layout, and signage (Lam, 2001), while online experiences involve technology-related features (Keiningham et al., 2017).

Redesigning Customer Experience Journey with AI and GAI

With the impact of AI on customer experience journey, several enhancements can be noted: during the awareness stage, predictive analytics (a subfield of AI) identifies key interests for customers and offers product or service recommendations. In the consideration phase, AI helps websites integrate vast amounts of data, enabling customers to learn about and compare similar products. During the purchasing process, AI analyzes customer data patterns to understand unique buying behaviors and provides tailored recommendations. In the support phase, AI monitors customer behavior and detects signs of dissatisfaction, allowing for timely and appropriate responses. These strategies result in highly personalized customer service, with AI facilitating a two-way conversation to deliver effective customer support (Daqar & Smoudy, 2019).

The iterative and dynamic Customer Experience (CX) process moves through the pre-purchase stage, the purchase stage, and the post-purchase stage. This process also includes past experiences and external influences (climate, economy, culture, unemployment). Throughout these stages, customers encounter numerous touchpoints.

The pre-purchase stage involves all interactions a customer has with a brand, category, and environment before making a transaction. Previous experiences influence current ones by shaping expectations and creating persistent evaluations (Lervik-Olsen, Van Oest, & Verhoef, 2015). Traditionally, pre-purchase behavior includes need recognition, search, and consideration, encompassing the entire customer experience prior to purchase. Generative Artificial Intelligence (GAI) can anticipate customer needs and suggest alternative products or services through tools like ChatGPT, chatbots, and virtual assistants. The purchase stage includes all customer interactions with the brand and its environment during the transaction, such as choosing, ordering, and payment. GAI significantly contributes by creating dynamic marketing content, producing unique and engaging material for emails, and social media posts (Verma & Kumari, 2023). The post-purchase stage involves customer interactions after the purchase, including usage, consumption, post-purchase requests. GAI can track customers for future needs and gather feedback.

Touchpoints refer to any instance where a customer interacts with a brand or company (Jain, Aagja & Bagdare, 2017). These interactions, such as visiting a website, making a phone call, or exchanging emails, are considered touchpoints, representing individual occasions that overlook the broader context of the customer's journey (Bansal, 2023). It's essential for organizations to reassess each customer touchpoint (Sadhorta & Gupta, 2023). Generative Artificial

Intelligence (GAI) and AI can assist in consolidating information from various touchpoints along the CX journey by harmonizing the derived data. Moreover, AI can manage customer personal information for experiential marketing (Schmitt, 1999), covering aspects like sensory experiences, emotions, thoughts, actions, and relationships.

Customers navigate a journey utilizing numerous touchpoints, with these touchpoints influencing one another. As customers progress through the journey towards making a purchase, they encounter and interact with various touchpoints, each of which has both direct and indirect impacts on purchase decisions and other customer behaviors. Despite the complexity involved, it is crucial to identify pivotal touchpoints, often referred to as "moments of truth" that wield the greatest influence on key customer outcomes. Generative Artificial Intelligence (GAI) can aid in tracking each touchpoint meticulously, treating them as well-planned and strategically executed activities aimed at engaging customers through multisensory cues, thus eliciting positive cognitive, emotional, and behavioral responses (Jain, Aagja & Bagdare, 2017).

A focus on outcomes involves minimizing customers' transaction costs, including efforts related to finding and vetting new providers, underscoring the significance of goal-oriented experiences in consumer behavior (Huffman & Houston, 1993). The concept of peace-of-mind pertains to the customer's evaluation of all interactions with the service provider before, during, and after the purchase of a service. GAI can track and manage touchpoints for each customer, leading to a more personalized experience and service.

Depending on the specific characteristics of the product/service or the unique journey undertaken by the customer, the significance or effectiveness of each touchpoint category may vary across different stages of the purchasing process. AI can play a role in orchestrating various touchpoints by establishing seamless connections and gathering feedback to ensure a cohesive customer experience. Customer experiences are comprehensive, encompassing a series of interconnected activities spanning from pre-purchase to post-purchase stages (Jain, Aagja & Bagdare, 2017). AI and Generative Artificial Intelligence (GAI) can contribute to achieving this holistic perspective.

AI can be leveraged to track website visitors and their areas of interest. Serving as a significant touchpoint empowered by AI, a website can provide valuable information or facilitate orders for customers at any stage of their journey. Technologies such as Non-Fungible Tokens (NFTs), virtual assistants, Virtual Reality (VR) websites, web Augmented Reality (AR) applications, and AI-driven chatbots can enhance the customer experience. The integration of various "phygital" touchpoints and brand-oriented experiential websites is not only essential but also a fundamental requirement for integrating customers' offline and online experiences (Koronaki, Vlachvei & Panopoulos, 2023). AI or GAI can be utilized at any stage and touchpoint throughout the customer journey.

Irrespective of whether it occurs online or offline, fostering a positive Customer Experience (CX) can cultivate an emotional connection between a company's brand and its customers, thereby bolstering customer loyalty (Gentile, Spiller & Noci, 2007). GAI driven technologies can monitor this emotional connection through methods such as text analysis of written feedback, customer blogs, or questionnaires. Consequently, it is highly likely that the CX journey will encompass omnichannel interactions, as customers may transition between different channels, potentially enhancing customer satisfaction and service quality (Bansal, 2023). GAI can play a crucial role in this endeavor by analyzing customer feedback across various online channels. Utilizing GAI, businesses can further refine their CX efforts and offer personalized services (Mondal, Das & Vrana, 2023).

Generative Artificial Intelligence (GAI) can monitor multiple channels for customers and provide options and information whenever needed. It can guide and assist customers in finding the most suitable channels. Customer experience management (CEM) process involves numerous functions, including service operations, IT, analytics, and marketing. CEM necessitates a multidisciplinary approach, wherein various functions such as IT, marketing, operations, customer service, and human resources collaborate to deliver a seamless customer experience. Generative Artificial Intelligence (GAI) and Artificial Intelligence (AI) can assist in managing the customer experience by orchestrating the different functions within or outside the organization.

Researchers should capitalize on the growing availability of "big data" by integrating survey data with transactional, channel, and operational data, both in aggregate and at the individual level (Bolton, Lemon, & Bramlett, 2006; Bolton, Lemon & Verhoef, 2008; Gijsenberg, Van Heerde & Verhoef, 2015). AI and GAI can aid in analysis through data mining. GAI can generate customer journey maps for individual customers as well as customer segments.

Effectively managing each customer's experience stands out as a crucial factor in fostering customer loyalty. The concept of the loyalty loop underscores the idea that customer loyalty strengthens with each instance where the service meets or exceeds customer expectations (Court et al., 2009). GAI and AI can assist companies in cultivating loyal customers. Furthermore, customer value is shaped by the overall experience encompassing various elements (Gronroos, 2008). From this perspective, the process of creating customer experience involves three stages of value creation: value before use, value during use, and value after use (Jain, Aagja & Bagdare, 2017). GAI and AI can aid in developing metrics to assess the value generated at each stage of the CX journey. By sharing their consumption experiences, customers offer valuable insights to firms, enabling them to refine their processes and redesign their offerings (Kristensson et al., 2008). AI-enabled technologies like service robots, e-kiosks, chatbots, metaverse, machine learning, NLP, and GAI can significantly contribute to enhancing effectiveness in value co-creation (Solakis et al., 2024).

GAI has emerged as a significant force in service industries, primarily due to its capacity to produce content resembling human-generated material, engage in customer interactions, and streamline operational processes. Its introduction has sparked a revolution in the service sector by enabling automated content creation, tailored recommendations, and interactive chatbot interactions (Sadhorta and Gupta, 2023). Notably, GAI has captured the attention of major tech companies like Google, Microsoft, Baidu, and Apple, with tools like OpenAI's Codex, DALL-E, and ChatGPT being among the prominent solutions utilized in this domain (Mondal, Das, and Vrana, 2023). Additionally, GAI has revolutionized the customer-business communication landscape through the implementation of chatbots and virtual assistants. Empowered by this technology, these GAI-driven assistants offer a more natural and efficient mode of interaction, adeptly handling a wide array of tasks ranging from addressing common inquiries to delivering personalized product recommendations (Verma and Kumari, 2023).

Generative AI in Hospitality and Tourism

There is significant potential for utilizing Generative AI (GAI) to discern a customer's intent and anticipate future impactful interactions (Thukral et al., 2023). GAI offers numerous applications across nearly all functional areas of Hospitality and Tourism (HT) firms, which can enhance operational efficiency, boost workplace productivity, and improve the customer experience (Dogru et al., 2023). Beyond ChatGPT, other tools like GPT-4, DALL-E2, Meta's AI Sandbox, Jasper. AI, Midjourney, Peppertype.AI, and Copy.AI demonstrate diverse applications in hospitality and tourism marketing (Kshetri et al., 2023). Text-generating AI systems, such as ChatGPT, are based on large language models (LLMs) that are trained on extensive datasets to perform tasks or answer questions based on statistical probabilities (Databricks). This variety of Generative AI solutions offers numerous benefits for the HT industry.

GAI can help hospitality and tourism businesses create personalized and relevant marketing campaigns, offers, recommendations, and feedback tailored to each customer's preferences, needs, and behaviors. For instance, Persado is a platform that uses AI to generate and optimize marketing messages that resonate with the audience and encourage desired actions. Additionally, AI can help develop conversational and interactive customer service agents, such as chatbots and voice assistants, which can naturally respond to inquiries, provide information, resolve issues, thereby enhancing the overall customer experience (Khan & Khan, 2024). Another example is GPT-4, the latest version of OpenAI's GPT model, which excels in generating high-quality, multilingual content for various platforms, including blogs, social media, emails, ads, and websites (Robbert et al., 2023).

GAI can assist HT businesses in generating personalized marketing content on a large scale with minimal human intervention. For instance, hotels can use GAI to create customized social media posts for guests based on their preferences and past behaviors. These systems can analyze data such as previous stays, and booked activities to generate social media posts that showcase new offers and experiences (Dogru et al., 2023).

GAI can enhance service recovery, customer satisfaction, loyalty, and positive word-of-mouth by crafting personalized responses to negative online reviews. Beyond inspiring travelers and creating customized trip plans at the pre-trip stage, GAI can be integrated into the destination experience by providing navigation routes tailored to visitors' interests and preferences during their trip. Additionally, GAI can reduce the need for human labor in the hospitality and tourism industry by handling repetitive and time-consuming tasks. The insights generated by GAI regarding new products, branding, promotions, pricing strategies, business models, and design can also stimulate the creativity and innovation of HT managers, thereby improving their marketing and operational processes, competitiveness, revenue, and profitability. GAI can offer travelers instant assistance, real-time travel updates, personalized travel suggestions, and language translation (Master of Code).

Opportunities for Customer Experience

AI technologies, when combined with augmented reality, computer vision-based image recognition, and predictive inventory management, offer promising opportunities to enhance customer experiences (Saponaro et al., 2018). GAI opens up personalized experience possibilities such as tailored emails, chat interactions, and product recommendations (Verma & Kumari, 2023). Personalization can manifest in various forms, including content, interaction, or strategy. Content personalization involves tailoring information based on individual user profiles, encompassing product or service offerings and pricing (Zanker et al., 2019).

Interaction personalization pertains to AI algorithms autonomously determining when and how to engage with users (Chen, Chiang & Storey, 2012; Zanker et al., 2019). Personalization strategies that yield high positive perceptions and low negative perceptions are likely to reinforce customer brand loyalty (Shen & Ball, 2009).

The utilization of predictive analytics and "Big Data" stands out as a pivotal opportunity facilitated by AI and/or GAI. Through the application of deep learning algorithms, it becomes feasible to predict customer behavior and preferences more effectively (Gartner, 2020). AI continuously enhances the ability to anticipate consumer needs and desires, often even before they are explicitly expressed through conventional search options embedded in company websites and apps (Batra, 2019). Leveraging customer data to forecast which aspects of the company are likely to influence the customer enables sales personnel to concentrate their efforts accordingly (Daqar & Smoudy, 2019). These data streams are processed to generate valuable leads, opportunities, and foster long-term customer relationships (Daqar & Smoudy, 2019).

Content creation presents another opportunity where GAI offers substantial opportunities. Marketers now have access to specific AI tools that facilitate the generation of unique, high-quality content for online promotions and advertisements, resulting in significant market impressions and viewership (Nwachukwu & Affen, 2023). GAI-generated content has notably enhanced productivity and cost-effectiveness within marketing endeavors. Marketers can capitalize on AI applications and tools, such as online virtual writing assistants, to assist them in crafting premium content. An illustrative example of such an application is Rytr.me, an AI writing assistant that empowers the creation of top-tier advertising messages online (Nwachukwu & Affen, 2023). Recent trends in content sharing also include videos and live-streaming platforms. Video platforms are increasingly integrating shopping features into their services, exemplified by recent developments (Batra, 2019). Additionally, AI holds the potential to boost workflow efficiency and effectiveness, affording marketers more time to focus on strategy, creativity, and optimizing outcomes (Daqar & Smoudy, 2019).

Furthermore, AI-enabled technologies and GAI introduce novel customer experiences that encompass hedonic and recognition elements. These advancements in AI-driven services have the capacity to enhance both the hedonic (pleasure-oriented) and recognition aspects of the customer experience, delivering improvements in terms of time, efficiency, enjoyment, and personalization (Saponaro et al., 2018). GAI takes this a step further by providing a wider array of text, video, or audio information tailored to assist the customer. GAI can contribute to fostering valued relationships with customers by facilitating personalized and amiable communication tailored to their specific needs.

Post-purchase customer support presents another avenue for enhancing the customer experience. Providing comprehensive assistance with product-related inquiries post-purchase is a pivotal aspect of the customer journey. This encompasses aiding customers with product installation, usage, and maintenance (Daqar & Smoudy, 2019). GAI plays a role in post-purchase customer support by delivering personalized content and interactions to address individual customer needs effectively. Utilizing chatbots or other automation technologies to reduce customer waiting time can alleviate pressure on employees. Ultimately, GAI has the potential to enhance the productivity and retention of individual workers (Brynjolfsson, Li & Raymond, 2023).

Photo generation presents a significant opportunity. Apps that generate photos can assist marketers in creating unique faces for advertising campaigns, eliminating the need to pay large sums to celebrities or individuals for their likeness. An example of such apps is Generated Photos, which utilizes GAI algorithms to produce human faces that are entirely unique and unavailable elsewhere online (Nwachukwu & Affen, 2023). This provides the opportunity to select the most suitable face for advertising or promotional campaigns.

Speed is another advantage of AI and GAI, accelerating content delivery and enhancing the customer experience. The proliferation of digital media, particularly social media, facilitates the rapid and widespread dissemination of consumer trends and fashions. Machine learning can streamline the data digestion process from various sources to identify emerging trends (Batra, 2019).

Dynamic pricing algorithms facilitate automated demand-driven pricing, a strategy widely employed by various travel sites like Priceline, airlines, and hotels (Batra, 2019). Modern technological advancements and applications, such as online shoppable content, cross-device communication, predictive customer search, chatbots, emotion recognition through machine learning, smart virtual reality, and dynamic pricing, are integral components of the contemporary customer purchase journey (Batra, 2019).

Video creation applications, such as Invideo.io, cater to the rising trend of content marketing through video in the digital realm. These GAI tools facilitate the quick and efficient creation of video content without the need for advanced editing skills. Considering these various opportunities, GAI holds significant potential to enhance customer experience by delivering personalized content, recommendations, support, and insights into customer behavior and preferences (Nwachukwu & Affen, 2023).

Threats for Customer Experience

Data security and trust represent significant concerns for customers, with the analysis process often perceived as a "black box" operation, leaving customers unaware of who accesses their information and for what purposes (Tuo, Ning & Zhu, 2021). This lack of transparency raises several ethical issues, including concerns about biased information, privacy infringement, transparency, fairness, accountability, and trust in AI-generated content (Burlacu, 2023; Turel & Cavarretta, 2020; Li, Wang & Zheng, 2019). Additionally, there are worries about the absence of emotional comprehension and personalization in AI-generated content.

The effectiveness of AI-driven services heavily relies on the quantity and quality of personal data gathered by a brand. Ethical issues, biased information, and the absence of emotional understanding and personalization in AI-generated content have also been highlighted (Burlacu, 2023). There's a risk that text-to-voice technology could mimic customers' information and be exploited. An example of such technology is Play.ht, an AI-powered text-to-speech software that generates lifelike human voices from entered text, including male, female, or child-like tones (Nwachukwu & Affen, 2023). Excessive personalization can lead to privacy concerns and mistrust, with consumers favoring a moderate level of personalization that respects their privacy and autonomy (Wu et al., 2021). To address these challenges, companies must be transparent about their AI usage and ensure the deployment of ethical and unbiased algorithms (Nwachukwu & Affen, 2023).

Another significant concern involves the proliferation of fake information and the lack of authenticity and originality. There are ethical apprehensions surrounding the potential for GAI tools to generate and disseminate fabricated videos or images that could be humiliating or negative. Generative AI solutions may contain inherent biases derived from the data used to train the foundational model. These biases could lead to unfair treatment of certain customers when models are fine-tuned based on specific contexts (Sadhorta & Gupta, 2023).

While obtaining a service may necessitate both monetary and non-monetary sacrifices, evaluating the potential implications of many non-monetary sacrifices can be challenging. Examples of such sacrifices include loss of control, loss of privacy, potential financial losses, required effort and time, and negative emotions (De Kerviler, Demoulin & Zidda, 2016; Merisavo et al., 2007).

Machine learning poses challenges to the accuracy and truthfulness of information. For example, large language models like ChatGPT may not keep pace with changes in the world around them. Therefore, individuals and organizations employing GAI tools must verify the accuracy of the information they generate before implementation. While machine learning models can be trained using generated data, determining their compliance with fair use doctrine remains unclear (Mondal, Das & Vrana, 2023).

Some of the risks faced by customers include misinformation (stemming from virtual reviews and deceptive practices), a diminished sense of human connection and interaction, apprehension towards technology, increased susceptibility to financial fraud when using credit cards online, potential exclusion or limitations for businesses or individuals lacking direct or continuous access to internet and information technologies, as well as for companies unable to invest in technology due to financial constraints. Additionally, concerns include privacy breaches through continuous data collection from online applications, inadequate assurance of personal confidentiality, and potential encroachment on personal boundaries through biometric tests (Bagdan, 2013; Buhalis et al., 2019).

Results

A customer journey encompasses every interaction point and the overall customer experience, taking into account customers' pain points and areas for potential growth. Achieving an integrated customer journey can foster customer loyalty and profitability (Bansal, 2023). However, the advent of GAI and AI technologies has transformed today's customer experience landscape, necessitating a different approach to customer experience management compared to traditional methods. Increasingly, customers are becoming familiar with and embracing technologies, alongside the brands they interact with. This chapter advocates for a data-driven, dynamic customer experience journey that comprehensively covers all touchpoints and provides thorough oversight of the customer experience using GAI and AI technologies. Companies must leverage distributed technology, artificial intelligence (AI), and extended reality technologies to enhance their brand and customer experience to foster positive digital relationships with their customers (Bansal, 2023), alongside the evolving applications of GAI.

Utilizing Generative AI (GAI) and Artificial Intelligence (AI) in the customer experience journey offers the potential for a more exhaustive oversight of the entire journey. Presently, AI can facilitate attaining a 360-degree perspective of the customer (Bansal, 2023), essential for effective Customer Relationship Management (CRM) and Customer Experience Management (CEM). Consequently, businesses will require data scientists capable of devising a comprehensive data strategy aimed at enhancing consumer experiences and fortifying customer experience journeys (Batra, 2019). In the future, the customer experience journey might undergo a transformation, potentially evolving into what could be termed Virtual Contact Centers (VCC). However, Bolton et al. (2018) argue that while these technologies hold promise as effective VCC mechanisms, their implementation still poses challenges, including customer acceptance. Customers' prior exposure to AI and automation can equip them with the necessary skills, knowledge, and readiness to engage with the VCC process (Solakis et al., 2024). Personalized and rapid feedback can then be provided to customers based on their history with the company, preferences, or objectives. This approach allows companies to gain deeper insights into their customers and tailor experiences accordingly (Bansal, 2023). Furthermore, understanding customers and customizing touchpoints according to their preferences is crucial (Tyrvainen & Karjaluoto, 2019).

The integration of Generative AI (GAI) and Artificial Intelligence (AI) into customer experiences presents a spectrum of opportunities and challenges. These technologies offer prospects such as personalized experiences, predictive analytics, content generation, leveraging "Big Data" cost reduction, enhanced flexibility, streamlined interactions with customers, novel customer experiences, and convenient services. Additionally, they facilitate seamless communication, foster valued relationships, provide after-sales customer support, empower human agents, reduce customer uncertainty and anxiety, and deliver personalized content.

Furthermore, GAI and AI bring benefits like photo generation capabilities, identifying pain points through sentiment analysis, enabling voice shopping, facilitating video creation, implementing dynamic pricing strategies, and ensuring swift service delivery. AI-powered chatbots and virtual assistants play a pivotal role in delivering instantaneous and tailored responses to customer queries, thereby enhancing satisfaction levels and reducing response times.

Ethical considerations, particularly surrounding data privacy and potential job displacement, underscore the need for transparent and regulated AI integration. Companies must prioritize transparency and ethical considerations in their AI usage to ensure a positive impact on customer experience. A framework for ethical AI design, emphasizing transparency, fairness, and accountability, has been proposed to address these concerns. Furthermore, companies should carefully weigh the potential risks and benefits of AI marketing, placing the interests and needs of customers above their business objectives. Trust emerges as a critical factor in fostering acceptance and advancement of AI technology. Establishing trust and confidence among customers through transparent practices is essential for building satisfying customer experiences in the realm of GAI, AI, and related technologies.

Current literature emphasizes the importance of marketers revamping their customer-service processes to leverage AI for enhanced operational flexibility, speed, scalability, decision-making, and personalized product and service offerings. Investing in AI-driven personalized experiences, integrating chatbots for improved customer service, incorporating virtual reality, and blending human touch with automation are highlighted as strategies to provide customers with personalized experiences across various activities and options.

The replacement of human labor with automation is anticipated to have substantial social implications for both the workforce and employers, as argued by Ivanov and Webster (2019). Despite the increasing role of technologies like chatbots, human intervention and interaction remain integral to the customer experience journey. While chatbots and similar tools excel at handling straightforward inquiries, customers still prefer human assistance for more complex questions, as noted by Samala et al. (2022). However, Generative AI (GAI) holds the promise of delivering a more personalized and human-like customer experience. Consequently, it is anticipated that the future of the customer experience journey will be characterized by more satisfied customers. Companies may think of sharing this cost with their partner companies whom they cooperate like deliver, production or market research leading to a more thorough and complete customer experience journey.

In addition to the above mentioned discussion, GAI can be developed and used for gamification to enrich CX journey in terms of hedonic customer values. Gamification can be used on different stages of customer experience journey and may offer bonus and promotions leading to more customer engagement.

Maintaining a human element throughout the process proves to be the most effective approach in overcoming these challenges. According to Sadhorta and Gupta (2023), a human review is essential before AI-generated content reaches consumers. However, the extent to which Generative AI (GAI) can eventually replace human review remains uncertain. Transitioning from face-to-face interactions to utilizing generative AI-based chatbots for accessing services would present a challenging behavioral shift to facilitate (Dwivedi et al., 2023).

The potential for Generative Artificial Intelligence (GAI) and Artificial Intelligence (AI) to revolutionize the customer experience journey appears significant. As customers become more accustomed to the benefits of AI and GAI, they may develop greater empathy towards their usage, potentially leading to increased customer advocacy for these new technologies (Cobos et al., 2016). However, the successful implementation of new technologies like GAI in enhancing customer experience relies heavily on maintaining a customer-centric approach. It's essential that the system or customer journey remains attentive to the customer at every stage, continuously learning and adapting to improve service and overall experience. Reliability is another concern in data governance, as Learning Language Models (LLMs) lack the ability to differentiate between true and false information, potentially memorizing inaccuracies or outdated data. However, the benefits of GAI are notable, including time savings, enhanced personalization, increased efficiency, heightened productivity, and improved marketing performance, particularly in the dynamic landscape of hospitality and tourism marketing.

Future Implications

In the future, advertisements may dynamically adjust based on customer mood, predicting potential needs and tailoring advertisements accordingly. Text or voice chatbots could interpret emotional states through voice analysis and respond with emotionally resonant messages. Certain businesses are currently exploring facial recognition technology to gauge customer moods and subsequently offer suitable product recommendations (Yang et al., 2021). As Generative Artificial Intelligence (GAI) systems advance, they are likely to foresee consumer behavior and engage with individuals proactively, potentially even before customers themselves are aware of their own needs or desires, as these systems acquire deeper insights into a business's offerings and operations (Sadhorta & Gupta, 2023).

In the future, the Visual Positioning System (VPS) might supersede the Global Positioning System (GPS). VPS is an advanced technology that provides users with a real-time view of the physical environment, including landmarks like stores, hotels, malls, and restaurants, enabling location-based experiences (Dumitru, 2023). Additionally, the metaverse could become an integral part of the evolving customer experience journey, working in tandem with Generative Artificial Intelligence (GAI). GAI can further bring new forms of marketing. GAI can further produce new promotion tools in addition to other applications for a better customer experience. Marketers need more creativity than before as the success of tomorrow can be hidden in creative tools by using these growing technologies.

GAI has the potential to orchestrate a more interconnected network of touchpoints, diverging from the traditional linear customer experience journey described as "fluid" and lacking a strict chronological order (Grewal & Roggeveen, 2020). In this context, digital touchpoints offer customers enhanced control, granting them instant access to information and the freedom to shop at their convenience, thus fostering a more personalized experience (Tyrvainen & Karjaluoto, 2019). However, it's crucial to manage irrelevant touchpoints effectively, as they may be perceived as intrusive to customers' lives (Del Bucchia et al., 2020).

Future research and advancements should continue to prioritize trust, alongside addressing other identified threats across all stages of the customer experience journey. GAI is anticipated to contribute significantly to personalization, insight generation, and content creation, enabling marketers to cultivate a compelling customer experience and engagement (Kshetri et al., 2023). GAI holds the potential to benefit marketers by reducing costs and enhancing efficiency and productivity in almost every stage of the customer journey especially in hospitality and tourism industry although human touch will always be the key factor for customer satisfaction.

References

Ameen, N., Tarhini, A., Reppel, A., & Anand, A. (2021). Customer experiences in the age of artificial intelligence. *Computers in Human Behavior, 114*, 106548.

Bagdan, P. (2013). *Guest Service in the Hospitality Industry*. Hoboken, NewJersey: John Wiley & Sons Inc.

Bansal, K. (2023). Customer experience: Creating value through transforming customer journeys. *International Journal of Advances in Engineering and Management (IJAEM)*, 5(9), 715–725, www.ijacm.net ISSN: 2395-5252

Batra, M. M. (2019). Strengthening customer experience through artificial intelligence: An upcoming trend. In Competition forum. *American Society for Competitiveness*, *17*(2), 223–231.

Bolton, R. N., Mccoll-Kennedy, J. R., Cheung, L., Gallan, A., Orsingher, C., Witell, L. & Zaki, M. (2018). Customer experience challenges: bringing together digital, physical and social realms. *Journal of Service Management*, *29*, 776–808.

Bolton, R. N., Lemon, K. N., & Verhoef, P. C. (2008). Expanding business-to-business customer relationships: Modeling the customer's upgrade decision. *Journal of Marketing, 72*(1), 46–64.

Bolton, R. N., Lemon, K. N., & Bramlett, M. D. (2006). The effect of service experiences over time on a supplier's retention of business customers. *Management Science, 52*(12), 1811–1823.

Brynjolfsson, E., Li, D., & Raymond, L. R. (2023). *Generative AI at work* (No. w31161). National Bureau of Economic Research.

Buhalis, D., Harwood, T., Bogicevic, V., Viglia, G., Beldona, S. & Hofacker, C. (2019). Technological disruptions in services: lessons from tourism and hospitality. *Journal of Service Management*, *30*, 484–506.

Burlacu, C. (2023). *The Impact of Ai-powered content generation on customer experience.* Bachelor's thesis, University of Twente.

Chen, H., Chiang, R. H., & Storey, V. C. (2012). Business intelligence and analytics: From big data to big impact. *MIS Quarterly*, 1165–1188.

Claveria, O. &Torra, S. (2014). Forecasting Tourism Demand To Catalonia: NeuralNetworks Vs. Time Series Models. *Economic Modelling, 36*, 220–228.

Cobos, L. M., Mejia, C., Ozturk, A. B., & Wang, Y. (2016). A technology adoption and implementation process in an independent hotel chain. *International Journal of Hospitality Management, 57*, 93–105.

Court, D., Elzinga, D., Mulder, S., Vetvik, O. J. (2009). The consumer decision journey. *McKinsey Quarterly, 3*, 96–107.

Daqar, M. A. A., & Smoudy, A. K. (2019). The role of artificial intelligence on enhancing customer experience. *International Review of Management and Marketing, 9*(4), 22.

Databricks, "The great acceleration: CIO perspectives on generative AI",MIT Technology Review Insight, https://www.databricks.com/sites/default/files/2023-07/ebook_mit-cio-generative-ai-report.pdf (Accessed 18.05.2024)

De Kerviler, G., Demoulin, N. T., & Zidda, P. (2016). Adoption of in-store mobile payment: Are perceived risk and convenience the only drivers?. *Journal of Retailing and Consumer Services, 31*, 334–344.

Del Bucchia, C., Miltgen, C. L., Russell, C. A., & Burlat, C. (2020). Empowerment as latent vulnerability in techno-mediated consumption journeys. *Journal of Business Research, 124*,629–651. https://doi.org/10.1016/j.jbusres.2020.03.014.

Dogru, T., Line, N., Mody, M., Hanks, L., Abbott, J., Acikgoz, F., Assaf, A., Bakir, S., Berbekova, A., & Bilgihan, A. (2023). Generative artificial intelligence in the hospitality and tourism industry: Developing a framework for future research. *Journal of Hospitality & Tourism Research*, 1–19. Doi: 10.1177/10963480231188663.

Dumitru, D. (2023). A fundamental insight over the consequences of artificial intelligence on the tourism industry. *Revista de turism-studii si cercetari in turism*, 35.

Dwivedi, Y. K., Pandey, N., Currie, W., & Micu, A. (2023). Leveraging ChatGPT and other generative artificial intelligence (AI)-based applications in the hospitality and tourism industry: practices, challenges and research agenda. *International Journal of Contemporary Hospitality Management, 36*(1),1–12.

Feuerriegel, S., Hartmann, J., Janiesch, C., & Zschech, P. (2024). Generative Aİ. *Business & Information Systems Engineering, 66*(1), 111–126.

Gartner. (2020). The future of personalization: Maximizing customer lifetime value. https://www.gartner.com/en/documents/3985104 (Accessed 20.05.2024).

Gentile, C., Spiller, N., & Noci, G. (2007). How to sustain the customer experience:: An overview of experience components that co-create value with the customer. *European Management Journal, 25*(5), 395–410.

Gijsenberg, M. J., Van Heerde, H. J., & Verhoef, P. C. (2015). Losses loom longer than gains: Modeling the impact of service crises on perceived service quality over time. *Journal of Marketing Research, 52*(5), 642–656.

Grewal, D., & Roggeveen, A. L. (2020). Understanding retail experiences and customer journey management. *Journal of Retailing, 96*(1), 3–8.

Gronroos, C. (2008). Service logic revisited: who creates value? And who co-creates?. *European Business Review, 20*, 298–314.

Groumpos, P. P. (2019). *Artificial intelligence: Issues, challenges, opportunities and threats*. In Creativity in Intelligent Technologies and Data Science: Third Conference, CIT&DS 2019, Volgograd, Russia, September 16–19, 2019, Proceedings, Part I 3 (pp. 19–33). Springer International Publishing.

Huffman, C., & Houston, M. J. (1993). Goal-oriented experiences and the development of knowledge. *Journal of Consumer Research, 20*(2), 190–207.

Iliadi, S. (2023). *Adoption of Artificial Intelligence in Tourism Industry: A Systematic Literature Review*, Master Thesis, International Hellenic University.

Ivanov, S., & Webster, C. (2019). Economic fundamentals of the use of robots, artificial intelligence, and service automation in travel, tourism, and hospitality. In Robots, artificial intelligence, and service automation in travel, tourism and hospitality (pp. 39–55). Emerald Publishing limited. Leeds, pp. i–xxii. https://doi.org/10.1108/978-1-78756-687-320191014.

Jain, R., Aagja, J., & Bagdare, S. (2016). Customer experience–a review and research agenda. *Journal of Service Theory and Practice, 27*(3), 642–662.

Keiningham, T., Ball, J., Benoit, S., Bruce, H. L., Buoye, A., Dzenkovska, J., et al. (2017). The interplay of customer experience and commitment. *Journal of Services Marketing, 31*(2), 148–160.

Khan, U., & Khan, K. A. (2024). Generative artificial intelligence (GAI) in hospitality and tourism marketing: Perceptions, risks, benefits, and policy implications. *Journal of Global Hospitality and Tourism, 3*(1), 269–284.

Kılıçhan, R., & Yilmaz, M. (2020). Artificial intelligence and robotic technologies in tourism and hospitality industry. *Erciyes Üniversitesi Sosyal Bilimler Enstitüsü Dergisi, 50*, 353–380.

Koronaki, E., Vlachvei, A., & Panopoulos, A. (2023). Managing the online customer experience and subsequent consumer responses across the customer journey: A review and future research agenda. *Electronic Commerce Research and Applications*, *58*, 101242.

Kristensson, P., Matthing, J. and Johansson, N. (2008). Key strategies for the successful involvement of customers in the co-creation of new technology-based services. *International Journal of Service Industry Management*, *19*, 474–491.

Kshetri, N., Dwivedi, Y. K., Davenport, T. H., & Panteli, N. (2023). Generative artificial intelligence in marketing: Applications, opportunities, challenges, and research agenda. *International Journal of Information Management*, *75*, 102716.

Kumar, A., & Anjaly, B. (2017). How to measure post-purchase customer experience in online retailing? A scale development study. *International Journal of Retail & Distribution Management*, *45*(12), 1277–1297.

Lai, W., C. & Hung, W., H. (2018). A framework of cloud and AI based intelligent hotel. In: Proceedings of the 18th international conference on electronic business, ICEB, Guilin, 2–6 Dec, 36–43.

Lam, S. Y. (2001). The effects of store environment on shopping behaviors: A critical review. In M. C. Gilly, & J. Meyers-Levy (Eds.), *Advances in Consumer Research*, *28*, 90–197. Valdosta, GA: Association for Consumer Research

Lemon, K. N., & Verhoef, P. C. (2016). Understanding customer experience throughout the customer journey. *Journal of Marketing*, *80*(6), 69–96.

Lervik-Olsen, L., van Oest, R., & Verhoef, P. C. (2015). When is customer satisfaction 'locked'. A longitudinal analysis of satisfaction stickiness. *BI Norwegian Business School.*

Li, X., Wang, D., & Zheng, X. (2019). Personalized marketing with privacy protection in the era of big data. *Journal of Business Research*, *98*, 265–279.

Margas, S. (2023). *Artificial intelligence and the industry of hospitality and tourism threats and opportunities*, Dissertation Thesis, International Hellenic University

Master of Code, "Generative AI in Hospitality and Travel: How to Boost Guest Experiences and Multiply Revenue" https://masterofcode.com/blog/generative-ai-chatbots-in-the-travel-and-hospitality-industry-use-cases#Language_Translation (Accessed 13.05.2024)

McColl-Kennedy, J. R., Zaki, M., Lemon, K. N., Urmetzer, F., & Neely, A. (2019). Gaining customer experience insights that matter. *Journal of Service Research*, *22*(1), 8–26.

Merisavo, M., Kajalo, S., Karjaluoto, H., Virtanen, V., Salmenkivi, S., Raulas, M., & Leppäniemi, M. (2007). An empirical study of the drivers of consumer acceptance of mobile advertising. *Journal of Interactive Advertising*, *7*(2), 41–50.

Meyer, C., & Schwager, A. (2007). Understanding customer experience. *Harvard Business Review, 85*(2), 116.

Mondal, S., Das, S. and Vrana, V. G. (2023). How to bell the cat? A theoretical review of generative artificial intelligence towards digital disruption in all walks of life, *Technologies, 11*(2), 1–17.

Naude, W. (2021). Artificial intelligence: neither Utopian nor apocalyptic impacts soon. *Economics of Innovation and New Technology, 30*(1), 1–23

Nwachukwu, D., & Affen, M. P. (2023). Artificial intelligence marketing practices: The way forward to better customer experience management in Africa (Systematic Literature Review). *International Academy Journal of Management, Marketing and Entrepreneurial Studies, 9*(2), 44–62.

Pine, B. J., & Gilmore, J. H. (1999). *The experience economy.* Harvard Business Press.

Robbert, K., Penn, C., & Wall, J. (2023). Use cases of large language models in marketing analytics. *Applied Marketing Analytics, 9*(3), 249–269.

Russell, S. and Norvig, P. (2016). *Artifficial Intelligence: A Modern Approach,* NY: Pearson.

Sadhotra, N., & Gupta, N. (2023) Generative AI in customer service. *Digital Paradigm Shift: Unravelling Technological Disruption in Business,* 31.

Samala, N., Katkam, B. S., Bellamkonda, R. S. and Rodriguez, R. V. (2022). Impact of AI and robotics in the tourism sector: a critical insight, *Journal of Tourism Futures, 8*(1), 73–87. https://doi.org/10.1108/JTF-07-2019-0065

Saponaro, M., Le Gal, D., Gao, M., Guisiano, M., & Maniere, I. C. (2018). Challenges and opportunities of artificial intelligence in the fashion world. In 2018 international conference on intelligent and innovative computing applications (ICONIC) (pp. 1–5). IEEE.

SAS, Experience 2030, https://www.sas.com/content/dam/SAS/documents/marketing-whitepapers-ebooks/ebooks/en/customer-experience-2030-111008.pdf (Accessed 14.03.24)

Schmitt, B. (1999). Experiential marketing. *Journal of Marketing Management, 15*(1–3), 53–67.

Shen, A., & Ball, A. D. (2009). Is personalization of services always a good thing? Exploring the role of technology- mediated personalization (TMP) in service relationships. *Journal of Services Marketing, 23*(2), 80–92.

Solakis, K., Katsoni, V., Mahmoud, A. B., & Grigoriou, N. (2024). Factors affecting value co-creation through artificial intelligence in tourism: A general literature review. *Journal of Tourism Futures,* 10(1), 116–130.

Thukral, V., Latvala, L., Swenson, M., & Horn, J. (2023). Customer journey optimisation using large language models: Best practices and pitfalls in generative AI. *Applied Marketing Analytics*, *9*(3), 281–292.

Tuo, Y., Ning, L., & Zhu, A. (2021). How artificial intelligence will change the future of tourism industry: The practice in China. In Information and Communication Technologies in Tourism 2021: Proceedings of the ENTER 2021 eTourism Conference, January 19–22, 2021 (pp. 83–94). Springer International Publishing.

Turel, O., & Cavarretta, F. (2020). Ethical implications of artificial intelligence in marketing. *Journal of Business Research*, *119*, 560–569.

Verhoef, P. C., Lemon, K. N., Parasuraman, A., Roggeveen, A., Tsiros, M., & Schlesinger, L. A. (2009). Customer experience creation: Determinants, dynamics and management strategies. *Journal of Retailing*, *85*(1), 31–41.

Verma, R. K., & Kumari, N. (2023). Generative AI as a Tool for Enhancing Customer Relationship Management Automation and Personalization Techniques. *International Journal of Responsible Artificial Intelligence*, *13*(9), 1–8.

Wu, Y., Wu, L., & Zhang, H. (2021). How much personalization is too much? A privacy calculus perspective. *Journal of Business Research*, *135*, 26–35.

Yang, X., Li, H., Ni, L., & Li, T. (2021). Application of artificial intelligence in precision marketing. *Journal of Organizational and End User Computing*, *33*(4), 209–219.

Zanker, M., Rook, L., & Jannach, D. (2019). Measuring the impact of online personalisation: Past, present and future. *International Journal of Human-Computer Studies*, *131*, 160–168.

Zhang, K., Chen, Y. & Li, C. (2019). Discovering The Tourists' Behaviors And Perceptions In a Tourism Destination By Analyzing Photos' Visual Content With A Computer Deep Learning model: The Case Of Beijing. *Tourism Management*, *75*, 595–608.

Tuncay Çanakçi*

Chapter 3 Api-Tourism in the World and in Turkiye: Evaluation of Api-Tourism Potential of the City of Kars

Introduction

Nowadays, both the intense work tempo and the monotony of crowded city life push people to seek holidays where they can spend alone with nature (Aydın, 2012). On the other hand, healthy nutrition and increased awareness of the negative consequences of mass tourism on society and ecology increase the tendency towards alternative tourism types (Gülbahar, 2009). People's growing and increasingly important awareness of healthy nutrition and protecting nature has begun to highlight nature-based tourism types among alternative tourism types (Çoban & Karakuş, 2019). Tourism types such as rural tourism, eco-tourism, farm tourism, and plateau tourism are generally listed as nature-based alternative tourism types. These types of tourism share the following general characteristics: they are carried out in authentic environments where construction and population density are low, production and consumption are largely carried out with traditional methods, and nature and the natural are integrated (Albayrak, 2013).

Nature-based alternative tourism types are extremely important for the regions where they are developed, as they protect natural resources and the environment, provide economic benefits to local people from tourism, and support the realization of sustainability in tourism (Hacıoğlu & Avcıkurt, 2011). Api-tourism, which has started to draw attention in the last decade, is considered a niche tourism type among nature-based tourism types (Wos, 2014). In addition to the consideration of this type of tourism as an opportunity to diversify alternative tourism types, it is also seen as an important opportunity to ensure sustainability in tourism and contribute to regional development (Çukur & Çukur, 2021).

The Tuscany and Alto Adige regions in Italy, the Alsace region in France, the Lake Balaton region in Hungary, and the Andalusia region in Spain stand out among the regions where api-tourism comes to the fore (Akça, 2004; Aydın,

* Ass.Prof., Kafkas University, Sarıkamış Tourism Faculty, Department of Gastronomy and Culinary Arts, tuncaycanakci@gmail.com

2012). In addition to its natural beauty, Türkiye is a suitable destination for api-tourism, and scientific studies are needed to contribute to the development of this type of tourism. Especially the Eastern Black Sea Region, the Mediterranean, and the Aegean regions have rich potential in terms of api-tourism. (Wos, 2014). This type of tourism has the potential to contribute to sustainable tourism and regional development, as well as increasing the diversity of alternative tourism (Çukur & Çukur, 2021).

The Concept of Api-Tourism

The word "Api," which means bee, is derived from the Latin name of the honeybee, "Apis mellifera," meaning "honey collector" (Abou-Shaara, 2014). Api-tourism has great potential to be an additional opportunity to develop and consolidate the position of beekeepers and api-therapy specialists in the niche tourism market (Korosec, 2016). A niche market is defined as a narrow, small market segment determined to better meet the needs of a more limited customer group whose needs cannot be truly met (Garda & Temizel, 2016). Api-tourism is defined as bringing together travel and educational experiences, starting with bee culture and awareness of responsibility for the environment and human life. Api-tourism, a new concept in the travel and sightseeing industry, has emerged and developed as a significant part of the green economy (Grigorova, 2015).

Figure 1 shows the position of api-tourism in relation to other forms of tourism, including sustainable, alternative, green, responsible, rural, and soft tourism (Suna, 2018).

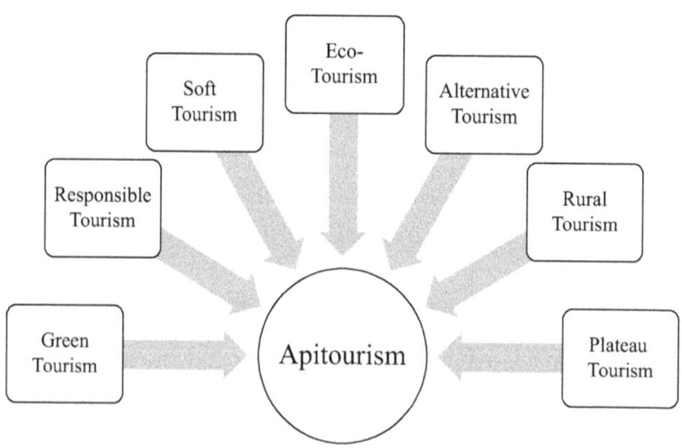

Figure 1. The position of api-tourism

Activities and apitherapy practices within the scope of api-tourism are listed in table 1.

Table 1. Api-tourism activities and api-therapy practices

Api-tourism Activities	Api-therapy practices
• Polen collection • Wax making • Honey harvest in beekeeping suits • Honey tasting • Visiting beehives • Seeing bee colonies and beekeepers' working environments on site • Visiting bee museums • Hiking on api-tourism routes	• Honey massage • Api spa • Listening to bee sounds • Api yoga • Cosmetic masks • Beehive air inhalation • Api-therapeutic treatments

Basic Functions of Api-Tourism

Api-tourism, a form of tourism that involves culture and traditions, is defined as not only a type of tourism focused on bees or honey, but also a type of tourism that combines many different sectors (Aliyeva et al., 2019). Api-tourism unites beekeepers, food and beverage businesses, health businesses, cosmetics and care product manufacturers, and collectors at a certain point. Api-tourism, which has a mostly positive impact on many sectors, generally has four basic functions: health, social, travel, and education.

Health function encourages treatment with api-therapy practices and natural methods, along with bee products such as pollen, propolis, and royal jelly (Wos & Bien, 2013). The **social function** of api-tourism reveals the communication between local people and tourists. It is inevitable for tourists who want to benefit from the knowledge and experience of beekeepers to establish social relations with the local people. The **travel function** includes tourists tasting different types of honey, visiting hives, buying honey, seeing natural beauties, and participating in various activities (Wos, 2014). The **education function** is the most important function of api-tourism. In addition to dissemination of all information about beekeeping, training sessions convey to tourists the richness that natural life brings to people (Arih & Korosec, 2015).

The Importance and Purpose of Api-Tourism

The benefits of bees to biodiversity and ecosystems are numerous. The use of bee products produced by beekeeping activities in the treatment of people, increasing the income of local people, and their place in human nutrition are very important (Beigi, 2018). While api-tourism is considered an important niche in terms of meeting the needs of a specific customer segment, it is also seen as a part of green tourism and therefore the green economy by integrating travel and educational experiences with a sense of responsibility for the environment and human life (Arih & Korosec, 2015). The purposes of api-tourism include:

- Examining beekeeping activities in the context of tourism,
- Increasing environmental awareness, especially in rural areas, and contributing to the development of environmental protection awareness,
- Understanding its contribution to social development, economic growth, and employment,
- Supporting the development of a generation that respects nature,
- Enabling people who prioritize a healthy life to interact with nature and experience this lifestyle (Arih and Korosec, 2015).

Api-Tourism in the World and Türkiye

The beginning of api-tourism in the world dates back to the 1700s. Slovenia is renowned for laying the initial foundations of api-tourism. The empress of the period, Maria Theresa, laid the foundations for today's Slovenian beekeeping and api-tourism when she assigned Anton Jansa to teach beekeeping (Liourta, 2018). Although the beginnings of api-tourism date back to the 1700s, its current use in the tourism sector has emerged in more recent years. It is claimed that the interest of Slovenian beekeepers, who played an important role in the emergence of api-tourism, first started at the International Beekeeper Congress hosted by Slovenia in 2003. Since 2007, api-tourism in Slovenia has begun to develop as a new type of tourism for beekeepers to develop products for tourists who are interested (Arih & Korosec, 2015). Conceptually, api-tourism was first introduced by ApiRoutes, the brand project of the Slovenian Beekeepers Association, in 2013 (Gandhy, Chang, & Rahmi, 2019). Many countries, particularly the Czech Republic, Poland, Lithuania, Spain, Ukraine, and Germany, have used this project, which began in Slovenia, in the years that followed. According to Wos (2014), there are currently seventy-five api-tourism destination centers in total, with forty in Poland, six in the Czech Republic and Germany, seven in Slovenia, four in Spain, two in Lithuania, Ukraine, Portugal, Slovakia, and Hungary, and one in Greece. According to Cesur (2021), Poland is another European country where api-tourism is developing. While there were initially beekeeping museums in the country, in recent years, various api-tourism services have also been provided in beekeeping farms. Sadecki Bartnik in Strozye is one of the farms with both national and international recognition where these services are provided. It is known that this farm also includes more than a hundred different beehive exhibits, a restaurant serving honey dishes, and an exhibition displaying equipment used for 2000 years. The tourist route 'Malopolska Miodowa Kraina', also located in the Malopolska city of Poland, attracts attention as an important api-tourism initiative. There are agricultural farms along this road, and beekeeping activities are carried out on all farms. Api-tourism offers tourists the chance to engage in beekeeping activities and sample honey.

According to Cesur's (2021) study, Tenerife in the Canary Islands and Thailand are two of the most prominent api-tourism centers in the world. Unfortunately, Türkiye does not have any api-tourism destinations designed in that manner. There is only one organization in Türkiye that fits the current definition of api-tourism. It is a non-profit organization called Balyolu. This organization was organized by American Cat Jaffee, who came to Türkiye in 2008–2010 to conduct research on internal migration. Jaffee, who had the opportunity to become more acquainted with beekeeping during their travels to eastern Türkiye (Kars and its surroundings), worked with women in a region famous for its honey and brought many American tourists to Türkiye for a honey-focused dining experience that included traditional cuisine. However, as will be explained in detail in the following sections, Türkiye has considerable api-tourism potential with its vast knowledge and cultural background on the history of beekeeping, its capacity in the number of hives and honey production, and its unique natural beauties, as well as the diversity of flora and fauna. Especially bee farms in the cities of Muğla, Ordu, and Adana, and traditional fixed beekeeping practices that still exist in some parts of the Eastern Black Sea Region, can be turned into api-tourism destinations in a short time.

Api-Tourism Potential of the City of Kars

The history of beekeeping in Kars dates back to ancient times and is one of the region's primary sources of income. Kars Honey is famous due to the region's special flora, natural conditions, and climate. There is information in written sources that "Kars Honey," with its aroma coming from thousands of wildflowers and its light scent and taste characteristics, was on the Ottoman palace tables and was among the products exported from the province during the Ottoman period. Beekeeping around Kars was done in long baskets until the beginning of the 20th century, and then new types of hives were used. The city of Kars was the first place in Türkiye to use new types of hives during World War I. Later, the new types of hives spread throughout Türkiye (Demir, 2014).

In terms of beekeeping, Kars, a border city where bee movements are easier to control than in other cities, is considered an isolated region. As a result of calculating the honey, pollen, and nectar potential of various plants reflecting the characteristics of the Kars region, it was determined that many of the honey plants growing in the region carry high amounts of honey, pollen, and nectar during periods when natural conditions are suitable (Varalan & Çevrimli, 2023).

The characteristics of Kars regarding beekeeping activities reveal the need for the development of api-tourism activities in the region. We can assert that the economic gain of the local people from beekeeping activities will increase significantly with api-tourism.

Conclusion and Evaluation

Türkiye ranks among the world's top countries in both colony count and honey production, showcasing significant potential for api-tourism. With its diverse flora and fauna across all four seasons, Turkey offers a rich tapestry of natural beauty. However, it remains uncertain whether this potential is adequately tapped. While viewed as a shortfall, this situation also presents a crucial opportunity (Ayaz Dönmez, 2023).

Beekeepers, bee farms, beekeeping museums, honey forests, bee hives, and bee rocks required for api-tourism are already waiting for api-tourism enthusiasts. All that needs to be done is for stakeholders to come together and act for sustainable and effective programs and projects without wasting time. Since api-tourism enthusiasts prefer being alone with nature and in natural environments over comfort and luxury, it does not require huge investments for other types of tourism. In this sense, what needs to be done includes (Şimşek, 2023):

- Determining Türkiye's beekeeping and api-tourism potential on a regional basis.
- Mapping suitable locations in the determined regions for possible api-tourism routes.
- Making plans to facilitate access to those routes,
- Building natural and simple accommodation and food and beverage venues in the residential areas along the routes, or supporting entrepreneurs in this regard,

- Providing training to beekeepers who are or will be on the determined routes so that they are at a level to provide training on honey and honey products,
- Taking precautions to ensure the safety of api-tourism participants (bee stings, bee allergies, etc.) during the api-tourism process,
- Training business owners, managers, and employees on api-tourism routes in human relations, communication, and customer relations, as well as tourism services,
- Implementing a comprehensive, sustainable, and effective promotion and marketing activity for api-tourism both nationally and internationally.

References

Abou-Shaara, H. F. (2014). The foraging behaviour of honey bees, Apis mellifera: A review. *Veterinarni Medicina, 59*(1), 1–10. doi:10.17221/7240-VETMED

Akça, H. (2004). Dünyada ve Türkiye'de Kırsal Turizm. *Ekonomik ve Teknik Dergi Standard, 43*(513), 67–70.

Albayrak, A. (2013). *Alternatif Turizm*. Ankara: Detay yayıncılık.

Aliyeva, Z. N., Baiburiyev, R. M., Lorant, D. D., Shagyrbay, A. S. & Kaliaskarova, Z. K. (2019). Problems And Prospects Of Development Of Apitourism in Kazakhstan. *Bulletin Of Naiıonal Academy of Sciences Of The Republic of Kazakhstan, 6*(382), 45–53. doi:10.32014/2019.2518-1467.144

Arih, K. & Korosec, T. (2015). Api-tourism: transforming Slovenia's apicultural traditions into a unique travel experience. *Sustainable Development and Planning VII, 193*, 963–974. doi:10.2495/sdp150811

Ayaz Dönmez, N. (2023). Apiturizm Araştırmaları: Sistematik Literatür İncelemesi. *Türk Turizm Araştırmaları Dergisi*, 7(4), 685–699.

Aydın, O. (2012). AB'de Kırsal Turizmde İlk 5 Ülke ve Türkiye'de Kırsal Turizm. *Karamanoğlu Mehmetbey Üniversitesi Sosyal ve Ekonomik Araştırmalar Dergisi, 14*(23), 39–46.

Beigi, H. (2018). Api-Tourism, Bees Tourism. *Honeybee Science Journal*, 9(16), 12–18.

Cesur, E. (2021). A Creative Approach in Creative Tourism: Apitourism. In V. Krystev, S. Çelik Uğuz, R. Efe & E. Kapluhan (Ed.), *Tourism Studies and Social Science* (pp. 75–91). Sofia: St. Kliment Ohridski University Press

Çoban, O. & Karakuş, Y. (2019). Doğa Temelli Alternatif Turim Türleri. In Ş. Aydın ve D. Eren (Ed.), *Alternatif Turizm* (pp. 101–126). Ankara: Detay Yayıncılık.

Çukur, F. & Çukur, T. (2021). Muğla İli Arıcılık Potansiyelinin Apiturizm Açısından Değerlendirilmesi. *Tarım Ekonomisi Dergisi, 27*(2), 113–122. doi:10.24181/tarekoder.882927

Demir, M. (2014). Kars İlinin arıcılık potansiyeli ve değerlendirme durumu. Doğu Coğrafya Dergisi, 19(32), 209–230.

Liourta, A. (2018). *Apitourism: Παρούσα Κατάσταση, Τάσεις και Προοπτικές μελλοντικής ανάπτυξης (Apiturizm: mevcut durum, eğilimler ve gelecekteki beklentileri gelişimi).* yayımlanmamış Lisans tezi, Sosyal Bilimler Yüksekokulu, Selanik.

Gandhy, A., Chang, M. ve Rahmi, S. (2019). Service Quality and Customer Satisfaction in the Apiculture-Based Agritourism. *Jurnal Organisasi dan Manajemen, 15*(2), 153–165. doi:10.33830/jom.v15i2.734.2019

Garda, B. & Temizel, M. (2016). Sürdürülebilir Turizm Çeşitleri, *Selçuk Üniversitesi Sosyal ve Teknik Araştırmalar Dergisi, 12*, 2016, pp. 83–103.

Grigorova, Z. (2015). Apitourism. *UARD Yearbook, 3*, 83–125.

Gülbahar, O. (2009). 1990'lardan Günümüze Türkiye'de Kitle Turizminin Gelişimi ve Alternatif Yönelimler. *Süleyman Demirel Üniversitesi İktisadi ve İdari Bilimler Fakültesi Dergisi, 14(*1), 151–177.

Hacıoğlu, N. & Avcıkurt, C. (2011). *Turistik Ürün Çeşitlendirmesi* (2. bs.). Ankara: Nobel Yayınları.

Korosec, T. A. (2016). *Api Turizmi, Api Sağlık, Api Terapi*, Muğla Arıcılık ve Çam Balı Kongresi, Muğla Sıtkı Koçman Üniversitesi.

Suna, B. (2018). Api Turizmin Türkiyedeki Yeri ve Önemi. *Uludağ Arıcılık Dergisi, 18*(1), 42–51.

Şimşek, M. (2023). Apiturizm. In M. Öksüz (Ed.), *Turizm Alanında Çok Yönlü Araştırmalar* (pp. 75–89). Özgür Yayın Dağıtım Ltd. Şti.

Varalan, A., & Çevrimli, M. B. (2023). Kars İlinde Arıcılık İşletmelerinin Sosyo-Ekonomik Yapısı Üzerine Bir Araştırma. *Dicle Üniversitesi Veteriner Fakültesi Dergisi, 16*(2), 102–107.

Wos, B. & Bien, W. (2013). Apitourism as a form of sustainable development. *Research Bulletin Leisure and Tourism, 1*(11), 83–90.

Wos, B. (2014). Api-tourism in Europe. *Journal of Environmental and Tourism Analyses, 2*(1), 66–74

Funda Yorulmaz[†]

Chapter 4 The Contribution of the Filyos Valley Project to the City Branding Process of Zonguldak Province and Its Potential Impact on Tourism Management in the Region[*]

Introduction

Today, competition exists not only between products and companies but also between countries and cities. People travel worldwide owing to many factors, such as the rapid development of communication and transportation facilities, the rapid spread of information, and increased income. Countries and cities face fierce competition to capture the largest share of global tourism revenue. Cities and countries must differentiate themselves to achieve superiority. Cities can be differentiated by branding (Çelik et al., 2020).

The concept of a brand is not limited to products and services but also includes countries, cities, and people. Over time, cities have been marketed and branded by evaluating them as products (Fırat & Kömürcüoğlu, 2015). Creating brand value for cities using modern marketing methods aims to create added value for the city, differentiate it from similar cities, ensure that it is preferred, and consequently, realize more investments. This is because a city or city branding is all the activities carried out to transform a particular place into a center of attraction where people want to live, work, invest, study, and visit instead of perceiving it as an ordinary settlement in the minds of the target audience using brand strategies and techniques (Avcılar & Kara, 2015).

City branding is an activity carried out to transform a city into a center of attraction where people want to live, work, invest, study, and visit instead of perceiving it as an ordinary settlement in the target audience's minds. City branding is the process of differentiating and diversifying cities. It brings together local tourism facilities, cultural and art organizations, museums, and historical sites to create a common image for the city. In addition, this branding process attracts

---

* 1st Filyos Congress, 15–16 Ekim 2021, Zonguldak Bülent Ecevit Üniversitesi, Türkiye

† Gökçebey Mithat-Mehmet Çanakçı Vocational School, Department of Foreign Trade and Marketing, Public Relations and Publicity Program, Zonguldak Bülent Ecevit University, Türkiye

the interests of consumers and investors and helps establish tourism businesses. There are two types of city branding; the first is the promotion of an existing city, and the second is the structuring of a new city. The media plays an important role here (Jojic, 2018).

According to Aaker, city branding is an approach to conceptualizing a city as a brand, and a brand is a multidimensional structure consisting of functional, emotional, relational, and strategic elements that create a set of associations in the public mind. City branding is performed to differentiate a city from other cities and to change the negative or neutral image of the city into a positive and desirable one. Branding a city is more complicated and complex than branding products. This is because the target audience and stakeholders are much more numerous and diverse (Görkemli, 2012).

City branding is the symbolic shaping of all information about a city for a specific purpose and is a multidisciplinary field involving city planners, marketers, architects, and tourism professionals. City branding aims to increase investment and tourism revenue in cities and develop a community, which is a tool that explains the city's competitiveness, spatial quality, history, culture, and lifestyle. Therefore, cities tend to brand themselves and aim to position themselves as brands in the minds of their target audiences (Akturan & Oğuztimur, 2016).

City branding provides essential information about a city's competitiveness, spatial quality, history, lifestyle, and culture and describes a more complex structure than product or service branding. From this perspective, city branching

- Acting to implement the branding strategies of products or services for cities,
- Create a positive perception in people's minds by adding value to all kinds of outputs related to the city and its citizens,
- Realize the claim of becoming a center of attraction for existing or potential guests/customers (Erim, 2017).

In particular, when a country or region is not well known or has a negative image, branding this city or region should be a priority for recognition and consideration by others (Mabillard, 2023).

The Process of City Branding

The branding efforts of city centers are primarily economic. When creating branding strategies for cities, the elements of the city's environment (natural, human, and artificial) should first be analyzed appropriately. City branding is a strategy to create an image that emphasizes the strengths of a city that are different from other cities and adds cultural, economic, and social value to the city.

In other words, the branding of a city is to bring together its cultural, historical, natural, and social characteristics and to develop them with a symbol that will reveal its differences from other cities, to promote itself in the national and international arena, and to gain an image. The promotion of cities with these characteristics is also easier (İçellioğlu Şarkaya, 2014). A positive image and identity of a city are essential for a competitive advantage, mainly because there is competition between cities and countries.

The competition between countries and cities is not new. Since the 1860s, products have competed in technological innovation, creativity, craftsmanship, values, and skills through fairs and exhibitions worldwide (Mabillard, 2023).

City branding is a process that depends primarily on the answer that cities give to the question: "What feature(s) of our city is/can be differentiated from other cities, taking into account the target audience and the competition?". This involves creating an image and promoting the city by revealing the unique qualities of the city structure. The city-branding process consists of beliefs that include many components and even the city's emotions (Özkaya & Deniz, 2019). A city's image is the sum of people's beliefs, ideas, and impressions. For a city image to be effective (Yükselen & Güler, 2009),

- It must be valid,
- It must be convincing,
- It must be simple,
- It must be attractive,
- Thus, they should have distinct features.

By turning the city into a center of attraction, it aims to attract foreign investors, tourists, and developed skilled populations to the city. Local governments have a great responsibility. Local governments should evaluate the potential of a city through various studies and arrangements in the city branding process. For this purpose, strategic planning should be made, and integration should be ensured by involving other stakeholders in the decision-making process. Additionally, the private sector's branding efforts can influence the city's branding process—for example, L'Oreal's branding of the city of Paris. In addition, the needs and expectations, mission, vision, and goals of the city should be addressed using a strategic approach to the branding process of the city. In strategic planning, we must look for answers to the question,' How can we achieve our goals? A city vision is essential for this purpose. City vision answers the question, "What kind of city should the city be in the future?" (Özkaya & Deniz, 2019).

On the other hand, the brand city concept consists of elements such as city identity, image, and positioning, which are considered interrelated in the formation process. In addition, the city's physical, social, cultural, economic, and historical elements, such as buildings, landscapes, natural and geographical qualities, people, architecture, streets and cityscape, universities, and transportation facilities, affect the formation of city identity. City identity is the basis for the formation of city images and positions, as it includes all the tangible and intangible values of the city. City image is a subjective personal judgment, and people can perceive cities differently. On the other hand, city positioning creates a value compatible with the city's tangible and intangible identity elements. It presents this value to the target audience to ensure it occurs in their minds (Beyaz & Boyraz, 2018).

City branding is a strategy for creating an image that emphasizes a city's strengths, which makes it different from other cities and adds cultural, economic, and social value to the city. If we expand this issue further, the branding of a city will bring together its cultural, historical, natural, and social characteristics and develop them with a symbol that will reveal its differences from other cities to promote itself in the national and international arena and gain an image. It is easier to promote cities with these characteristics. Visual images and symbols play essential roles in branding. Accordingly, for a city to be considered a brand, the needs and expectations of many segments, such as locals, foreign tourists, students, homemakers, institutions and organizations operating in the city, investors, and tourists, should be identified. To become a brand city, it must attract the attention of foreign investors, buyers, and tourists (Uygurtürk & Yılmaz, 2019). Undoubtedly, being a city that starts with "World Famous" and creating such a brand is a tremendous competitive advantage.

Since it is known that branded products are more likely to be preferred by consumers, cities' characteristics are used in creating a city brand. The critical point is to create a brand perception of the city and the power of attraction by contributing to the economic and social improvement of the city and region. One of the stages of becoming a "brand city" is to realize that it competes with all other cities to offer better quality to residents and external customers/visitors (Bekar & Şengel, 2021).

Cities on the road to branding are taking necessary steps toward achieving high economic and social standards in their environment. Thus, increasing economic attractiveness is a prerequisite for branding. Economically attractive cities attract investors, buyers, and tourists from all over the world and from their own countries. This led to economic growth. It also provides residents with the opportunity to live at higher standards. The work carried out for the brand-city process is the most critical investment in the future of the city and the country. Branded cities create tourist value by being centers of attraction, and this value can pave the way for sustainable development. Another important aspect of city branding is promoting a city's opportunities, potential, and attraction to the outside world. Graphic design also plays a vital role in creating branded cities. It is a universal way to convey a society's culture, values, and potential. In addition to design elements, such as symbols, logos, emblems, graphics, characters, typography, signage, and colors, graphic design effectively conveys a city's spirit, energy, history, and geography (Davras & Uslu, 2019).

The Relationship Between City Branding and Tourism and Its Impact on Tourism Management

City branding offers several significant benefits. Efforts to create a city brand represent a significant investment in the city's future and its inhabitants. Today, branded cities are attracting people worldwide, thus ensuring sustainable development by providing high economic and social standards for people living in the city. It is well known that tourism contributes to the national economy by increasing the inflow of foreign currency and providing employment and industry. It has a significant share in protecting world peace with its integrative and unifying effect on international cultural and social communication.

While cities' development attracts tourists, increasing tourist mobility positively affects city development. Many cities that want to develop their economies have developed tourism products based on their historical and contemporary heritage. A city's history, monumental or industrial heritage, gastronomy, art, culture, events, popular traditions, and entertainment offer tourists various attractions. Cities attempt to increase their potential to attract investors and tourists through their features and services. A city with brand value has a high potential to attract foreign tourists and investors. Tourists do not want to visit cities with a negative image. In addition, cities that have high potential but do not use their resources effectively and do not make the right and sufficient investments may not be branded (İçellioğlu Şarkaya, 2014).

Destination branding should be discussed in this context. The origin of the destination concept comes from the French language, and in the Turkish Language Association Dictionary, it is expressed as "place to arrive"; in literature, it means a place of attraction, a tourist place, and a tourist region. In a broader sense, a destination is defined as a region that is smaller than a country but larger than many cities within the country, a branded national area with a specific image in the memory of individuals, a region with significant tourist attractions, a good transportation network, many events such as festivals and carnivals, and a sufficient geographical area for the development of tourism enterprises. Factors such as natural and historical attractions, quality service and product diversification provided by qualified personnel, and natural and historical attractions in tourist destinations satisfy tourists, and satisfied tourists increase tourism revenue (Davras & Uslu, 2019).

As a destination becomes a brand, the brand identity of cities in the destination develops, and this situation causes the destination to have a positive image of people. Cities with a brand image have the advantage of being perceived as a city where it is convenient to live, where cultural and tourist activities predominate, and a city with a place in people's minds has the advantage of increasing awareness with its unique images. The number of visitors to cities with a high destination brand image has increased, and these cities desire to be revisited at a higher level by the target audience (Giritlioğlu & Öksüz, 2016).

The difference in destination can only be communicated to customers through a brand. Each destination has unique characteristics and is branded. Branding is possible by effectively using and promoting the resources of the destination. Attracting tourists to a destination is essential (Aslan et al., 2014). The tourist value of a city is created when its tourist supply capacity is perceived and interpreted as a tourist attraction by the target audience. The stakeholders want to create a tourism brand city and aim to improve the city's and its managers' image—non-governmental organizations. Branding studies are very important for investors of tourism enterprises, such as hotels, to see which richness of the city can be evaluated as elements of attraction (Çelik & Sandıkçı, 2015).

Exploration of Zonguldak Province as City Branding

Today, with the realization of the power that branding brings and an understanding of its importance, almost everything is branded. However, branding is time-consuming and requires long-term effort. The branching of cities offers several benefits. Creating a city brand is a significant investment in the city's future and its inhabitants. Today, branded cities attract people and ensure sustainable development by providing high economic and social standards for people living in the city (Avcılar & Kara, 2015).

In the town of Filyos in the Çaycuma district of Zonguldak, the Filyos Valley Project, which consists of a port, a free zone, and an industrial zone, was planned by Ottoman Sultan Abdülhamit II approximately 100 years ago and was scheduled to be completed in 1909. However, the construction of this project only started in 2016 (Ceyhan et al., 2017).

With the Filyos Valley Project, known as a "mega project," Zonguldak is expected to become a new image and center of attraction. This project will support the region's economic development and commercial growth. In addition to this project, it is also interesting that the Çaycuma Filyos region was a similar port and trade city, even in the years before Christ. Accordingly, discovering ancient artifacts in this region will contribute to tourism in both regions and countries. With such historical, cultural, natural, and other features, the Zonguldak Province can become a positive city brand. The aim is to create a new and positive city image using the Filyos Valley Project of Zonguldak Province. Since international recognition will increase with a branded city, it can convey different messages to the world by using the strengths of the city and provide economic benefits by attracting investment (İçellioğlu Şarkaya, 2014).

This regional project, which includes many private sector investments and public sector investment projects, will impact beyond the borders of Çaycuma District. With this project, there will be significant developments in the economic structure of the Zonguldak, Bartın, and Karabük provinces, unemployment and migration will decrease, and new business lines will open. Thus, this region will become a substantial industrial and logistics base (Çetinkaya, 2012). Thus, many new industrial facilities are located in the region within the Filyos Valley Project, creating new employment opportunities. As a result of the development of industrial activities, residential areas will form on the Filyos and Çaycuma sides. Since the Filyos-Çaycuma region will become more attractive to long-term residents, area planning should now be conducted.

In addition, Filyos has the potential to be a major port connecting the Western Black Sea and Central Anatolia to countries that share a familiar coast with the Black Sea and even to Central Asia and the oceans by land. The fact that there are maritime tourism destinations in the region where travel can be planned and that there is a port that can be docked by ship gives local and foreign tourists an alternative holiday experience. Cruise ship voyages, which are considered alternative tourism, and their effect on national income per capita has been seen to increase by 3 % in ten other countries with significant tourism income volume (Antalya, Aydın, Bartın, Çanakkkale, İstanbul, İzmir, Muğla, Samsun, Sinop, Trabzon) (Ceyhan et al., 2017). Even considering this aspect, it can be estimated that Filyos Port can significantly contribute to the country's welfare due to research. Investments in this region should include not only industrial but also tourism enterprises. The fact that there are ruins from ancient antiquities in the region and swimmable beaches increases this region's attractiveness.

In this context, the ancient city of Tios in Çaycuma District of Zonguldak Province on the Western Black Sea coast can be a strategic approach in city branding. The remains of the castle, coastal walls, aqueduct, vaulted gallery, theater, defense tower, and tombs from the Roman, Byzantine, and Medieval periods can be seen as the remains of the ancient city of Tios. The city also has a harbor and two completely underwater breakwaters. Besides, excavations have revealed partially paved roads, architectural structures considered bathhouses and nymphaeum, and many small finds. Kale Hill (Acropolis): It is located just east of today's Filyos, on a natural hill perpendicular to the sea. With the opening of this region to tourism, it will become more active in industry, trade, and tourism. With these features of the Zonguldak Province, it will be highly beneficial to the development of the country and the region to increase its recognition worldwide with a new city brand.

Today, tourism is recognized as necessary for national and regional development worldwide. As it is known, the tourism sector has a significant place in countries' economies thanks to its relationship with many sub-sectors, its contribution to employment, and the foreign currency inflow it generates. Therefore, countries and regions with tourism potential strive to accelerate national and regional development and economic growth by focusing on domestic and international tourism activities.

Recognizing the power of tourism, countries are looking for new ways to increase their existing tourism potential and strengthen the competitiveness of their destinations. Although each country's destinations' natural and cultural characteristics differ, the resources that make a destination stand out are tourist attractions, activities, enterprises, and infrastructure. In addition, tourism enterprises are more environmentally friendly than industrial enterprises.

Turkiye's urban branding strategy in the 2007–2013 Action Plan is adopted to brand our cities with rich cultural and natural values and turn them into points of attraction for tourists. In this plan, various arrangements such as architectural arrangements, arrangements related to the transportation system, arrangement of cultural axes, and physical and social arrangements are envisaged for the realization of branding on an urban scale in Turkiye. The action plans for urban branding include the creation of branded cultural cities by revitalizing cultural tourism; the establishment of city museums; the restoration, repair, and landscaping of historical, cultural, and architectural ruins; the establishment of food and beverages and daily commercial centers around historical attraction points; the establishment of qualified accommodation facilities around airports; the establishment of a fast, safe, and comfortable system to provide transportation between airports and city centers; the development of marine tourism infrastructure; the development of fair and congress centers; and the creation of art villages. A SWOT analysis should highlight the city's strengths and weaknesses compared to other cities and examine its opportunities and threats. The recommendations presented at the end of the analysis will be helpful in terms of the development of the city, revealing the vision, branding, and increasing tourism investments (İçellioğlu Şarkaya, 2014).

Conclusion

Cities are marketed products. A city is a product that can be marketed with its physical structures, infrastructure, dimensions from the present to the past and future, the people who live in it, and the values that it creates individually or collectively. The satisfaction of customers who benefit from a product is essential, and the same applies to city brands. In other words, the satisfaction of those visiting a city is essential. Just as managers are responsible for maximizing the benefits customers can derive from a product; city managers are also responsible for making the city a brand. Ensuring visitor satisfaction is possible only if visitor expectations are satisfied. Unfulfilled expectations hurt a city's image, as visitors may negatively approach the city. The city's stakeholders, especially the local authorities, are responsible for knowing what visitors expect, taking the necessary measures, preparing the infrastructure, etc. It should not be forgotten that creating a city brand is a long-term, conscious, patient, planned, and programmed process. This requires effective communication with the target audience.

However, cities' high expectations also make them more valuable. Increasing the value of a city as a product will be a factor that will increase the revenues of the city's stakeholders. The point is that the expectations promised to customers must be fulfilled. The benefits expected of tourists visiting this location are essential. In other words, the city brand and perceived city brand should be the same. For this reason, each city's stakeholders should participate in efforts to increase their value and meet visitors' expectations simultaneously. As a brand represents a certain quality, care should be taken to ensure the city's image is positive, improvable, and sustainable. Another critical problem is insufficient or limited budget allocation.

It is well known that tourism contributes to the national economy by increasing the inflow of foreign currency and providing employment and industry. Its integrative effect on international cultural and social communication contributes significantly to maintaining world peace. A city's historical and cultural heritage is as much a part of society as the city itself. Transforming these values into economic value in tourism plays a role in developing the local region and nation. Because the tourist elements of a city are used in city branding, the cultural values, immovable assets, and symbols that reveal the city's identity should be studied in detail and collected as written and visual data.

In this regard, with the Filyos Valley Project, which will contribute to the branding of Zonguldak Province, the region will not only have a large port. However, it will also become a trade, industry, and tourism region. As a result, the number of companies producing products and goods in the region will increase, the trade will become widespread, and with the increase of these and visitors, accommodation, restaurants, etc., will cause the need for and establishment of many tourism companies in the service sector. In addition, foreign investors will be attracted to this region, as the port of Filyos is essential in the Black Sea region and is well-known in neighboring countries. In this regard, it is believed that the realization of the city's branding will play an essential role in making the region a center of attraction.

When the above action plan is examined, it can be concluded that Zonguldak does not have sufficient branding infrastructure. In addition to the fact that there are deficiencies in the transportation of passengers landing at Çaycuma Airport, there are no comfortable accommodation facilities near the airport. The lack of a direct connection between airports and high-speed trains adds a new transportation problem. There is also a need for large museums to play a role in promoting the city. Historical buildings must be protected and restored to their original state. Service areas such as cultural and art organizations and fairs should be created.

Neither Zonguldak Central Municipality, Çaycuma Municipality, nor Filyos Municipality has any studies on how urban development and transformation will occur if the project is realized. A new urban project can be created based on good examples worldwide. Considering the distorted urbanization problems of the Zonguldak Centre, the Filyos Project is an essential opportunity in the context of new urban development and transformation. However, suppose the necessary planning is not performed. In that case, it will be inevitable to repeat past mistakes, that is, unplanned urbanization, environmental pollution, and destruction of the ecological system and historical sites. Local and central governments perform a significant amount of work.

As it is known, Zonguldak Province has been known for its coal and mining activities for many years. Thanks to the Filyos project, it can play an active role in the city branding of Zonguldak Province by eliminating its un positive city image and gaining a positive city image. In particular, it is essential to revive tourism, increase the number of tourists visiting the city, and make it a center of attraction by encouraging investments in this direction. In this way, stopping emigration from the city and attracting qualified people and workforce is possible.

On the other hand, with the realization of the Filyos project, other visitors, including those who eat, drink, and stay in accommodations, etc., will be attracted apart from industrial investments, primarily commercial and industrialists. Considering the need for activities that will occur, tourism companies will be needed mainly for new investments. The fact that Filyos and Zonguldak are resort areas and that one can swim in the sea in summer from many parts of the region makes it a place of attraction in terms of tourism management. Again, the local culinary culture can be included in city branding.

In addition to the natural, historical, cultural, and social features that will ensure regional and urban branding, amusement parks, such as Disneyland-like entertainment places, can also be established. A successful and positive city brand can be created by designing new residential areas with modern buildings, including various activities, that will be built in the future. For example, the ports of Hamburg or Rotterdam and their surroundings can be built with smooth and modern urbanization. What is important here is that, by creating a thriving city brand, Zonguldak Province will be transformed into a city with a more positive urban identity, beyond the perception of being identified only with hard coal and workers' cities. At the same time, it is possible to travel, have fun, live, invest, etc., by adopting a new attractive city image.

In this context, the public and private sectors should support and help each other with complete coordination. It should be remembered that creating and promoting a new place brand will be a patient endeavor, as it will take time. The success of the strategies depends more on the consistency of the decisions. The infrastructure and superstructure of the city, the urban consciousness, the local people's ownership of their city, and the bond between the public and local administration play an essential role. The more positive the brand image created in a city, the more successful it will be more successful. This will increase the city's attractiveness to tourists, and the added value tourism brings to the region. In addition, with increased visitors, tourism activities will increase, and the city will develop economically and socially. It also provides economic attractions for various sectors. As a result, sustainable economic development will be ensured for the region, city, and country.

References

Akturan U., Oğuztimur S. (2016). Kent Markalaşması Kavramının İçeriği ve Gelişimi: Farklı Disiplinler Farklı Yaklaşımlar, *TMMOB Şehir Plancıları Odası Planlama*, Sayı: 26(2), ss. 117–129.

Aslan Z., Güngeren E., Çoban G. (2014). Destinasyon Markalaşma Sürecinde Yöresel Mutfağın Rolü: Nevşehir Örneği, *Journal of Tourism and Gastronomy Studies* Cilt:2, No:4, ss. 3–13.

Avcılar M. Y., Kara E. (2015). Şehir Markası Kavramı ve Marka Şehir Yaratma Stratejilerine Yönelik Literatür İncelemesi, *Sosyal ve Beşeri Bilimler Araştırmaları Dergisi*, Bahar, Sayı 34.

Bekar A., Şengel Ş. (2021). Eskişehir'in Markalaşmasında Grafik Tasarımın Rolü ve Eskişehir İçin Tasarım Önerileri, *Pamukkale Üniversitesi Sosyal Bilimler Enstitüsü Dergisi*, Sayı:45, ss. 401–418.

Ceyhan S., Kamacı A., Peçe M. A. (2017). Filyos Liman Projesinin Kruvaziyer Gemi Turizmi açısından Değerlendirilmesi ve Bölge Ekonomisine Katkısı, *Uluslararası Yönetim İktisat ve İşletme Dergisi*, ICMEB17, Cilt: 13, Sayı:13, ss. 271–283.

Çelik S., Sandıkçı M.(2015).Türkiye'de Kentler İçin Turizm Destinasyonu Değer Kıstaslarının Belirlenmesi ve Destinasyon İmajı, *Journal of Tourism and Gastronomy Studies* 3/4, ss. 45–57.

Davras Ö., Uslu A. (2019). Destinasyon Seçimini Belirlemeyen Faktörlerin Destinasyon Memnuniyeti Üzerindeki Etkisi: Fethiye'de İngiliz Turistler Üzerine Bir Araştırma, *MANAS Sosyal Araştırma Dergisi*, Cilt:8, Sayı:1, ss. 679–696.

Edit. Çelik D., Kayahan K.; Yazarlar: Uygur S. M., Çelik D., Köylü P., vd. (2020). Marka, Kimlik, Turizm ve Bartın Kent Kimliği Çalıştayı, *Bartın Üniversitesi Yayınları*, No:35, MYO Yayın No:3

Erim G. (2017). Kent Markalaşması Yolunda Konya, *Konya Ticaret Odası*, http://www.kto.org.tr/d/file/kent-markalasmasi-yolunda-konya-gizem-erim.pdf

İçellioğlu Şarkaya C. (2014). Kent Turizmi ve Marka Kentler: Turizm Potansiyeli Açısından İstanbul'un Swot Analizi, *İstanbul Üniversitesi Dergisi*, Sayı:2014/1, 37–55.

Fırat A., Kömürcüoğlu F. (2015). Muğla Şehir Markası ve İmaj Algısı; Muğla Sıtkı Koçman Üniversitesi Öğrencileri Üzerine Bir Alan Araştırması, *Yönetim Bilimleri Dergisi*, Cilt: 13, Sayı: 26, ss. 285–304.

Giritlioğlu İ., Öksüz E. N. (2016). Turizmde Destinasyon Aidiyeti: Gaziantep İli Üzerine Bir Alan Araştırması, https://dergipark.org.tr/tr/download/article-file/273720

Görkemli N. H. (2012). Kent İmajı ve Markalaşan Kentler, *Akdeniz Üniversitesi İletişim Fakültesi Dergisi*, Sayı:17, ss. 141–155.

Jojic S. (2018). City Branding and the Tourist Gaze: City Branding for Tourism Development, *European Journal of Social Science Education and Research*, Vol: 5, No: 3.

Özkaya B., Deniz E. (2019). Kent Markalaşması Faaliyetleri Üzerine Bir Araştırma: Karşıyaka/ İzmir Örneği, *Business and Economics Research Journal* Vol.10, No:2, ss. 557–574.

Uygurtürk H., Yılmaz A. (2019). Şehir Pazarlamasında Marka Değeri Algısının Önemi: Sivas İli Örneği, C.Ü., *İktisadi ve İdari Bilimler Dergisi*, Cilt:20, Sayı:1, ss. 394–408

Yüksel C., Güler E. (2009). Antakya Marka Kent Görüş ve Öneriler, Detay Yayınları, Ankara.

Fidel Toker*

Chapter 5 The Impact of Participation in Outdoor Recreation on Healthy Aging

Introduction

Recreation plays a crucial role in maintaining and improving health across all ages. It includes a wide range of activities such as sports, arts, social activities, travel, outdoor pursuits that contribute to physical, mental, and social well-being. The importance of recreation in health promotion is increasingly recognized by health professionals and policy makers alike. The way to explain the vital link between recreation and health is to point out that participating in fun activities is a critical bridge in improving the quality of life. Recreation encompasses a diverse range of activities pursued for pleasure, relaxation, and stress relief. From sports and physical exercises to arts, crafts, and social gatherings, recreational activities vary but share common objectives: enjoyment, personal fulfillment, and health enhancement.

This chapter examines various forms and benefits of recreation, illustrating that whether through vigorous physical activity or serene leisure pursuits, recreation is a crucial component of a healthy lifestyle. Studies have consistently shown that engaging in recreational activities can significantly improve health outcomes across various populations. Recreational activities that involve strength training, such as weightlifting or resistance exercises, play a critical role in musculoskeletal health. These activities enhance muscle strength, flexibility, and bone density. This is particularly important as one ages, helping to combat the natural decline in muscle mass and bone density a condition known as osteoporosis.

Beyond the physical benefits, recreation greatly impacts mental health. Engaging in enjoyable activities can mitigate stress and anxiety, acting as a natural antidepressant. Recreational activities have a therapeutic effect, improve mood and cognitive functions, and provide a sense of achievement and satisfaction (Stebbins, 2007). Whether it involves the mental challenge of a strategic game or the creative outlet of the arts, recreational activities help maintain

* PhD., Council of Higher Education (CoHE) 06539 Bilkent/Ankara, fidelcann@gmail.com

mental agility and emotional balance. Establishing a connection through recreation, it strengthens social aspects and supports social health, whether through team sports and group activities or by participating in activities that gather like-minded individuals. Recreational activities improve social interactions and strengthen community ties (Coalter, 2015). By interacting with others, individuals improve their social skills, develop supportive networks and increase their overall social well-being. Recreation is important in maintaining social connections, mental health, especially for those prone to isolation, such as the elderly or remote workers. Why is participation in outdoor recreation more desirable and why does it have a positive impact on healthy aging?

Aging and Being Healthy

Aging is a multifaceted process characterized by gradual physiological, cognitive, and social changes that occur as individuals grow older, and decline of physiological functions, leading to increased vulnerability to disease and death (López-Otín et al. (2023)). This process is influenced by a combination of genetic, environmental, and lifestyle factors. According to World Health Organization (WHO) reports, the global population aged 60 years and over is expected to reach 2 billion by 2050, up from 900 million in 2015 (WHO, 2018).

With the advancement of medical sciences and improvements in public health, life expectancy has significantly increased, leading to a higher proportion of older adults in the population. This demographic shift presents both challenges and opportunities for societies worldwide. As the aging population grows, it becomes increasingly important to understand the biological mechanisms of aging and to develop strategies to support healthy aging and improve the quality of life for older adults.

Recent scientific research continues to uncover the multifaceted mechanisms underlying aging, with studies spanning genetics, cellular biology, and lifestyle factors. In their article, López-Otín et al. (2023) explore various biological mechanisms, including genomic instability, telomere attrition, and mitochondrial dysfunction that contribute to the aging process and influence the health span and lifespan of organisms. He and Sharpless (2017) discuss in their review cellular senescence, a state where cells cease to divide in response to stress, contributing to aging and age-related diseases. The authors explore the dual role of senescence in promoting both health and disease, depending on the context.

Other research includes: The contribution of dietary practices such as fasting and time-restricted eating to a healthy lifespan by influencing metabolic and cellular pathways involved in aging (Longo & Panda, 2016); genetic factors contributing to exceptional longevity and healthy lifespan (Milman & Barzilai, 2016); the role of epigenetics in aging, risk of developing age-related diseases (Raj et al., 2020); the emerging field of geroscience, which aims to understand the biological links between aging and chronic diseases (Kennedy et al., 2014); how emotion regulation changes with age, how it affects psychological well-being in older adults, and the importance of adaptive emotional strategies to age well (Charles, 2020).

Healthy aging is about promoting and maintaining the functional ability that enables well-being in older age. Functional ability is the capability of individuals to be and do what they value throughout their lives. It encompasses factors such as mobility, cognitive health, emotional well-being, and social connectedness. The concept of healthy aging emphasizes the need for societies to adapt to the demographic shift by ensuring that older adults can live not only longer but also healthier lives.

Rowe and Kahn's (1997) seminal work on successful aging highlights the importance of maintaining physical health, cognitive functioning, and active engagement with life as key components of healthy aging. Additionally, the WHO's framework on healthy aging focuses on creating environments that foster the abilities of older people, suggesting interventions ranging from promoting physical activity to ensuring social participation and security (WHO, 2015). Encouraging active aging through outdoor recreation is increasingly recognized as an essential strategy for enhancing the overall health and quality of life among the elderly. This chapter explores the multidimensional benefits of outdoor activities for older adults, integrating physical, mental, and social health aspects, and discusses the implications for policy and community planning.

Health Benefits of Outdoor Activities

Outdoor recreation provides numerous benefits in various aspects of health and well-being, making it a valuable component of a healthy lifestyle for individuals of all ages. Freedom of movement and the freshness of open spaces positively affect human health. The feeling of freedom and fresh air created by open spaces supports socialization also. Participating in outdoor physical recreation activities such as walking, cycling, swimming or team sports helps to improve cardiovascular health, increases muscle strength, flexibility and overall endurance, maintains muscle function and overall physical flexibility. This section examines both structured and unstructured physical recreation and their contribution to maintaining an active, healthy body.

A study by Pedersen and Saltin (2015) shows that regular muscle-strengthening activities, such as swimming, walking, reduce the risk of chronic diseases such as heart disease, type 2 diabetes and obesity. In addition to these benefits of outdoor recreation on physical health, it is also necessary to indicate that it develops motor skills and coordination, which are necessary to maintain independence as the age progresses. Outdoor activities increase exposure to sunlight, which is crucial for the body to produce Vitamin D. Adequate Vitamin D levels are essential for bone health and are linked to a lower risk of chronic disease (Holick, 2007). The outdoors improves mental health and leads to a reduction in stress and anxiety. Natural environments have been shown to reduce stress and anxiety levels. The presence of green spaces is associated with lower levels of cortisol, a stress-related hormone (Hartig et al. 2014).

As for the cognitive benefits of open spaces; regular interaction with nature is linked to improved cognitive functions, especially attention and memory capacities (Berman et al. 2008). Open spaces are conducive to enhanced social interactions, which are vital for mental health and well-being. Community gardens, parks and recreational trails provide opportunities for socialising and can help strengthen community ties (Kuo, 2003). Stodolska and Yi-Kook (2005), state that outdoor recreation promotes social integration and equality by providing inclusive opportunities for people of all ages and abilities. Regular exposure to natural environments increases environmental sensitivity, which leads to environmentally responsible behavior by increasing people's awareness and appreciation of the natural world (Chawla, 2007). The popularity of outdoor activities helps to increased efforts to protect and maintain natural landscapes and parks, which are essential for biodiversity and ecological health, ensuring the preservation of natural areas (Godbey et al., 2005). Activities in natural environments increase mood and general sense of happiness. Because nature has a unique ability to improve emotional health by providing a sense of peace and fulfilment (Ulrich et al., 1991).

Facing challenges during outdoor activities like hiking or rock climbing is effective in resilience, self-esteem, and personal development. Indeed, these activities often require overcoming obstacles, which builds confidence and coping skills (Ommundsen et al., 2011). In summary, outdoor recreation not only supports physical and mental health but also enhances social connections, encourages environmental stewardship, and contributes to emotional and psychological well-being. The wide-ranging benefits make outdoor recreation essential for the development of lifelong health and prevention of diseases.

Open Spaces: Vital for Aging Adults

Outdoor recreation, which includes a variety of physical activities and leisure pursuits in natural settings and fresh air, plays an important role in promoting health and well-being among older adults. Outdoor activities, such as hiking, bird-watching, or even walking in a park, offer numerous health benefits that are particularly substantial for older adults. These benefits extend beyond the physical, encompassing mental health improvements due to exposure to nature and social interaction opportunities. Regular outdoor activities help maintain mobility, flexibility, and balance, which are crucial for aging adults.

The natural resistance provided by walking on uneven surfaces, for example, strengthens the musculoskeletal system, reducing the risk of falls and fractures (Hunter-Jones, 2003). According to Warburton et al. (2006), individuals who engage in regular physical activity have approximately a 30 % lower risk of cardiovascular disease compared to those who are inactive. A study by Lacharité-Lemieux et al. (2015) highlights the importance of outdoor physical activity in reducing the risk of several chronic conditions prevalent among older adults, enhancing overall life expectancy and quality of life.

Social isolation and loneliness are significant risk factors for adverse health outcomes in older adults (Woolley et al., 2016) and, outdoor recreation, which offers substantial psychological and social advantages as well as physical activity, reduces this risk. Social interactions during these activities help reduce feelings of loneliness and isolation among seniors. Group activities in natural settings may increase feelings of belonging and purpose. For example educational programs about local flora and fauna, guided walks, and bird-watching tours can also provide cognitive stimulation and new learning opportunities, which are vital for maintaining mental acuity in older adults (Ommundsen et al., 2011). The walking for health initiative in the UK exemplifies how structured group walks in natural settings can significantly enhance social ties among participants, leading to improved social well-being and reduced loneliness (Hanson & Jones, 2015).

Regular exercise not only improves physical fitness but also enhances brain health, suggesting a critical role for physical activity in promoting healthy aging. In term of mental health, nature has a profound calming effect, helping to reduce stress levels and depression. Activities in green settings help decrease cortisol (a stress hormone) levels and improve mood by increasing endorphin production and serotonin flow (Gilbert & Abdullah, 2004). Research by Thompson Coon et al. (2011) found that activities in green spaces have a high potential for improving mental health, particularly in reducing symptoms of psychological distress among older adults. Physical activity, particularly of moderate to vigorous intensity, has been consistently linked to better cognitive health and a reduced risk of developing cognitive disorders like Alzheimer's disease (Srivastava, 2017). Engaging in physical activity in stimulating environments, like those offered outdoors, help enhance cognitive function in older adults. The combination of physical exertion and environmental complexity can lead to better concentration and memory retention (Steptoe et al., 2015).

Results and Suggestions

Recreation is not just an optional extra in human life, but an integral part of achieving and maintaining optimal health. The physical benefits of recreational activities are extensive and vital for maintaining a healthy lifestyle. Not only do these activities improve the obvious aspects of physical health, such as cardiovascular fitness and muscle strength, but they also enhance more subtle elements like immune system function, coordination, and balance. To maximize these benefits, it is recommended that individuals incorporate a variety of recreational activities into their regular routine, tailored to their personal health needs and preferences.

Engaging older adults in outdoor recreation emerging as a key approach to developing public health strategies that will promote health, longevity and quality of life. This chapter examines the diverse benefits of outdoor activities, highlighting their impact on physical, mental, and social health, while also discussing policy implications and the need for accessible outdoor environments. It includes case studies that showcase the transformative power of recreation on individual health. It also integrates research findings that provide empirical evidence of the health benefits associated with regular recreational engagement (Caldwell, 2005). Outdoor recreation, such as walking, swimming, and cycling, is crucial in maintaining cardiovascular health, muscle strength, and overall mobility, which are essential for particularly older adults' independence and well-being. Activities in natural settings not only aid in weight management but also help in mitigating chronic diseases such as diabetes, hypertension, and arthritis. The examples and studies cited support the claims made and inspire readers to incorporate more recreational activities into their own lives.

Research shows that travel and recreational preferences among seniors often include quieter, nature-based destinations, especially those that offer restorative experiences. Preferences for travel timing, type of travel, and companionship vary, with many opting for off-peak times to avoid crowds and reduce costs (Patterson et al., 2021). While many aging adults are interested in outdoor activities, several barriers reduce their participation. These include physical limitations, accessibility issues; lack of transportation, and sometimes, financial constraints. Addressing these barriers by providing more accessible outdoor spaces, offering senior-friendly travel packages, and improving public transportation can enhance engagement (Fleischer & Pizam, 2002). For policymakers and community planners, understanding the importance of outdoor recreation for aging populations is crucial. Enhancing access to parks and recreational areas and considering the needs of the elderly in urban planning can significantly impact public health (Patterson et al., 2021). By recognizing the specific needs and preferences of aging adults regarding outdoor recreation, communities can develop more effective programs that support healthy aging. Integrating these insights with broader health and social policies will help ensure that the benefits of outdoor recreation are accessible to all seniors, fostering a healthier, more active, and engaged aging population.

References

Berman, M. G., Jonides, J., & Kaplan, S. (2008). The cognitive benefits of interacting with nature. *Psychological Science*, 19(12), 1207–1212. DOI: 10.1111/j.1467-9280.2008.02225.x.

Caldwell, L. L. (2005). Leisure and health: Why is leisure therapeutic? *British Journal of Guidance & Counselling*, 33(1), 7–26.

Charles, S. T. (2020). Emotion regulation and aging. *Annual Review of Gerontology and Geriatrics*, 40(1), 15–28. https://doi.org/10.1891/0198-8794.40.15

Chawla, L. (2007). Childhood experiences associated with care for the natural world: A theoretical framework for empirical results. *Children, Youth and Environments*, 17(4), 144–170.

Coalter, F. (2015). The social benefits of sport: Where's the proof? In E. Sherry & K. S. Gelsthorpe (Eds.), *Sport and Social Capital*, 24–3, Elsevier.

Fleischer, A., & Pizam, A. (2002). Tourism constraints among Israeli seniors. *Annals of Tourism Research*, 29(1), 106–123.

Gilbert, D., & Abdullah, J. (2004). Holidaytaking and the sense of well-being. *Annals of Tourism Research*, 31(1), 103–121.

Godbey, G., Caldwell, L., Floyd, M., & Payne, L. (2005). Contributions of leisure studies and recreation and park management research to the active living agenda. *American Journal of Preventive Medicine*, 28(2), 150–158. DOI: 10.1016/j.amepre.2004.10.027.

Hanson, S., & Jones, A. (2015). Is There Evidence That Walking Groups Have Health Benefits? A Systematic Review and Meta-Analysis. British Journal of Sports Medicine, 49, 710–715. https://doi.org/10.1136/bjsports-2014-094157.

Hartig, T., Mitchell, R., de Vries, S., & Frumkin, H. (2014). Nature and health. *Annual Review of Public Health*, 35, 207–228. DOI: 10.1146/annurev-publhealth-032013-182443.

He, S., & Sharpless, N. E. (2017). Senescence in health and disease. *Cell*, 169(6), 1000–1011. https://doi.org/10.1016/j.cell.2017.05.015

Holick, M. F. (2007). Vitamin D deficiency. *New England Journal of Medicine*, 357, 266–281. DOI: 10.1056/NEJMra070553.

Hunter-Jones, P. (2003). Health and pleasure from the mature tourist's perspective. *Tourism Management*, 24(6), 709–716.

Kennedy, B. K., Berger, S. L., Brunet, A., Campisi, J., Cuervo, A. M., Epel, E. S., … & Rando, T. A. (2014). Geroscience: linking aging to chronic disease. *Cell*, 159(4), 709–713. https://doi.org/10.1016/j.cell.2014.10.039

Kuo, F. E. (2003). The role of arboriculture in a healthy social ecology. *Journal of Arboriculture*, 29(3), 148–155.

Lacharité-Lemieux, M., Brunelle, J. P., & Dionne, I. J. (2015). Adherence to physical activity recommendations and the influence of socio-demographic correlates - a population-based cross-sectional study. *BMC Public Health*, 15, 1140.

Longo, V. D., & Panda, S. (2016). Fasting, circadian rhythms, and time-restricted feeding in healthy lifespan. *Cell Metabolism*, 23(6), 1048–1059. https://doi.org/10.1016/j.cmet.2016.06.001

López-Otín C, Blasco MA, Partridge L, Serrano M, Kroemer G. Hallmarks of aging: An expanding universe. Cell. 2023 Jan 19;186(2):243–278. doi: 10.1016/j.cell.2022.11.001. Epub 2023 Jan 3. PMID: 36599349

López-Otín, C., Galluzzi, L., Freije, J. M. P., Madeo, F., & Kroemer, G. (2023). Metabolic control of longevity. *Cell*, 166(4), 802–821. https://doi.org/10.1016/j.cell.2016.07.031

Milman, S., & Barzilai, N. (2016). Dissecting the mechanisms underlying unusually successful human health span and life span. *Cold Spring Harbor Perspectives in Medicine*, 6(1), a025098. https://doi.org/10.1101/cshperspect.a025098

Ommundsen, Y., Klasson-Heggebø, L., Anderssen, Patterson, I., & Pegg, S. (2011). Marketing to the ageing consumer: The secrets to building an age-friendly business. *Palgrave Macmillan*, 3(1), 32. DOI: 10.1186/1479-5868-3-32.

Patterson, I., Balderas-Cejudo, A., & Pegg, S. (2021). Tourism preferences of seniors and their impact on healthy ageing. *Anatolia*, 32(4), 553–564.

Pedersen, B. K., & Saltin, B. (2015). Exercise as medicine - evidence for prescribing exercise as therapy in 26 different chronic diseases. *Scandinavian Journal of Medicine & Science in Sports*, 25(S3), 1–72.

Raj K, Horvath S. Current perspectives on the cellular and molecular features of epigenetic ageing. Experimental Biology and Medicine. 2020;245(17):1532–1542. doi:10.1177/1535370220918329

Rowe, J. W., & Kahn, R. L. (1997). Successful aging. *The Gerontologist, 37*(4), 433–440.

Srivastava, S. (2017). "The Mitochondrial Basis of Aging and Age-Related Disorders" *Genes* 8, no. 12: 398. https://doi.org/10.3390/genes8120398

Stebbins, R. A. (2007). Serious leisure: A perspective for our time. New Brunswick, *NJ: Transaction Publishers.*

Steptoe, A., Deaton, A., & Stone, A. A. (2015). Psychological wellbeing, health and ageing. *The Lancet*, 385(9968), 640–648.

Stodolska, M., & Yi-Kook, J. (2005). Ethnicity, immigration, and constraints. *Journal of Leisure Research,* 37(4), 466–489.

Thompson Coon, J., Boddy, K., Stein, K., Whear, R., Barton, J., & Depledge, M. H. (2011). Does participating in physical activity in outdoor natural environments have a greater effect on physical and mental wellbeing than physical activity indoors? *Environmental Science & Technology*, 45(5), 1761–1772.

Ulrich, R. S., Simons, R. F., Losito, B. D., Fiorito, E., Miles, M. A., & Zelson, M. (1991). Stress recovery during exposure to natural and urban environments. *Journal of Environmental Psychology*, 11(3), 201–230. DOI: 10.1016/S0272-4944(05)80184-7.

Warburton, D. E., Nicol, C. W., & Bredin, S. S. (2006). Health benefits of physical activity: the evidence. *Canadian Medical Association Journal*, 801–809. DOI: 10.1503/cmaj.051351.

Woolley, H., Rose, R., & Carmona, M. (2016). The Value of Public Space: How High Quality Parks and Public Spaces Create Economic, Social and Environmental Value. *CABE Space.*

World Health Organization. (2015). *World report on ageing and health.* WHO. Retrieved from https://www.who.int/ageing/events/world-report-2015-launch/en/

World Health Organization. (2024). *Ageing.* https://www.who.int/health-topics/ageing#tab=tab_1

H. Tanju Ipek[*] and Çağla Özer[†]

Chapter 6 From the Perspective of Viticulture and Wine in Anatolia, Terroir, Local Wine Grapes: Wine Tourism in Turkey

Introduction

Turkey is one of the world's leading countries in terms of viticulture and grape production. Unlike world trends, Turkey uses half of its grape production for fresh and table use, 39 % for drying, and about 11 % to produce wine and other alcoholic spirit (TEPGE, 2021). Anatolia is home to many important grape varieties with a long history of viticulture and winemaking. Grapes bear the contributions of Anatolian civilizations in production, processing techniques, storage and ripening techniques, trade, and even the naming of products.

Among the grapes produced and utilized in various ways today, Turkey has important varieties used in wine production and the wines have recently been receiving successful results in competitions abroad and have production regions and regions that attract the attention of domestic and foreign wine tourists. In this study, the historical development of viticulture and winemaking in Turkey, Old World and New World approach in wine, terroir in vineyard and wine, wine tourism issues are examined and the possibilities of developing wine tourism in Turkey are evaluated.

Viticulture and Winemaking in Anatolia in Ancient Times

It is estimated that the lands where wild vines were cultivated and wine was first produced were the Southern Caucasus, the southern coasts of the Caspian Sea and Anatolia (Aktan & Kalkan, 2000). The research reveals the existence of findings showing the prevalence and development of viticulture and wine culture within the triangle of Georgia, Armenia, and Eastern Anatolia.

* PhD., İstinye University, Faculty of Fine Arts, Design and Architecture, Department of Gastronomy and Culinary Arts, tanju.ipek@istinye.edu.tr

† Associate Professor, İstinye University, Faculty of Fine Arts, Design and Architecture, Department of Gastronomy and Culinary Arts, cozer@istinye.edu.tr

Biomolecular Archaeologist Patrick E. McGovern and his colleagues, during excavations in 1968 at the Neolithic settlement of Haji Firuz Tepe near Lake Urmia (Iran), found a layer of yellowish and reddish residues of tartaric acid and calcium tartrate in terracotta jars dated to 5400–5000 BC (Doğer, 2004). These substances, which are found only in grapes in nature, are the oldest evidence of wine culture, indicating that wine was present and stored in these vessels.

The Hattians and then the Hittites, one of the oldest known civilizations of Anatolia, represent a very advanced civilization in agriculture, viticulture, and mining within the Kızılırmak arc since 2500 BC. Golden bowls, beaklike jugs and stemmed glasses unearthed during the excavations in Alacahöyük are evidence of a sophisticated wine civilization (Doğer, 2004). There are articles about agriculture and viticulture in the famous Hittite laws. The 100-article 2nd Series of the Hittite laws, starting with the phrase "If a vineyard", includes actions that harm viticulture and the penalties to be given to those who do this (Reyhan, 2010). Around the Sakarya River in western Central Anatolia, tribes coming from Thrace settled in the region starting around 1200 BC. In Gordion, the capital of this civilization known as the Phrygians, McGovern et. al (1996) have revealed that the ruins in the banquet vessels found during excavations in the tumulus, the tomb of King Midas, were a mixture of wine, beer, and honey (Kocabaşoğlu, 2007). The homeland of the Lydians, neighbors of the Phrygians in western Anatolia, was a land revered by the Anatolian god Dionysus. Dionysus is the god of vineyards, grapes, vintage, other fruits, and virgin nature.

The people who came to Western Anatolia during the Aegean migrations (1150–1050 BC) and settled on the Aegean coasts and the Aegean islands established the oldest vineyards of the ancient period in the lands from Çanakkale to Antalya. These people, who were engaged in agriculture in the fertile lands of Western Anatolia, prospered with the grape, wine and olive oil products they produced and exported.

During the Roman Republican and Roman Imperial periods, wines produced in the Aegean islands and Anatolia were imported to meet the increasing consumption of wine. The settlement that exported the highest amount of wine in Anatolia was Klazomenai (Urla İskele). In addition, wines from Katakekaumene (Burnt Country), Telmessus (Fethiye), Mesogis Mountains (Aydın), Knidos (Datça Peninsula) were exported (Strabon, n.d.)

Viticulture and Winemaking in Anatolia in the Middle Ages

At the beginning of the Middle Ages, viticulture and winemaking were powerful and effective economic activities in the Balkans, Thrace, Aegean Islands,

Anatolia, and the Middle East. A significant portion of the wines produced were exported as prestigious products to European countries with growing demand for wine. The grapes grown at that time and today on the islands of Samos (Samos) and Limni produce one of the Aegean's famous sweet wines, vinum moscatos (Dalby, 2014). This grape, Bornova Misketi, is still produced today in the vineyards of Izmir and the Aegean Region and a wine of the same name is made with rose and floral aromas.

Famous for their wines in antiquity, the wines of the islands of Chios (Sakız) and Lesbos (Midilli) were still acclaimed by travelers during the Byzantine period and even during the Ottoman Empire (Dalby, 2014). Likewise, the wines from of Crete, Samos (Sisam) and ganitic wines from Ganos Mountain (Tekirdag) are the best wines (Dalby, 2014). The region between the Ganos Mountains and the Sea of Marmara is still one of the most important viticultural and wine-making centers of Thrace, with Kumbağ in the east and Gaziköy, Hoşköy, Mürefte and Şarköy in the west. In Southern Marmara, Triglia (Tirilye) and Nikea (Iznik) produce important wines of Bithynia.

Viticulture and Wine Production in the Ottoman Empire

During the establishment and expansion period of the Ottoman Empire, viticulture and wine production were important economic indicators of the people living in Anatolia and the Balkans. The grapes produced by the Muslim population were consumed fresh for the table or utilized in various ways, especially dried. Non-Muslim populations, on the other hand, cultivated wine grape varieties in their vineyards and produced wine from them. Viticulture and wine production were particularly important in the Balkans, Thrace, and Western Anatolia. Some of the wines produced were consumed domestically, while others were exported (Yıldırım, 2021).

In the second half of the 19th century, phylloxera (*Viteus vitifolii*),an important grapevine weevil, was first seen in France in the early 1860s and spread to Europe in the face of the rapid spread of the disease in vineyards and the magnitude of the disaster, European countries organized a congress on the control of phylloxera in Lausanne in 1877. The Ottoman Empire was not a member of this congress and participated with an observer delegation and the Phylloxera Regulation was issued, determining the measures to be taken in 1880 (Keskin, 2015). In the early 1880s, phylloxera spread from Europe to Ottoman lands and was first seen in the Thrace region and then in Istanbul vineyards. From there, it spread rapidly to Izmit, Bursa, Eskişehir, Kütahya, İzmir, Manisa, Balıkesir, Denizli, Afyonkarahisar and Ankara vineyards (Özgün, 2019). Between 1885 and 1902, 550,000 acres of the 762,000 acres of vineyards in 20 provinces of Aydın Province were destroyed by phylloxera. (Keskin, 2015). Efforts to prevent the phylloxera disease began to give positive results in 1887. Due to the persistence of the phylloxera disease, measures to prevent the spread of phylloxera were adopted with the Phylloxera Law enacted in 1911, including the distribution of American grapevine rods to producers free and the provision of financial support by the Ziraat Bank to damaged growers (Keskin, 2015).

Viticulture and Wine in Anatolia in the Republican Era

In 1923, after the establishment of the Republic, the founders of the young republic started a new struggle to solve many economic and social problems of the country. The phylloxera in that viticulture, which is an important economic activity of the country, faced in the last quarter of the 19th century was not yet completely over. To find scientific and practical solutions to the problems and to produce American grapevine rootstocks, Viticulture Research Institutes were established in Manisa and Tekirdağ in 1930. During the Ottoman Empire, small-scale wine production units, especially those established or operated by non-Muslims, where in the regions of wine grapes were traditionally produced. With the Republic, Turkish entrepreneurs also began to establish wine-producing businesses.

The Ottoman-era İnhisarlar Administration was reorganized in 1941 as Tekel Genel Directorate in 1946. Until 1983, it continued to produce alcoholic beverages along with wine, among other duties. In 1987, the structural changes in Tekel entered a new phase with the privatization of Tekel in 2001. In the same years, new entrepreneurs started to invest in vineyard and wine production facilities in addition to the existing organizations of the private sector. In early 2004, Tekel was sold to a consortium and the state withdrew from alcoholic beverage production.

Wine Production in Turkey and the World Today

As it is known, viticulture is a productive and economically important sector in the world and is practiced in temperate climates (average annual temperature of 10–20°C). The regions in the world that meet these conditions are located between 30th-50th latitude in the Northern and Southern hemispheres. Undoubtedly, the duration of sunlight utilization, protection from frost, air flow and well-drained slopes have an impact on yield and quality. In addition, climate change and environmental problems are issues that may cause significant changes in viticulture regions.

According to OIV (International Organisation of Vine and Wine) statistics (2019), total vineyard area in the world is 7.4 million ha and total grape production is 78 million tons. Turkey ranks 5th in the world in terms of vineyard area size with 431 thousand ha and 6th in the world in terms of grape production with 4.1 million tons. While 36 % of the total grapes produced in the world are consumed as table and fresh, 57 % is used in wine production and 7 % is used for drying. The types and utilization of grapes produced in Turkey are different from global trends; 50 % is used for table and fresh, 39 % for drying and 11 % for wine and other alcoholic beverages (TEPGE, 2021).

By 2023, the amount of wine (including sparkling wine) produced in Turkey is 66.6 million liters and the amount of wine (including sparkling wine) imported is 5.9 million liters. Interestingly, since 2021, the amount of wine produced in Turkey has decreased, while the amount of imported wine has increased.

Old World and New World in Wine

Cultivated between the south of the Caucasus and Anatolia, the vine, viticulture techniques and the first wine produced developed over thousands of years in the hands of the peoples of Anatolia and the Middle East and reached European lands, first with the colonies of the Ancient Greek civilization and then with the contributions of the Roman Empire. European countries, under favorable climatic conditions, have maintained and developed viticulture and wine production as an important economic activity. They established and implemented techniques and rules for the development of viticulture and winemaking, such as the establishment of vineyards, the breeding and development of wine grape varieties, the production of new varieties, the development of processing techniques and equipment in wine production, maturation and maturation containers, wine styles and new products, origin control and development in winemaking. The church and the monasteries played an important role in this development, especially from the beginning. Thus, European countries transferred their knowledge and experience in various ways to the countries of the world that started viticulture and winemaking later. Due to this pioneering role in viticulture and wine techniques, these countries are referred to as "Old World Winemaking" countries. The fact that the recently discovered lands were called "New World" during the period of geographical discoveries must also have contributed to this naming. These countries are mainly France, Italy, Spain, Portugal, Germany, Italy, Italy, Spain, Spain, Portugal, Germany, and other European countries engaged in viticulture and winemaking. Some sources add the countries of the Middle East and the Mediterranean Region, which have a significant viticulture and winemaking history.

On the other hand, when we examine the tasting characteristics of old-world wines in general, we can say that these wines have a lighter body, acidity and minerality and lower alcohol and fruit flavors (Puckette, n.d.) We should consider that these characteristics are defined in comparison with new world wine and are a general assessment.

In developing their own viticulture and winemaking, the countries that created "New World" winemaking undoubtedly added to their winemaking techniques and traditions what they had learned from the Old World. They brought grape varieties that they did not have, adapted them to their own soils and developed local varieties. The European immigrants who settled in California, the leading winegrowing and winemaking state in the USA, brought new grape varieties and their experience. California developed its own wine industry by taking advantage of the knowledge of European countries, which were advanced in vineyard and wine techniques, and achieved many innovations. The proof of its success was an interesting competition.

Steven Spurrier, a British wine expert and writer, owner of a wine shop and wine school in Paris, persuades California winemakers and French winemakers to organize a competition[‡] in Paris in 1976. At the end of the competition, California wines be the first both in the white wine and in the red wine category. The results of this competition led to important developments in world winemaking. We can consider the results of the competition and the worldwide interest it generated as one of the leading signs of globalization in winemaking. Considering the characteristics of the wines, we can make an overall assessment of Old World and New World winemaking in comparative terms;

- The Old World generally offers elegant wines from cooler climates, while New World wines are full-bodied with a strong fruit flavor from warmer regions.
- Old World wines are characterized by region and terroir; New World wines are characterized by grape varieties.
- Old World wines are characterized by tradition and traditional methods; New World wines are characterized by technological development.

‡ For the jury of the Paris Wine Tasting and the results of the competition see: Kırım (2006) Paris Wine Tasting and the Globalization of Wine (Hürriyet Gazetesi 01.04.2006). The story of the competition was filmed in 2008 by director Ranndall Miller, actor Alan Rickman playing the role of Steven Spurrier.

- The Old World considers that environmental factors, broadly defined as terroir, lead to the production of different wines; in the New World the winemaker is important.

Terroir in Vineyard and Wine

Terroir, which comes from the French root terre (earth), is used as terroir (teruar). A terroir can be defined as the part of a region's soil, landforms (topography), climate, traditions and culture that gives a product its characteristic features. Although the size of a terroir varies, it is the part of a region that bears the characteristic features of a region. In this way, there can be more than one terroir within a region, or a whole region can be a terroir. Since our subject is viticulture and wine production, the terroir is characterized by the type and permeability of the soil, its height above sea level, direction and slope, summer-winter temperature difference, day-night temperature difference, number of sunny days, annual rainfall, and other vegetation. In addition, terroir is very important in old world wine producing countries, as traditions, technology and equipment are also important for the expected characteristics of the wine produced.

In contrast, New World wine-producing countries have achieved successful results, initially by applying the knowledge and techniques they learned from Old World winemakers, and later by developing modern techniques against the rules and regulations of the Old World in the winemaking process, giving themselves more freedom in the winemaking process. For example, Malbec, Carménéré and Sauvignon blanc from the Bordeaux region have gained an important place in world winemaking with new styles and characteristics in Argentina, Chile, and New Zealand respectively.

Wine Grape Varieties in the World and in Turkey

Internationally recognized grape varieties that are widely produced, Sauvignon Blanc, Chardonnay and Riesling among the white varieties, Cabernet Sauvignon, Merlot, Syrah, and Pinot Noir among the red varieties. Undoubtedly, internationally recognized wine grape varieties are not limited to these. Robinson & Hardings (2015) has identified 1368 wine grapes around the world with their flavors, characteristics, and origins. Table 1 is a summary chart prepared with the information provided in this source, where 26 wine grapes from Turkey are introduced.

In Turkey, public and private sector organizations, researchers, working groups, scientific and practical studies are carried out on the wine characteristics of the existing grapes and the wine grapes that are facing extinction today, and the results are discussed in many congresses and symposiums[§]. As a result of these and similar studies Acıkara (Elmalı, Antalya Region), Barburi (Hatay Region), Patkara and Gök Üzüm (Toros Mountains, Mersin Region), Beylerce (Bilecik Region), Ten Göynek (Beyşehir Region). Kerküş, Mazrona, Midin Karası (Southeastern Anatolia Region) are wine grape varieties that existed in the past and are facing the threat of extinction.

Wine Tourism (Oenotourism)

Towards the end of the 20th century, the explosion of demand and intensity due to package tours organized by tour operators caused reactions due to their negative impacts on the environmental, ecological, social and cultural environment (Kozak & Bahçe, 2012). In the same period, we also see that tourists who are tired of participating in coastal tourism, which consists of sun, sand and sea, are looking for different activities and adventures (Holloway & Humphreys, 2016). The concept of sustainability, which entered the world's agenda with the 1992 Rio Conference, has brought sustainable and alternative tourism issues in tourism. The "sustainable tourism" approach brought to the agenda by the World Tourism Organization (UNWTO) has led to an increase in alternative tourism types and widespread awareness on the subject.

Wine tourism, which is among alternative tourism types, is evaluated within special interest tourism. It is associated with gastronomy tourism, rural tourism, agro-tourism, eco-tourism, and cultural tourism considering its various characteristics, interests and connections. On the other hand, wine tourism is a concept and product that is undergoing a process of development and is one of the interrelationships between the tourism industry and the wine industry (Hall et al., 2011).

A comprehensive definition of wine tourism is "an important member of the of agrotourism, located at the intersection of culture and gastronomy tourism, which allows the discovery of vineyard products (that can include direct product sales), nature, history and heritage including accommodation, natural activities and varied gastronomic experiences that is a sustainable, high value-added, boutique and human-scale form of tourism" (Hall et al., 2011). Robinson & Hardings

§ The first Heritage Grapes of Anatolia–Root, Origin Soil Conference was held in 2019 The aim of the organizers and participants of the symposium which was held with the participation of academicians, wine professionals and producers on the subject is to protect local grapes with thousands of years of history and value.

(2015) attributes the increasing importance of wine tourism today primarily to the growing interest in wine and travel, the appeal of aesthetically attractive vineyards and winemaking, the suitability of the climate for tourism when grapes are grown and wine is produced, the need for urban dwellers to connect with nature, the privilege of gaining specialized knowledge about wine, and the opportunity to taste and purchase affordable and unique wines

Thomas Jefferson, the 3rd President of the United States of America, traveled to the famous wine regions of France during his term as the Ambassador of the United States of America to France (1785–1789), inspected vineyards and famous chateaus, conducted tastings in these establishments and purchased wines for his cellar (Wallace, 2013). Considering the definitions of wine tourism, we can argue that Jefferson was one of the first wine tourists.

Positive developments in Europe in the second half of the 19th century, increased prosperity and safer travel conditions encouraged increased travel. During the 1855 Paris International Exhibition, at the request of Napoleon III, the "Bordeaux Wine Classification" was officially announced for the first time, reinforcing the concept of terroir in wine, and helping to use it as a marketing tool for the region (Hall et al. 2011). This classification forms the basis of the Appellation d'origine Côntrolée (AOC) applied today in France.

In parallel with the development of tourism after 1950s, thematic cultural routes were established. For example, the German Wine Route (Deutsche Weinstrasse), which was created to promote wine tourism, is about 85 km long and the centers on this route host many events, including vineyards. In 2009, the Council of Europe registered the Iter Vitis Route as a Council of Europe Cultural Route. This route, which starts in the UK and goes all the way to Azerbaijan, covers a large part of the European countries (22 countries).

Wine Tourism in Turkey and Its Development

In Turkey Doluca Inc.Co.(1926) and Kavaklıdere Şarapları Inc.Co. (1929), which were founded in the early years of the Republic, continue their production and marketing activities. Recently, two separate investments made by Kavaklıdere Inc.Co. in Bordeaux/France (2016) are very important for the development of the national winemaking and for to access the global market. In addition to these two reputable companies, Turkey have valuable sectoral organizations that contribute to the national economy, agriculture, and industry. Their success in the national and international markets, despite all the difficulties they face, is a source of pride. In these lands that cultivated the grapevine, if one can mention about Turkish winemaking in the future, they will undoubtedly have the biggest share in it. In 1986, the changes initiated in the Tekel Administration were completed with the sale of the alcoholic beverage division to the private sector in 2004. These were the years when the private sector started to make new investments in viticulture and winemaking with modern and scientific methods.

Individual or group tours to vineyards and wineries and vintage festivals are popular activities. To develop these activities in a certain order, Thrace Vineyard Route, which was established in 2013 by 12 organizations operating in this region with the support of the Thrace Development Agency and the Thrace Tourism Operators Association, is the first initiative in Turkey in this sense. The purpose of the organization is to contribute the development of the tourism sector in the region and to ensure the sustainability of the social and economic foundations as well as the environment by determining the wine routes throughout the Thrace Region, mobilizing the tourism potential it has, increasing the number of domestic and foreign tourists. There are 12 organizations, 7 vineyard houses/hotels, 12 restaurants/cafes operating on the Trakya Vineyard Route, whose awareness and popularity are increasing by the day, host their visitors and promote their products (Gündoğan & Yankı, 2021).

The Urla Vineyard Route Association, to promote the vineyards and winemaking establishments in the Urla district of Izmir, created the Urla Vineyard Route in 2016, which includes 10 wine producing and associated gastronomy and accommodation establishments in the region. In Denizli, which is an important viticulture and winemaking region of Turkey, the Çal Vineyard Route, which was established in 2022, includes 4 producers in the region. Producers have 2 restaurants and tasting areas. Considering the number of producers and potential in the region, the Çal Vineyard Route will attract more wine tourists in the process. Apart from these, viticulture and winemaking organizations operating in the Aegean, Mediterranean, Central, Southeastern and Eastern Anatolia regions continue their production activities, host visitors, and try to promote their activities and products.

Results

We have tried to briefly present the historical development of viticulture and winemaking in Turkey, where grapevine cultivation, viticulture, winemaking, and other viticultural products have been an important agricultural and economic activity throughout history. In this section, we try to evaluate the current situation and development opportunities of viticulture and winemaking in Turkey in terms of wine tourism as the land has a very special position in terms of the past and future of viticulture and winemaking.

First, Turkey is the 5th largest country in the world in terms of vineyard area and the 6th largest country in grape production but lost approximately 20 % of vineyard areas in the last 10 years (OIV, 2023). According to TADAB (Republic of Türkiye Ministry of Agriculture and Forestry Tobacco and Alcohol Directorate), data, wine production has been decreasing since 2021, while wine import have been increasing (TADAB, 2023). 1368 grape varieties used in wine production in the world and 26 known varieties are specific to Turkey. In addition to this, relevant public and private sector organizations, researchers and working groups have recently working on more than 10 varieties that are about to disappear. On the other hand, the important varieties, which are known and recognized in Turkey but not abroad, have started to achieve significant success in their promotion in recent years. The producers of these grapes and wines, the region and terroir where they grow, and the characteristics of the grapes and wines need to be promoted. Although the Vineyard Routes realized in recent years are important and useful activities but not sufficient when we consider the existing potential.

The importance of wine tourism is increasing in the world and in our country. Vineyards as an agricultural wealth, grape varieties produced, unspoiled, high quality production regions and terroirs attract the attention of wine tourists and visitors. They also participate in wine tours to see, learn, and taste wine making, wine varieties and styles, meet and converse with vineyard and wine producers, and learn about vineyard and wine culture. They want to buy wines they like or are interested in to put in their cellars. On the other hand, Article 20.1 and 4 of the regulation on the procedures and principles regarding the sale and presentation of tobacco products and alcoholic beverages stipulates that alcoholic beverages cannot be advertised or promoted or distributed to consumers in any way (TADAB, 2011). Considering this regulation what will the business officer who is prevented from providing information and promoting the products to wine tourists and visitors coming to vineyard and winery company do and how will he be able to turn away a guest who wants to taste the product in a winery. It would be appropriate to review the regulation on behalf of wine tourism, wine tourists and producers.

Many countries in the world that are in the wine-growing and winemaking zone utilize their vineyards and wine production facilities in tourism development activities. In Turkey, the existing vineyard regions attract wine tourists with their unique landscapes, valuable and rich grape varieties, and excellent wineries. In addition to these values, the natural, historical, and cultural assets and gastronomic richness will support the development of wine tourism.

Table 1. Wine Grape Varieties of Turkey (Alphabetical)

Row	Type	Color	Region
1	Adakarası	Red	Avşa Island, South Marmara
2	Boğazkere	Red	Diyarbakır, Southeastern Anatolia
3	Çalkarası	Red	Çal, Denizli
4	Çavuş	White	Bozcaada. Aegean Coasts
5	Dimrit	Red	Nevşehir, Kayseri, Niğde
6	Emir	White	Kapadokya
7	Foça Karası	Red	Foça, Izmir Region
8	Hasandede	White	Kırıkkale, Nevşehir-Kapadokya
9	Horozkarası	Red	Kilis, Gaziantep Region
10	İri Kara	Red	Hadim, Konya
11	Kalecik Karası	Red	Kalecik, Kızılırmak Coasts, Ankara.
12	Karalahna	Red	Bozcaada, Çanakkale
13	Karasakız (Kuntra)	Red	Bozcaada, Çanakkale
14	Kolorko	White	Uçmakdere-Şarköy Region
15	Köhnü	White	Arapgir, Malatya
16	Kösetevek	Red	Elazığ
17	Merzifon Karası	Red	Merzifon, Amasya Region
18	Narince	White	Tokat Region
19	Öküzgözü	Red	Elazığ Region, Manisa ve Denizli Region
20	Papazkarası	Red	Thrace Region
21	Sidalan	White	Bozcaada, Güney Marmara
22	Sultaniye	White	Manisa, İzmir, Denizli Region
23	Urla Karası	Red	Urla
24	Vasilaki	White	Bozcaada, Çanakkale
25	Yapıncak	White	Şarköy-Mürefte, Tekirdağ Region
26	Yediveren	Red	Şanlıurfa, Southeastern Anatolia

References

Aktan, N., & Kalkan, H. (2000). Şarap Teknolojisi, Kavaklıdere Eğitim Yayınları, No: 4, Ankara.

Dalby, A. (2014). Bizans'ın Damak Tadı, Çev. Ali Özdamar, Alfa/Tarih-Yemek No:40, İstanbul.

Doğer, E. (2004). Antik Çağda Bağ ve Şarap, İletişim Yayınları. İstanbul.

Gündoğan, G., & Yankı, M. (2021). Türkiye Önoturizm Rehberi. Alfa Basım Yayım. İstanbul.

Hall, C. M., Sharples, L., Cambourne, B., Macionis, N., Mitchell, R., & Johnson, G. (2011). Wine Tourism Around the World Development, Management and Markets. Routledge, Taylor & Francis Group. London.

Holloway, J. C., & Humpreys, C. (2016). The Business of Tourism (10th Ed.) Pearson Education, Caanada.

Keskin, Ö. (2015). Üzümün Bağı Asmanın Kurdu: Osmanlı İmparatorluğu'nda Filoksera ile Mücadele. *Tarih İncelemeleri Dergisi,* 30(2), 479–505.

Kırım, A. (2006). 1976 Paris Şarap Tadımı ve şarabın küreselleşmesi. Hürriyet Gazetesi (04.01.2006)

Kocabaşoğlu, Y. E. (2007). Eğitim Seminer Notları, Veritas Ltd. Şti.

Kozak, M., & Bahçe, A., S. (2012). Özel İlgi Turizmi (2. baskı) Detay Yayıncılık Ankara.

McGovern, P. E., Glusker, D. L., Exner, L. J. & Voigt, M. M. (1996). Neolithic resinated wine. *Nature* 381, no. 6582, 480–481.

OIV (2019). International Organisation of Vine and Wine, Statistical Report on World Vitiviniculture.

OIV (2023). International Organisation of Vine and Wine, Statistical Report on World Vitiviniculture

Özgün, C. (2019). Osmanlı'dan Cumhuriyet Türkiye'sinin İlk Yıllarına Filoksera Hastalığının Batı Anadolu'daki Ekonomik Sonuçları Üzerine Tespitler. 3.İktisat Tarihi Kongresi Bildirileri, İzmir Demokrasi Üniversitesi Fen Edebiyat Fakültesi.

Puckette, M. The Real Differences Between New World and Old World Wine. https://winefolly.com/deep-dive/new-world-vs-old-world-wine.

Reyhan, E. (2010). Hititlerde Gündelik Hayata Dair İktisadi İlişkiler. *Tarih Araştırmaları Dergisi* 29(47), 65–82.

Robinson, J., & Hardings, J. (2015). The Oxford Companion to Wine, Oxford Univesity Press 4th Ed, Oxford, UK.

Strabon Antik Anadolu Coğrafyası Kitap: XII-XIII-XIV (Adnan Pekman, Çev.) Arkeoloji ve Sanat Yayınları (3. Baskı) 1993.

TADAB (2023). Tütün ve Alkol Dairesi Başkanlığı, Alkol ve Alkollü İçkiler Daire Başkanlığı Piyasa İstatistikleri.

TADAB (2011). Tütün Mamulleri ve Alkollü İçkilerin Satışına ve Sunumuna İlişkin Usul ve Esaslar Hakkında Yönetmelik. Resmi Gazete Sayısı: 27808 (18.09.2013 Resmi Gazete Sayısı: 28769 yapılan değişiklik dahil).

TEPGE, Tarımsal Ekonomi Politika Geliştirme Enstitüsü (2021). Tarım Ürünleri Piyasaları Raporu.

Yıldırım, F. (2021). Mühimme Defterleri'ne Göre Osmanlı Devleti'nde Şarap. *Fırat Üniversitesi Sosyal Bilimler Dergisi,* 31(1), 495–514.

Wallace, B. (2013). Milyarderin Sirkesi, Dünyanın en pahalı şarap şişesinin esrarı. Kitabix Yayınları İstanbul.

Tuğçe Özoğul Balyali*

Chapter 7 Reflections of Project-Based Learning in Recreational Material Development

Introduction

Project-based learning is a student-centered form of teaching characterized by students acting independently, setting goals, researching, collaborating, communicating, and experiencing real-world implementations. The most important feature of project-based learning is the generation of creative thinking by students in classroom environments. This research aimed to determine student opinions about project-based learning implemented in the Recreational Material Development course.

The reason why the project-based learning method was chosen in the Recreational Material Development course is that this course must be practice-based. The researcher taught the course researcher and the entire course-created content. First of all, the course content was determined and weekly course plans were prepared. Afterward, the implementation phase started. The implementation continued for one semester (2022–2023 Spring semester). In implementing the course, theoretical information that can be used in developing recreational materials is first conveyed. Afterward, the students were given a project proposal form and asked to design their projects. Students were asked to design projects that would contribute to the development of recreational activities in the city they live in and to make models of their projects from waste materials. In the evaluation of the course, the presentation of the project proposal and the presentation of the completed projects in the final stage were used. In other words, a note was given for implementation. However, this evaluation is not within the scope of the research. The qualitative research method was used in the research. A semi-structured interview form was used to collect data. The sample of the research consisted of all students who took the Recreational Material Development course in the 2022–2023 Spring semester. Content analysis was used to analyze the data. Themes were created from the data obtained and discussed in line with

* Assoc. Dr., Van Yüzüncü Yıl University, Faculty of Tourism, Department of Tourism Guidance, tugceozogulbalyali@yyu.edu.tr

the relevant literature. With project-based learning, students can be motivated and encouraged to take interest in the course. Additionally, this method can contribute to the development of many skills. This research may offer ideas to educators and practitioners about using the project-based learning method in their courses.

Project-Based Learning

Project-based learning is an active student-centered form of teaching characterized by students' autonomy, collaboration, goal setting, communication, constructive investigations, and reflection on real-world implementations (Kokotsaki, Menzies, & Wiggins, 2016). This learning method was initially developed to solve real-life problems using various educational techniques such as the laboratory method. Instead of creating an education program based on disciplines, project-based learning is a teaching approach based on the search for solutions to real problems that exist in human life (Toci, 2000). While teaching courses with project-based learning, original thinking should be developed in the classroom. By associating the method with daily life, students can be motivated and encouraged to be interested in the course (Alacapınar, 2008).

Courses such as Recreational Material Development should be conducted in a student-centered manner. In such courses, rather than transferring theoretical knowledge, a method should be used in which students are encouraged to think creatively and can use real-life information and observations. Therefore, one of the methods that can create a suitable learning environment is the project-based learning method. In the research, the answer to the question "How does teaching a course with the project-based learning method affect students' views on the course?" was sought.

Methods

Course Design

The reason why the project-based learning method was chosen in the Recreational Material Development course is that this course must be practice-based. The researcher taught the course researcher and the entire course-created content. First of all, the course content was determined, and weekly course plans were prepared. Afterward, the implementation phase started. The practice continued for a period. The content of the course is given in Table 1.

Table 1. Content of the Recreational Material Development Course

COURSE CONTENT		
Week	**Topics**	**Course Duration**
1	Course introduction	3 hours
2	Need for Recreation, Benefits of Recreation, Classification of Recreational Activities	3 hours
3	Material Examples Used in Recreational Activities, Basic Functions of Materials, and Material Design Principles	3 hours
4	Creativity in Material Development	3 hours
5	Types of Materials: Study guides, brochures, photographs, drawings, charts and graphs, cartoons, real objects and models, audio, multimedia, videos.	3 hours
6	Material Development for Games, Sports Implementations and Animation	3 hours
7	Other materials that can be used in recreational activities: Storytelling and creative drama workshop	3 hours
9	Recreational Material Design and Preparation	3 hours
10	Recreational Material Design and Preparation	3 hours
11	Recreational Material Design and Preparation	3 hours
12	Recreational Material Design and Preparation	3 hours
13	Recreational Material Design and Preparation	3 hours
14	Presentation and Evaluation of Student Designs	3 hours
15	Presentation and Evaluation of Student Designs	3 hours

Implementation

In implementing the course, theoretical information that can be used in developing recreational materials is first conveyed. Students were given a project proposal form (Table 2) and asked to design their projects. While designing the projects, support was provided by brainstorming with the students.

Table 2. Project Proposal Form

Project name:
Name and Surname of the Project Owner:
The goal of the project:
Regional/national benefits of the project, its innovative and original aspects:
Target audience of the project:
Main activities of the project:
Estimated investment amount:

While designing the projects, it was stated that students should design projects to develop the tourism and recreational activities of the province they live in. Afterwards, the designed projects were evaluated and the implementation phase began. During the implementation phase, students were asked to make models of the projects they designed from waste materials. Images of project examples are as follows (Table 3).

Table 3. Images from Project Examples

Project Name: Return to Nature and Protect Nature

Project Name: Sled-Ski Center

Project Name: Sports Facility with Accommodation

Project Name: Barrier-Free Life Park

In the evaluation of the course, the project proposal presentation (Table 2) and the presentation of the completed projects in the final stage (Table 3) were used. In other words, a note was given for implementation. However, this evaluation is not within the scope of the research.

Research Design

The qualitative research method was used in the research. A semi-structured interview form was used to collect data. Alacapınar (2008) research was used to create the interview questions. The sample of the research consisted of all students (12 students) taking the Recreational Material Development course. 10 of the students participating in the research are female and 2 are male. All of the students have not taken courses with the project-based learning technique before. Data was collected at the end of the 2022–2023 Spring semester. The interviews lasted an average of 20 minutes. Interview questions are:

- What are your feelings about the project-based learning method applied in the course?
- In what ways is the project-based learning method different from other study/taking methods/techniques?
- In what ways has project-based learning been beneficial to you?

Content analysis was used to analyze the data. Codes and themes were created using the inductive approach. Opinions were presented without including the names of the students.

Results

Three themes emerged from the data obtained. These are the themes of 'skills', 'moods' and 'other outcomes'.

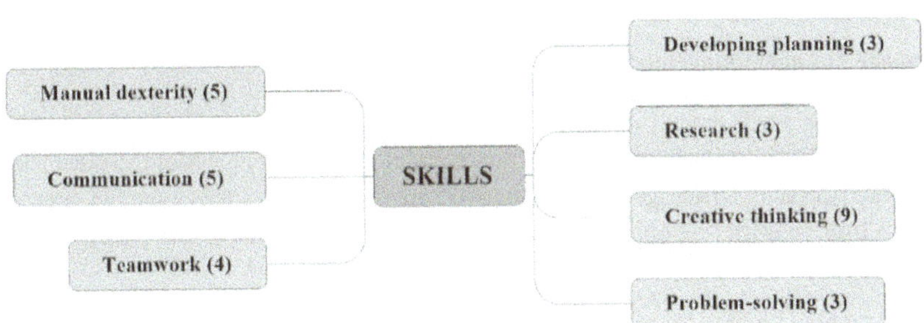

Figure 1. Skills Theme

Most of the students emphasized that the project-based learning method encourages creative thinking skills. Other skills gained by the method are planning, research, creative thinking, problem-solving, teamwork, communication, and manual dexterity (Figure 1).

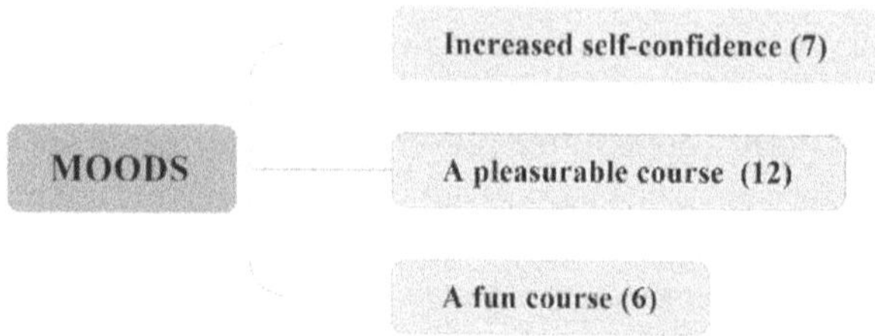

Figure 2. Moods Theme

Students stated that the course was a pleasurable course with the project-based learning method. They also stated that this method was fun and had a self-confidence-increasing effect (Figure 2).

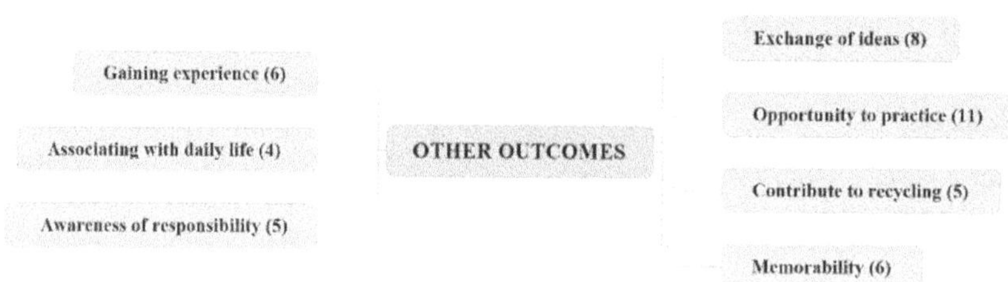

Figure 3. Other Outcomes Theme

It was mostly emphasized by the students that this method provides the opportunity to practice. Other gains of the method are exchanging ideas, contributing to recycling, making the course memorable, gaining experience, associating it with daily life, and awareness of responsibility (Figure 3).

Discussion and Conclusion

Despite the existence of many different approaches, the teaching paradigm in tourism research is moving towards the implementation of active and student-centered methods that ensure student participation in the teaching-learning process and increase their motivation and autonomy (Dinis, Melo, & Sousa, 2021).

The design principles of project-based learning emphasize the importance of the project as the central vehicle of instruction and students as active participants in structuring knowledge. The project-based learning approach is a method that supports and encourages students' self-control and lifelong learning.

Students in the field of recreation management associated with the tourism industry will need different skills to face the challenges of the highly competitive industry. The experiential framework that emerged through the project-based learning method built on the recreational material development course not only exposed students to concrete experiences but also encouraged them to think about industry-specific implementations. Previous research (Simpson, 2011; Aiedah and Audrey Lee, 2012; Pham, Nguyen, & Hoang, 2023) has also shown that project-based learning generally has a positive impact on the learning experience.

The results obtained in this study regarding teamwork, development of planning and problem-solving skills, and increased self-confidence are similar to the results obtained by Simpson (2011) from foreign language class students studying tourism at a large regional university in Thailand. Katz and Chard (2000) and Kraus (2009) confirmed that the project-based learning method encourages creativity, critical thinking, collaboration, individual study, and other study skills. The findings of this research regarding the implementation opportunity that enables learning by doing, the opportunity to exchange ideas, teamwork, and communication also coincide with the findings of Azanza, Fernández-Villarán, and Goytia (2022).

In this research, the student's views on the fact that the course taught with the project-based learning method is pleasurable and fun, that it contributes to recycling thanks to the use of waste materials, and that it raises awareness of responsibility differed from other research results. Project-based learning is a solution that will help students learn most effectively and improve their study capacity (Pham et al., 2023). At the same time, this method is seen as an alternative method that reflects contextual learning and allows students to learn through intense interaction and creativity (Suprayogi & Eko, 2020).

Crossley, Jamieson, and Brayley (2007) suggest that some personal characteristics and skills needed by professionals in the recreation sector are difficult to teach with traditional academic methods such as lectures. According to the authors, since recreation is predominantly a leisure activity, learning how to plan and implement activities or programs is not possible with theoretical courses alone. Therefore, students need courses that use methods such as project-based learning where they can learn by doing.

Suggestions

The biggest obstacle to integrating the project-based learning method, especially in departments related to the field of tourism, is that it requires simultaneous arrangements in assessment, teaching, and curriculum. These arrangements are often complicated foreign to students as well as teachers. Considering the complexity of recreation and tourism products and activities, it is important to develop a roadmap for interdisciplinary project-based learning.

The practical contribution of the research is that it provides ideas for teachers to develop their project-based learning method approaches. Courses using the project-based learning method can be prepared and research can be conducted for different courses related to tourism and recreation. In-service training can be given on how to teach courses using this method. This research was designed within the framework of the qualitative research method. In future studies, studies based on a pretest-posttest design can be conducted using a quasi-experimental design. Mixed research can be conducted using both quantitative and qualitative data collection tools.

References

Aiedah, A. K., & Audrey Lee, K. C. (2012). The application of project-based learning in enhancing students' engagement In Malaysian studies and English language at Taylors' College School Of Hospitality and Tourism, Malaysia. *Journal of interdisciplinary Research in Education,* 2(1), 37–46.

Alacapınar, F. (2008). Effectiveness of project-based learning. *Eurasian Journal of Educational Research,* 32(1), 17–34.

Azanza, G., Fernández-Villarán, A., & Goytia, A. (2022). Enhancing learning in tourism education by combining learning by doing and team coaching. *Education Sciences,* 12(8), 1–16.

Crossley, J. C., Jamieson, L. M., & Brayley, R. E. (2007). Introduction to commercial recreation and tourism: An entrepreneurial approach (5th ed.). Champaign, IL: Sagamore.

Dinis, G., Melo, C., & Sousa, J. (2021). Teaching and learning in tourism: The case of tourism laboratory. In Perspectives and Trends in Education and Technology: Selected Papers from ICITED 2021 (pp. 537–549). Singapore: Springer Singapore.

Katz, L. G., & Chard, C. (2000). Engaging children's minds: The project approach (3rd ed.). California: ERIC Press.

Kokotsaki, D., Menzies, V., & Wiggins, A. (2016). Project-based learning: A review of the literature. *Improving* schools, 19(3), 267–277.

Kraus, H. A. (2009). generating depth and engagement: The development of a project-based curriculum. *Asia CALL Online Journal*,4(1), 30–38.

Pham, T. H. Y., Nguyen, T. P. N., & Hoang, T. T. (2023). The effect of using project-based learning on ımproving self-regulated language learning among english-majored students at Ba Ria-Vung Tau University, Viet Nam. *International Journal of Current Science Research and Review*, 6(4), 2632–2637. |

Simpson, J. (2011). Integrating project-based learning in an English language tourism classroom in a Thai university (Doctoral dissertation, Australian Catholic University).

Suprayogi, S., & Budi Eko, P. (2020). The implementation of virtual exhibition project in English for tourism class for university students. *Academic Journal Perspective: Education, Language, and Literature*, 8(2), 87–97.

Toci, M. J. (2000). The effect of a technology-supported, project-based learning environment on intrinsic and extrinsic motivational orientation (Doctoral dissertation, The Pennsylvania University).

Filiz Arslan[†]

Chapter 8 Psychological Empowerment and Its Importance in Tourism Businesses[*]

Introduction

Psychological empowerment is a whole that increases a person's intrinsic task motivation by influencing the individual's tendency towards the role he/she assumes at work. Spreitzer (1995) defines psychological empowerment as a psychological state and process that occurs in four cognitive areas such as "meaning, self-determination, competence and impact". Psychologically competent people see themselves as capable of influencing their jobs and work environments in meaningful ways, displaying proactive behavior, taking initiative, and acting independently. When the elements that can be effective in making the individual feel empowered are introduced, the effectiveness of empowerment practices will increase. In today's conditions of international competition, the human factor is of great importance in the existence of tourism and service businesses. Employees who feel empowered will be more effective in resolving customer complaints and meeting customer requests, and their contribution to business success will also increase. Accordingly, in this study, the concepts of empowerment and psychological empowerment, factors affecting empowerment and the importance of psychological empowerment in tourism will be explained.

The Concept and Importance of Empowerment

The concept of employee empowerment has attracted increasing attention in management studies. There are various reasons for this. Firstly, it is suggested that practices that empower subordinates increase administrative and organizational effectiveness, secondly, that senior managers share their power and control with their employees, showing that the organization's power and effectiveness increases, and finally, empowerment techniques in the business play a very important role in improving working together (Conger and Kanungo, 1988:471).

---

* This study was expanded from a part of the author's doctoral thesis.

† Associate Prof., Akdeniz University, Göynük Culinary Arts School, filizarslan@akdeniz.edu.tr

When we look at the historical development of personnel empowerment, it can be seen that its history dates back to before the industrial revolution, that is, to the periods when all stages of production were carried out under the responsibility of craftsmen. In line with the idea of efficiency, which formed the basis of management under the influence of neo-classical view in the 1950s and 1960s, McGregor, Argyris and Likert, with their approaches, developed; He emphasized the importance of human resources in organizations and argued that the morale, motivation and commitment of employees to the organization will change in direct proportion to productivity. In the 20th century and later, with the modern management approach, the concept of personnel empowerment has penetrated into businesses (Yıldırım, 2015: 56–57).

Among the first to use the concept of empowerment are Harrison and Kanter in 1983, Bennis and Nanus in 1985, Burke and Nielsen in 1986, and Block in 1987. Empowerment gained its fame with Block's work, was accepted as a new form of management, and became a slogan after this date. In 1986, Bandura defined empowerment as employees feeling that they can create benefits, and in 1988, Conger and Kanungo argued that it should be defined with employees' "motivation processes" and defined empowerment as an increase in employees' effort-performance expectations (Temel, 2016: 64).

Power can emerge not only in the behavior of an individual or the performance results of a subunit, but also in the actions of others. An individual's relative power relative to others is a product of the individual's dependence on another individual. At the organizational level, the main sources of power are the individual's ability to provide resources or the ability to deal with important organizational problems. At the interpersonal level, the main sources of power over others can be listed as follows: a. Structural location, b. personal characteristics, c. expert knowledge and d. It is the power to access information. Theorists about empowerment state that individuals who have power are more likely to achieve the desired results, whereas those who do not have power are prevented by those who have power from achieving the desired results (Lodahl and Gordon, 1972; Salancık and Pfeffer, 1974, 1977; Kotter, 1977, 1979; Hills and Mahoney, 1978; Pfeffer, 1981). These theorists propose to focus on the source of power and the conditions that encourage commitment to that power. In addition, these theorists laid the foundation for the development of resource allocation strategies and tactics to increase the power of individuals who are less powerful than others and to reduce the power of those who are stronger than others (Conger and Kanungo, 1988: 472).

Empowerment can be described as an extension of delegation of authority. However, empowerment has a broader meaning than delegation. In delegation of authority, the leader, as the person responsible for the outcome of the work, delegates his powers to his subordinates with the assumption that he can get better results or because he deems it necessary. In empowerment; To enable the employee who personally performs the relevant job to see the expert knowledge, opportunities, and make the necessary decisions and to change his attitude towards the job, in short; It is possible to make the employee the owner of the job (Şimşek, 2003: 18). The concept of empowerment is seen more as an understanding shared by the manager who has power in an organization, and employees' participation is seen as equal only if they are empowered. Therefore, when the literature on this subject is examined, it is seen in management styles aimed at ensuring participation such as empowerment, management according to goals, delegation of authority, target-oriented management and quality (Conger and Kanungo, 1988: 473). In addition, in research, empowerment is also associated with the concept of job enrichment.

Although there is a common understanding between the concept of empowerment and job enrichment approaches, these two concepts differ from each other. Empowerment is based on the assumption that the individual can have a significant impact on shaping and influencing activities at the organizational level. Second, it is possible for individuals to become cognitively stronger even if their work does not enrich them. Third, empowerment focuses on the individual's level in relation to his or her work environment. However, job enrichment concepts focus more on both the individual and team levels of analysis (Speitzer, 1996: 484–485).

Empowerment gives employees the opportunity to make their own decisions or take responsibility for their own actions. Over the last 20 years, two complementary approaches to workplace empowerment have been developed in the literature. The first of these approaches is the empowerment type, which has a more macro perspective and focuses on socio-structural situations in workplaces. The other one has a more micro perspective and focuses on the empowerment perceptions of employees in the workplace or their perception that they are empowered. Macro and micro perspectives of empowerment can also be considered from "behavioral" and "cognitive" perspectives (Arslan, 2019: 76–77).

The focus of studies examining empowerment in the behavioral dimension is the role of managers in empowering employees. Authors who deal with empowerment in a cognitive dimension, on the other hand, consider this phenomenon from a "psychological" perspective and focus on employees' perceptions. In this perspective, it is stated that empowerment is fundamentally related to individuals' own existence and perceptions, and therefore, within this framework, the empowering behaviors of the top management or the working conditions they create are not sufficient alone (Aydoğmuş, et al., 2015: 256–257).

Psychological Empowerment

In the psychological empowerment approach, where empowerment is evaluated cognitively, the emphasis is not on what the top management should do in order for empowerment to occur, but on how what is done is perceived by the employees. Psychological empowerment is a whole consisting of four basic cognitive concepts: meaning, competence, autonomy and impact, which increases the individual's intrinsic task motivation by influencing the individual's tendency towards the role he/she assumes at work (Aydoğmuş, et al., 2015: 256–257).

In their studies, Thomas and Velthouse (1990) and Conger and Kanungo (1988) define empowerment as internal motivation that emerges in four cognitions that reflect the individual's orientation to the work role. The four cognitions are meaning, competence, self-determination and impact (Spreitzer, 1996: 484). This definition refers to psychological empowerment in the literature.

Conger and Kanungo (1988:474) define psychological empowerment as "the process of increasing employees' feelings of self-efficacy through the identification of conditions that increase vulnerability and through both formal organizational practices and informal techniques of providing effectiveness information." Spreitzer (1995) expanded this concept by stating the concepts of "meaningfulness, competence, choice and impact" from a comprehensive perspective, which are important in determining the task motivation of employees, based on the studies of Conger and Kanungo (1988) and Thomas and Velthouse (1990). He defines it as a process or psychological state that emerges in four cognitive dimensions such as "meaning", "competence", "self-determination" and "impact". In particular, "meaning" is related to the person's feeling that the work he does is important. It means that the person has the ability or self-efficacy to perform his/her duties successfully. "Self-determination" refers to the individual's perception that they have the freedom to choose in matters such as initiating and carrying out their task. "Impact" refers to the degree to which a person's behavior makes a difference in the process, from the work performed to the business results. Spreitzer (1995) suggests that these four dimensions "meaning, competence, self-determination, and impact" reflect a general structure of psychological empowerment, although they are different. Thus, psychological empowerment is seen as employees' involvement in an effective process to initiate a task and increase the permanence of this task (Zhang and Bartol, 2010: 110).

Psychological empowerment refers to a psychological state that exists within individuals and reflects an active orientation towards their work (Pieterse, et al., 2010: 612). It therefore differs from the concept of delegation as a set of managerial practices that focus on the delegation of authority. Psychological empowerment is a motivational construct that refers to an employee's perception of having choice in initiating and controlling actions, the ability to execute the job well (self-efficacy), and the ability to influence the environment (Spreitzer, 1995, Thomas and Velthouse, 1990). The topic of psychological empowerment has received remarkable attention in recent years. Psychological empowerment has antecedents such as the person, the environment or the organization in the environment, colleagues and many other resources (Pieterse, et al., 2010: 613).

Conger and Kanungo (1988) emphasize the motivating effect of empowerment and state that psychological empowerment will be achieved by increasing employees' sense of self-efficacy (Conger and Kanungo, 1988: 474). Likewise, Thomas and Velthouse (1990) emphasize the psychological aspect of empowerment and state that it is based on perceptions and that it is effective in motivating staff. suggests that it is an important factor (Thomas and Velthouse 1990: 666).

Psychologically competent people see themselves as capable of influencing their jobs and work environments in meaningful ways, facilitating proactive behavior, showing initiative and acting independently (Pieterse, et al., 2010: 613).

When the elements that can be effective in making the individual feel empowered are introduced, the effectiveness of empowerment practices will increase. Thomas and Velthouse (1990) enriched the research of Conger and Kanungo (1988) by developing a new theoretical framework and established the "Cognitive Enhancement Model" regarding empowerment. Based on the work of Conger and Kanungo, Thomas and Velthouse introduced a perceptual dimension to empowerment and suggested that empowerment is intrinsic task motivation that reflects the individual's orientation to the work role and is explained by four perceptual dimensions. Similar to the definitions provided by Spreitzer, Thomas and Velthouse, they defined empowerment as a motivational structure consisting of four dimensions: meaning, competence, autonomy and impact, and stated that each dimension should be considered independently of each other (Temel, 2016: 67).

Spreitzer (1995) defines these four dimensions as meaning, competence, autonomy and impact as follows:

- Meaning: It refers to a harmony between a job role and a person's beliefs, values and behaviors (Speitzer, 1996: 484). Meaningfulness, which is the value that the purpose of performing a task has for the employee; It is related to the level of employees caring about the work they do and thinking that their work is important (Sürgevil, et al., 2013:5374).

- Self-Efficacy / Competence: It is related to the level of confidence that employees have in themselves and their abilities to do their jobs well.
- Autonomy (Self-determination): It refers to employees' freedom to decide how to do their jobs and to have control over their work.
- Impact: It is about whether employees feel they have any impact on organizational results (strategic, managerial or operational). It shows the extent to which employees can leave a mark in their workplace and the organization takes their ideas seriously (Sürgevil, et al., 2013:5374). Put differently, influence is the extent to which a person can influence strategic, managerial or business outcomes in the workplace. The concept of affect has been implicitly examined in research on learned helplessness; Definitions of learned helplessness highlight the perceived lack of influence over workplace forces built on past experiences. These four cognitions expressed above reflect a passive, directive attitude to a work role. Thomas and Velthouse (1990) advocate the addition of four dimensions (Speitzer, 1996: 484). The relationship between social structure and empowerment may not be unidirectional. Over time, authority figures can also influence their environments through proactive behavior (Thomas &Velthouse, 1990). Thus, the relationship between individuals' perception of what can be an empowering environment and empowerment consciousness can be mutually reinforcing through a feedback loop between empowering behaviors and the work context (Speitzer, 1996: 4).

Conger and Kanungo (1988) state that the empowerment process is a process consisting of 5 stages. These; 1. Identifying psychological situations that cause a feeling of powerlessness within the organization, 2. Managers' inclusion of strategies and activities that provide empowerment, 3. Managers providing this information support to ensure the self-efficacy of subordinates. Thus, not only managers remove some external conditions, but also (and most importantly) the personal competence of subordinates is ensured, 4. Employees who achieve self-efficacy as a result of providing sufficient information feel empowered, 5. In this last stage, the behavioral effects of empowerment are realized (Conger and Kanungo,1988: 474). As Conger and Kanungo (1988) state, empowerment is a tool used by managers to motivate their employees. Empowerment refers to a process in which an individual's belief in his or her own self-efficacy is increased. In this respect, the behavioral dimensions of empowerment provide benefits to the organization on many levels.

Behavioral consequences are very important to organizational leaders. Empowerment processes can allow leaders to reduce the demoralizing impact of organizational changes or mobilize organizational members in the face of tough competitive challenges. These processes can enable leaders to set higher performance goals and help employees accept these goals. Empowerment practices may also be useful for motivating subordinates to persist despite difficult organizational/environmental obstacles (Conger and Kanungo,1988: 476).

For individuals to feel empowered, the role environment must be perceived as liberating rather than restrictive. For example, resources can be decentralized within an objective reality, but if employees are not informed that these resources are made available (access to resources will have little impact on their sense of empowerment). Thus, individuals' perceptions of their work environments shape empowerment rather than some objective reality. Moreover, since empowerment is defined as a set of cognition, it should also be evaluated through perceptions (Speitzer, 1996: 486). For these reasons, perceptions regarding empowerment and social structural features gained weight in Speitzer's (1996) study.

Factors Affecting Empowerment

An empowering business describes a structure that provides opportunities rather than restrictions on behavior (Torbert, 1991), has high conversion (Lawler, 1992), develops commitment-based (Walton, 1985) designs, and has a less challenging environment. Such organizations provide a wide range of information and resources for their employees' decision-making. Such a system creates cognition that employees can increase organizational effectiveness. In these systems where participation is high, it increases employees' trust in the organization and increases their commitment to the organization (Speitzer, 1996: 485). Conger and Kanungo (1988: 477) state that organizational factors, management model, reward system and a participatory organizational climate affect empowerment.

Speitzer (1996: 486–487) suggests that in high-involvement systems, six work unit social structural characteristics create a work context that facilitates empowerment: (1) span of control, (2) low role ambiguity, (3) sociopolitical support, (4) access to information, (5) access to resources and (6) participatory climate.

Span of control refers to the number of people a manager supervises. Narrow span of control is associated with centralized decision making. Decentralized control helps subordinates make decisions in their areas of responsibility and enhances their self-determination. Decentralized control makes employees feel like they are contributing to their organization's operations and increases their sense of making an impact. It may be noted that people working under conditions conducive to a narrow scope of control tend to feel that one or both of them do not have management confidence in their skills and abilities, feelings that result in feelings of personal inadequacy. Moreover, people working with auditors with narrow control areas are less intrinsically motivated (they have less personal experience) than people working with broad control areas (Speitzer, 1996: 487).

Role ambiguity can be defined as the person's lack of clear and consistent information about what is expected of him/her due to his/her role. Role can be defined as the set of duties or behaviors expected from a person within the group of which he is a member. In this context; It can be said that people have various roles in both social life and work environments. In addition, it is normal for people to want to know what behaviors are expected from them due to their roles in the organizations they work for (Walker et al., 1975:33). It is stated that employees with low role ambiguity feel more powerful.

Sociopolitical support is defined as the endorsement of legitimacy or given by organizational constituencies and is often gained from membership in organizational networks (Kanter, 1983). Social networks define the social fabric of an organization and provide members with important channels to get started. The relevant support network includes a person's boss, colleagues, subordinates, and members of the work group. Support network membership increases social exchange with important organizational constituencies and thus increases the sense of personal power. Such personal power can manifest itself in strong feelings of both self-determination and influence. In contrast, lack of support from core constituencies can lead to feelings of inadequacy (Speitzer, 1996: 488).

Access to information is related to the data, technical information and communication with experts that people need to perform their job. Getting support means accessing feedback and necessary guidance from superiors, subordinates and colleagues to increase effectiveness. *Access to resources* means the ability of the personnel to meet some of their needs, such as materials, equipment, and money necessary to achieve organizational goals. Access to information is related to the data, technical information and communication with experts that people need to perform their job. Receiving support means accessing feedback and necessary guidance services from superiors, subordinates and colleagues to increase effectiveness (Nedd, 2006: 14).

Climate is a phenomenon that forms the personality of an organization, shapes the attitudes and behaviors of its employees, and affects its characteristics. In participatory organizational climates with a democratic structure, employees participate in joint decisions, while in a non-democratic organizational climate, order, control and predictability are evaluated (Evered and Selman, 1989). In addition, in the participatory organizational climate, individual contribution and initiative are emphasized rather than a vertical hierarchy and a top-down chain of command and control (Speitzer, 1996: 489). Such an organizational climate is very important for business success. It is aware of the critical importance of human resources in success. It is aware of the need for its employees to show initiative in order to stand out in a competitive external environment (Speitzer, 1996: 490).

The Importance of Reinforcement in Tourism

Tourism has an important place in the macro and micro economies of countries. For this reason, countries attach increasing importance to tourism day by day. According to the definition of AIEST (International Association of Scientific Tourism Experts), tourism is defined as "a set of events and relationships arising from people's travels outside the places where they permanently reside and work and where they meet their usual needs, and their temporary accommodation, generally by demanding goods and services produced by tourism enterprises"(Arslan, 2019: 43–44).

The production and marketing of tourist products is different from industrial products. Tourism products are intangible, cannot be stored (consumed where they are produced), and require large capital. When a tourist spends his holiday in a holiday destination, he consumes not only the product of a supplier, but also the service package as a whole. Many different service providers are involved in creating a tourism experience. The product covers the experience from the moment he leaves home until he returns. This applies to all tourism services in accommodation, catering, travel agencies, transport systems and all other businesses related to the tourism value chain. A guest who books a particular holiday package does not know for sure what to expect and what it will be like (Weiermair, OECD, 2004: 3).

In today's conditions of international competition, the human factor is of great importance in the existence of tourism and service businesses. The fact that tourism is a service sector makes the impact of employees on business outputs important. It can be said that having personnel empowerment practices that increase employee behavior positively will increase outputs such as customer satisfaction and service quality in the tourism sector.

It is stated that people who feel empowered and psychologically competent are able to influence themselves, their jobs and work environments in meaningful ways, show proactive behavior, take initiative and see themselves as capable of acting independently (Spreitzer, 1995; Thomas and Velthouse, 1990). It is stated that empowered employees in tourism businesses can be more effective in resolving customer complaints and meeting customer requests. It can be stated that the presence of employees with these behaviors in the service-oriented tourism sector will contribute to the success of the businesses in the sector and will also increase the competitiveness of the businesses.

References

Arslan, F (2019) *Yiyecek-İçecek Çalışanlarında Dönüşümcü Liderlik ve Yenilikçi İş Davranışı İlişkisinde Psikolojik Güçlendirmenin Aracı Etkisi*, Zonguldak Bülent Ecevit Üniversitesi, Sosyal Bilimler Enstitüsü, İşletme Anabilim Dalı Doktora Tezi, Zonguldak.

Aydoğmus, Ceren, Azize Ergeneli, Selin Metin Camgöz (2015). The Role Of Psychologıcal Empowerment On The Relationship Between Personality and Job Satisfaction. *Research Journal of Business & Management*, 2, 251 276.

Conger, Jay A. and Kanungo, Rabindra N. (1988). The Empowerment Process: Integrating Theory and Practice. *The Academy of Management Review*, 13(3), 471–482.

Walker, O. C., Churchill, G. A. ve Ford, N. M. (1975). Organizational Determinants of the Industrial Salesman's Role Conflict and Ambiguity. *The Journal of Marketing*, 39 (1), 32–39.

Evered, R. D., & Selman, J. C. (1989) Coaching and the art of management *Organizational Dynamics*, 18(2): 16–32.

Hills, F. S., & Mahoney, T. A. (1978) University budgets and organizational decision making *Administrative Science Quarterly*, 23, 454–465.

Kanter, R. M. (1983) The change masters New York: Simon & Schuster.

Kotter, J. P. (1977) Power, dependence, and effective management *Harvard Business Review*, 55(4), 125–136.

Kotter, J. P. (1979) Power in management. New York: Ama-com.

Lawler, Edward E. (1992) Total Quality Management and Employee Empowerment and Involvement: Similarities, Differences, and Future Direction, CA: Ceo Publication.

Lodahl, J., & Gordon, G. (1972) The Structure Of Scientific Fields And The Functioning Of University Graduate Departments *American Sociological Review*, 37, 57–72.

Nedd, N. (2006). Perceptions of empowerment and intent to stay. *Nursing Economics.* 24 (1), 13–18.

Pfeffer, J. (1981) Power in Organizations Marshfield, MA: Pitman Publishing.

Torbert, W. (1991) The Power of Balance: Transforming self, society and social inquiry. Newbury Park, CA: Sage.

Thomas, Kenneth W. and Velthouse, Betty A. (1990). Cognitive Elements of Empowerment: An "Interpretive" Model of Intrinsic Task Motivation. *The Academy of Management Review*, 15(4), 666–681.

Salancik, G. R., & Pfeffer, J. (1974) The bases and use of power in organizational decision making: The case of a university *Administrative Science Quarterly*, 19, 453–473.

Salancik, G. R., & Pfeffer, J. (1977) Who gets power and how they hold on to it: A strategic-contingency model of power *Organizational Dynamics*, 5(3), 2–21.

Spreitzer, Gretchen M.(1995). Psychological Empowerment in The Workplace: Dimensions, Measurement, and Validation. Academy of Management Journal, 38(5), 1442–1465.

Spreitzer, Gretchen M (1995) Psychological Empowerment In The Workplace: Dimensions, Measurement, and Validation *Academy of Management Journal*, 38(5), 1442– 1465.

Şimşek, Necla (2003). *Personeli Güçlendirme ve Türk Sigorta Sektöründe Bir Uygulama*. Marmara Üniversitesi, Sosyal Bilimler Enstitüsü Doktora Tezi, İstanbul.

Temel, Eren (2016) *Dönüşümcü Liderlik ve Psikolojik Güçlendirme Arasındaki İlişkide Örgütsel Özdeşleşmenin Aracılık Rolü: Nazilli'deki Kamu Kurumlarında Bir Uygulama*, Adnan Menderes Üniversitesi Sosyal Bilimler Enstitüsü, İşletme Anabilim Dalı Yüksek Lisans Tezi, Aydın.

Walton, R.E. (1985) From control to commitment in the workplace, *Harvard Business Review*, 63 (2), 77–84.

Weiermair, K. (2004). Product improvement or innovation: what is the key to success in tourism?, OECD.

Yıldırım, Fatih (2015). *Örgüt Kültürünün Yeniliğe Etkisinde Personel GüçlendirmeninBiçimlendirici Rolü*, Atatürk Üniversitesi Sosyal Bilimler Enstitüsü, İşletme Anabilim Dalı Doktora Tezi, Erzurum.

Sürgevil, Olca, Ebru Tolay, Mert Topoyan (2013).Yapısal Güçlendirme ve Psikolojik Güçlendirme Ölçeklerinin Geçerlilik ve Güvenilirlik Analizleri. *Journal of Yasar University*, 8(31), 5371–5391.

Pieterse, Anne Nederveen, Daan Van Knippenberg (2010). Transformational and Transactional Leadership and İnnovative Behavior: The Moderating Role Of Psychological Empowerment. *Journal Of Organizational Behavior*, 31, 609 623

Zhang, Xiaomeng, Kathryn M. Bartol (2010). Linkıng Empowering Leadership And Employee Creativity: The Influence Of Psychological Empowerment, Intrinsic Motivation, And Creative Process Engagement. *Academy Of Management Journal*, 53(1), 107–128

Emre Ozan Aksöz* and Gamze Kayan Ürgün†

Chapter 9 The Role of Destination Anthropomorphism in Destination Promotion and Marketing

Introduction

The tourism industry is a field that constantly evolves and changes in the promotion and marketing of destinations. In recent years, increasing research on the promotion and perception of destinations with human-like characteristics has emphasized the importance of destination anthropomorphism. The concept of destination anthropomorphism refers to the tendency to attribute human qualities or characteristics to tourist destinations, relating not only to the physical characteristics of destinations, but also to their perception as a personality or living entity. Anthropomorphism is a cognitive process used by humans to understand and interact with nature and the environment. The concept of anthropomorphism has become important in the tourism sector through the characterization of destinations with human-like qualities and the establishment of emotional connections with them. In the tourism industry, destination anthropomorphism plays a significant role in destination promotion and marketing. By promoting destinations with anthropomorphized qualities, emotional connections are fostered among tourists. However, the effects of destination anthropomorphism on consumers are not limited to emotional bonds. Anthropomorphized destinations not only provide memorable experiences to visitors but also enhance the destination's image and increase its brand value. This study aims to examine the concept of destination anthropomorphism and discuss its effects on destination promotion and marketing.

* Prof., Anadolu University, Tourism Faculty, Department of Tourism Management, ozana@anadolu.edu.tr

† Res. Assist., Ondokuz Mayıs University, Tourism Faculty, Department of Tourism Management, gamze.kayan@omu.edu.tr

The Concept of Anthropomorphism

The term "anthropomorphism" is derived from the Greek words "Anthropos" (human) and "morphe" (shape or form), and it refers to attributing human characteristics to objects that do not have human identity, motivation, intention, or mental state (Epley et al., 2007: 865). Waytz et al. (2010: 220) explain anthropomorphism in two ways. First, it involves ascribing human physical features (such as face, hand, foot) to other entities; second, it entails transferring human mental attributes (such as mind, intention, consciousness) to other entities. From past to present, humans have tended to anthropomorphize religious figures, animals, the environment, brands, and even technological products (Salles, Evers & Farisco, 2020: 89).

Psychologically, anthropomorphism is considered a cognitive process in humans' interaction with the world. By humanizing complex or incomprehensible phenomena, people tend to make more meaningful and predictable explanations about them. This is observed in relationships with objects as well as in social relationships (Epley et al., 2007). Philosophically, anthropomorphism is often regarded as a universal and fundamental human trait. From ancient Greek philosophy to the present, anthropomorphism has led to profound reflections on human nature, the gods, and the relationship with nature. The debate on whether anthropomorphism reflects the self or an assumption about the external world still continues among philosophers (Boyer, 1996; Dacey, 2017).

Overall, it can be said that anthropomorphism represents a broad psychological process that can vary depending on specific situations. For example, some factors may lead individuals to a greater anthropomorphic tendency than others, some cultures may be more inclined to anthropomorphism compared to others, and according to past studies, children are generally observed to be more prone to anthropomorphic tendencies than adults (Han, Wang & Yang, 2023; Baskentli, Hadi & Lee, 2023).

In the past 20 years, the concept of anthropomorphism, which has attracted great interest from researchers, has been extensively examined and discussed by a range of disciplines. These disciplines include consumer behavior, advertising, brand development, robotics, psychology, social psychology, product design, hospitality and tourism, and artificial intelligence (Epley et al., 2007; Waytz et al., 2010; Miesler, 2012; Aggarwal & McGill, 2012; Wan, Chen & Jin, 2015; Lee & Oh, 2021; Blut et al., 2021; Murphy et al., 2021; Karimova & Goby, 2021; Ding et al., 2022).

Studies in the literature indicate that anthropomorphism has positive effects between brands and consumers. Consequently, anthropomorphizing products and brands has become a popular marketing strategy (Sheehan, Jin, & Gottlieb, 2020: 16). For instance, anthropomorphism can enhance attitudes toward a brand's or company's older products, increase the intention to visit hotels, boost trust in autonomous vehicles, and encourage sustainable behaviors (Ketron & Naletelich, 2019; Koo et al., 2019; Lee & Oh, 2021; Waytz et al., 2014).

In this context, the number of studies addressing anthropomorphism in the tourism sector is also increasing. Ding et al. (2022) reviewed the literature on anthropomorphism in the hospitality and tourism fields over the past twenty years and concluded that previous studies have primarily focused on themes of technology, brand, and product anthropomorphism. However, it was observed that studies on destination anthropomorphism are limited in the literature. In the following sections, the concept of anthropomorphism and consumer behavior will be discussed, and the concept of destination anthropomorphism in the tourism sector will be examined, along with its role in destination promotion and marketing.

Anthropomorphism and Consumer Behavior

Since the late 2000s, the concept of anthropomorphism has become a significant focus in the literature on consumer behavior, particularly in relation to brand, product, and more recently, technology anthropomorphism (Agarwal & Khandelwal, 2024). As a field of consumer psychology, anthropomorphism allows people to establish an emotional bond with products, brands, or services by attributing human-like qualities to them (Chen, 2017).

For example, previous studies have shown that human-like features, such as a smile on a car's grille, increase interest in the product, reduce the desire to replace it, and affect consumers' risk perceptions (Chandler & Schwarz, 2010; Aggarwal & McGill, 2012). These studies demonstrate that presenting products with human-like physical characteristics leads to positive responses from consumers.

As mentioned, consumers can form emotional bonds with products by attributing human-like qualities to them through anthropomorphism. These bonds have a significant impact on consumer loyalty and brand preferences. Aaker and Fournier (1995) suggested that when brands possess human characteristics, it leads consumers to develop positive attitudes towards these brands and remain loyal to them. This indicates that brands have a strong influence on consumer behavior, and anthropomorphism can enhance this effect.

In conclusion, existing literature supports that anthropomorphism has a significant impact on consumer behavior and is widely used in marketing strategies.

Marketers can appeal to consumers' emotional and social connections by adding human-like features to a brand or product, thereby increasing loyalty and positive brand evaluations.

The Concept of Destination Anthropomorphism and Its Role in Destination Promotion and Marketing

The concept of a destination is defined in various ways by different authors in the literature. Leiper (1990) described destinations as places where people stay for a certain period to experience specific travel experiences and choose based on the attractions of the destination. According to Buhalis (2000), a destination includes the core resources of tourism activities, where touristic goods and services are offered to consumers and where touristic experiences occur in an integrated manner. Kozak et al. (2012) defined destinations as tourism regions that create an integrated holiday package by offering touristic experiences to visitors.

As can be understood from these definitions, destinations share common features, such as creating an attraction with their values, being preferred by a specific group of visitors, and meeting travel needs with the various products they offer. To be a touristic destination and to develop in terms of tourism, a region must possess certain characteristics. These characteristics include a combination of attractions, image, transportation, accommodation, food and beverage, parks, museums, archaeological sites, and entertainment and recreation facilities (Ilban, 2007: 7). Due to their various features, destinations are among the most challenging touristic products to manage and market.

Destination anthropomorphism is defined as a marketing strategy used in tourism, where human characteristics or qualities are attributed to destinations. This concept is derived from the consumer behavior literature, particularly from brand and product anthropomorphism. Brand anthropomorphism refers to the tendency of consumers to associate brands with human characteristics. Product anthropomorphism refers to the tendency of consumers to interact with products by attributing human features to them. The aim of destination anthropomorphism is to help tourists form an emotional bond with destinations and increase their interest in them. By using personification and personalization techniques in the promotion and marketing of destinations, the goal is to make destinations more appealing to tourists (Ding et al., 2022).

Destination anthropomorphism is a commonly used strategy in promoting tourist sites. This strategy aims to create an emotional connection with visitors by associating destinations with human traits or qualities (Kwak et al., 2020). For example, destination mascots or characters often become symbols of destinations and play an effective role in promotional and marketing activities (Radomskaya and Pearce, 2021). In this field, Soltani et al. (2018) found that mascots with anthropomorphic features are an effective marketing tool that enhances tourists' emotional bonds and contributes to the success of city branding. Destination anthropomorphism is also used in advertising campaigns, especially with personalized messages (Letheren et al., 2017). In their study exploring the role of anthropomorphic communication in brands, Lee and Oh (2021) found that using first-person pronouns in their messages made anthropomorphic communication an effective strategy. Moreover, it is stated that anthropomorphic messages in destination marketing develop positive attitudes among tourists, increase their intention to travel, and discourage antisocial behaviors during travel (Letheren et al., 2017). These techniques have become important tools today for enhancing the impact of destinations on tourists and making destinations more attractive.

In today's tourism industry, environmental protection and sustainability are gaining increasing importance. Tourists' environmental awareness and attitudes are closely related to their attitudes towards the natural and cultural assets of the destinations they visit (Budeanu et al., 2016). Recent research has also shown that destination anthropomorphism positively affects tourists' environmental attitudes. When tourists start to see destinations not just as places to visit but as living entities, they feel a greater sense of responsibility for their protection and sustainability (Tam et al., 2013; Lin et al., 2024).

In conclusion, the concept of destination anthropomorphism parallels brand and product anthropomorphism and is recognized as an important tool for understanding and evaluating the impact of tourist destinations on consumers.

Results

This section examines the role of destination anthropomorphism in destination promotion and marketing. Anthropomorphism, the tendency to attribute human characteristics to non-human entities, has existed throughout human history and has been observed in many cultures and periods. For example, in Greek mythology, gods and goddesses were depicted with human-like traits. Similarly, in other ancient civilizations, gods and goddesses were portrayed with human emotions, desires, and even flaws. Anthropomorphism is known to have originated from humans' curiosity about nature and their environment, their search for meaning, and their cognitive processes (Epley et al., 2007).

Anthropomorphic elements are used in many areas today, including as an effective marketing strategy for destinations. Previous studies have shown that defining destinations with human-like characteristics allows tourists to form a deeper connection with them and makes their travel experiences more meaningful (Letheren et al., 2017; Kwak et al., 2020; Ding et al., 2022). Therefore, in the tourism sector, destination anthropomorphism emerges as a significant topic in the promotion and marketing of destinations.

Moreover, the use of anthropomorphic elements is said to strengthen the brand personality of destinations and provide a competitive advantage by making them distinguishable from others (Hankinson, 2004; Ekinci and Hosany, 2006). When tourists perceive destinations as characters, their feelings of interest and attachment to that destination are strengthened, making the destination's marketing efforts more effective. However, anthropomorphism must be used carefully. Defining destinations with human-like characteristics should not deviate from their real features and should not create a misleading perception for tourists. Additionally, promoting destinations with anthropomorphic features should align with the preferences, cultural characteristics, and expectations of the target audience (Kwak et al., 2020).

Research and applications in destination anthropomorphism are considered highly important and represent a new area in tourism studies. Work in this area will help destinations form stronger bonds with people and achieve more sustainable success in the tourism industry. In conclusion, destination anthropomorphism is seen as an effective strategy for destination promotion and marketing. It is necessary to increase research and applications on this topic. However, for this strategy to be successful, careful planning and appropriate implementation for the target audience are essential.

References

Aaker, J., & Fournier, S. (1995). A Brand as a Character, A Partner and a Person: Three Perspectives on the Question of Brand Personality. *Advances in Consumer Research*, 22(1), 391–395.

Agarwal, M., & Khandelwal, U. (2024). Research on anthropomorphism in the 21st century: a bibliometric study. International Journal of Internet Marketing and Advertising, 20(2), 188–215.

Aggarwal, P., & McGill, A. L. (2012). When brands seem human, do humans act like brands? Automatic behavioral priming effects of brand anthropomorphism. *Journal Of Consumer Research*, 39(2), 307–323.

Baskentli, S., Hadi, R., & Lee, L. (2023). How culture shapes consumer responses to anthropomorphic products. *International Journal of Research in Marketing*, 40(3), 495–512.

Blut, M., Wang, C., Wünderlich, N. V., & Brock, C. (2021). Understanding anthropomorphism in service provision: a meta-analysis of physical robots, chatbots, and other AI. *Journal of the Academy of Marketing Science*, 49, 632–658.

Boyer, P. (1996). What makes anthropomorphism natural: Intuitive ontology and cultural representations. *Journal of the Royal Anthropological Institute*, 83–97.

Budeanu, A., Miller, G., Moscardo, G., & Ooi, C. S. (2016). Sustainable tourism, progress, challenges and opportunities: an introduction. *Journal of Cleaner Production*, 111, 285–294.

Buhalis, D. (2000). Marketing the competitive destination of the future. *Tourism Management*, 21(1), 97–116.

Chandler, J., & Schwarz, N. (2010). Use does not wear ragged the fabric of friendship: Thinking of objects as alive makes people less willing to replace them. *Journal of Consumer Psychology*, 20(2), 138–145.

Chen, K. J. (2017). Humanizing brands: an examination of the psychological process of anthropomorphism and its effects on consumer responses. *Journal of Marketing Management*, 5(2), 75–87.

Dacey, M. (2017). Anthropomorphism as cognitive bias. *Philosophy of Science*, 84(5), 1152–1164.

Ding, A., Lee, R. H., Legendre, T. S., & Madera, J. (2022). Anthropomorphism in hospitality and tourism: A systematic review and agenda for future research. *Journal of Hospitality and Tourism Management*, 52, 404–415.

Ekinci, Y., & Hosany, S. (2006). Destination personality: An application of brand personality to tourism destinations. *Journal of Travel Research*, 45(2), 127–139.

Epley, N., Waytz, A., & Cacioppo, J. T. (2007). On seeing human: A three-factor theory of anthropomorphism. *Psychological Review*, 114(4), 864–886.

Han, J., Wang, D., & Yang, Z. (2023). Acting like an interpersonal relationship: Cobrand anthropomorphism increases product evaluation and purchase intention. *Journal of Business Research*, 167, 114194.

Hankinson, G. (2004). Relational network brands: Towards a conceptual model of place brands. *Journal of Vacation Marketing*, 10(2), 109–121.

Ilban, M. O. (2007). "*Destinasyon Pazarlamasında Marka İmajı ve Seyahat Acentalarında Bir Araştırma*", Doktora Tezi, Balıkesir Üniversitesi Sosyal Bilimler Enstitüsü, Balıkesir.

Karimova, G. Z., & Goby, V. P. (2021). The adaptation of anthropomorphism and archetypes for marketing artificial intelligence. *Journal of Consumer Marketing*, 38(2), 229–238.

Ketron, S., & Naletelich, K. (2019). Victim or beggar? Anthropomorphic messengers and the savior effect in consumer sustainability behavior. *Journal of Business Research*, 96, 73–84.

Koo, M., Oh, H., & Patrick, V. M. (2019). From oldie to goldie: Humanizing old produce enhances its appeal. *Journal of the Association for Consumer Research*, 4(4), 337–351.

Kozak, N., Kozak, M. A., ve Kozak, M. (2012). *Genel Turizm: İlkeler – Kavramlar.* (12. Baskı). Ankara: Detay Yayıncılık.

Kwak, H., Puzakova, M., Rocereto, J. F., & Moriguchi, T. (2020). When the unknown destination comes alive: The detrimental effects of destination anthropomorphism in tourism. *Journal of Advertising*, 49(5), 508–524.

Lee, S. A., & Oh, H. (2021). Anthropomorphism and its implications for advertising hotel brands. *Journal of Business Research*, 129, 455–464.

Lee, S. A., & Oh, H. (2021). Anthropomorphism and its implications for advertising hotel brands. *Journal of Business Research*, 129, 455–464.

Leiper, N. (1990). Tourist attraction systems. *Annals of Tourism Research*, 17(3), 367–384.

Letheren, K., Martin, B. A., & Jin, H. S. (2017). Effects of personification and anthropomorphic tendency on destination attitude and travel intentions. *Tourism Management*, 62, 65–75.

Lin, Z. C., Ruan, H., Zhang, H., & Xiao, X. S. (2024). Antecedents, manifestations, and environmental consequences of destination anthropomorphism: The case of nature-based destinations. *Journal of Hospitality and Tourism Management*, 58, 432–444.

Miesler, L. (2012). Product choice and anthropomorphic designs: Do consumption goals shape innate preferences for human-like forms?. *The Design Journal*, 15(3), 373–392.

Murphy, J., Gretzel, U., & Pesonen, J. (2021). Marketing robot services in hospitality and tourism: the role of anthropomorphism. *In Future of Tourism Marketing* (pp. 16–27). Routledge.

Radomskaya, V., & Pearce, P. L. (2021). Adding character: The role of destination mascots in tourism development. *Tourism Management*, 84, 104248.

Salles, A., Evers, K., & Farisco, M. (2020). Anthropomorphism in AI. *AJOB Neuroscience*, 11(2), 88–95.

Sheehan, B., Jin, H. S., & Gottlieb, U. (2020). Customer service chatbots: Anthropomorphism and adoption. *Journal of Business Research*, 115, 14–24.

Soltani, A., Pieters, J., Young, J., & Sun, Z. (2018). Exploring city branding strategies and their impacts on local tourism success, the case study of Kumamoto Prefecture, Japan. Asia Paci•c Journal of Tourism Research, 23(2), 158–169.

Tam, K. P., Lee, S. L., & Chao, M. M. (2013). Saving Mr. Nature: Anthropomorphism enhances connectedness to and protectiveness toward nature. *Journal of Experimental Social Psychology*, 49(3), 514–521.

Wan, E. W., Chen, R. P., & Jin, L. (2015). The Influence of Anthropomorphism on Product Attributes Processing and Consumer Preference. *Asia-Pacific Advances in Consumer Research*, 2015, 116.

Waytz, A., Cacioppo, J., & Epley, N. (2010). Who sees human? The stability and importance of individual differences in anthropomorphism. *Perspectives on Psychological Science*, 5(3), 219–232.

Waytz, A., Heafner, J., & Epley, N. (2014). The mind in the machine: Anthropomorphism increases trust in an autonomous vehicle. *Journal of Experimental Social Psychology*, 52, 113–117.

Ömer Ceyhun Apak* and Hami Ünlü†

Chapter 10 Factors Affecting Tourists' Sustainable Tourism Intention: A Parallel Mediation Analysis

Introduction

It is stated that approximately 375 million tourists travelled worldwide in 2023. In 2024, this number is expected to increase further (UNWTO, 2023). In terms of Turkey, the number of tourists travelling in 2023 is around 56 million. Similarly, it is estimated to increase even more in 2024 (Kaygısız, 2023). Therefore, the carbon footprint of these intense movements negatively affects the climate. At the same time, although various natural disasters (earthquakes, extreme weather conditions, volcanic activities, etc.), pandemics, and sea level rise slow down tourism activities, tourism movements are increasing day by day (Scott et al., 2011). In this framework, 2015 sustainable development goals have been proposed by the United Nations. Within these goals, there are also statements regarding the sustainability of tourism. Therefore, researchers are conducting studies to ensure sustainability in tourism. However, these studies mostly focus on regional development (Çeken, 2008), pollution in tourist destinations (Zhang et al., 2020), sustainable local foods (Süzer & Doğdubay, 2022), sustainable destination management (Zengin & Kiper, 2020), sustainable purchasing (Özkök & Gümüş, 2009; Uçgun & Narcı, 2023), sustainability of alternative tourism types (Gürbüz et al., 2019).

Studies have focused on the environmental awareness of various stakeholders towards sustainable tourism. However, attitudes, motivations, intentions and behaviours are ignored (Juvan and Dolnicar, 2014). There are a limited number of studies that reveal the behavioural intentions of stakeholders in sustainable tourism (Nguyen et al., 2023; Luongo et al., 2024). In this context, this study attempts to estimate the impact of tourists' attitudes towards sustainable tourism on behavioural intention. In addition, the mediating role of travel motivation

* PhD., Bayburt University, Social Sciences Vocational School, Department of Hotel, Restaurant and Catering Services, ceyhun.apak@hotmail.com
† PhD., Bayburt University, Social Sciences Vocational School, Department of Hotel, Restaurant and Catering Services, hamiunlu@bayburt.edu.tr

and moral reflection in the effect of attitude towards sustainable tourism on behavioural intention is examined. This study was carried out from the perspective of domestic tourists visiting the Eastern Black Sea Region as a region where sustainable tourism activities are considered to be the most feasible. Especially highland tourism attracts the attention of visitors. Therefore, the sustainability of tourism in this region is important for the protection of nature.

Theoritical Backround and Hypothesis Development

Tourist attitude and motivation are two research topics that tourism researchers have been working on (Huang & Hsu, 2009). Studies have shown that there is a relationship between these two issues (Hsu et al., 2010). When tourists want to overcome negative moods, their motivation may affect their intention to participate in tourism (Shien et al., 2022). Hsu et al. (2010) found that there is a relationship and effect between attitude and motivation in visiting a destination. Considering this relationship and its effect on the literature, the following hypothesis was developed.

H1. Attitude towards sustainable tourism positively affects travel motivation.

There are many studies examining tourists' attitudes, behaviours, values and concerns towards sustainable tourism (Barber, 2014; Imran et al., 2014; Hedlund et al., 2012). In these studies, tourists' ethical values, sensitivity to environmental problems, and tendency to support green initiatives of tourism enterprises have been examined, and it has been observed that people with these values support sustainable tourism (Miao & Wei, 2016; Dief & Font, 2012; Puhakka, 2011; Andereck, 2009). Tourists' attitudes towards sustainable tourism vary depending on their ethical values and moral behaviours (Weeden, 2011). Based on the relationship between attitudes towards sustainable tourism, ethical values and moral behaviour, the following hypothesis was developed.

H2. Attitude towards sustainable tourism positively affects moral reflection.

When the studies conducted in the field of tourism are examined, it is seen that there is a relationship between attitude and behavioural intention (Kim & Hall, 2019; Safshekan et al., 2020; Sujood et al., 2023). When individuals have a positive attitude towards a specific behaviour, they are more inclined to perform that behaviour. It is seen that people's environmental concerns stem from their interest in issues such as recycling, green consumption and energy saving (Cheng et al., 2006; Han et al., 2023). It has been found that tourists who care about sustainable tourism participate in short-distance travelling, have high environmental protection awareness and behave more responsibly in this regard (Fraçois, Lecompte & Prim Allaz, 2009). Based on the relationship between attitude towards sustainable tourism and intention to participate in tourism activities, the following hypothesis was developed:

H3. Attitude towards sustainable tourism positively affects sustainable tourism intention.

It is stated that travel motivation is one of the main determinants of behavioural intentions in destination choice (Chang et al., 2014). However, it is seen that travel motivation not only explains the behaviour of tourists but also predicts the intention to visit (Huang & Hsu, 2009). Considering this situation, the following hypothesis is proposed:

H4. Travel motivation positively affects sustainable tourism intention.

Moral reflection can motivate people to leave a breathable environment for future generations rather than personal interests (Afsar & Umrani, 2020). Environmentally friendly concepts such as environmental awareness and sustainability affect people's green hotel visits (Han, 2015) and travel preferences (Doran & Larsen, 2016). People's ethical norms may be effective in these preferences. In short, related studies show that when moral reflection is developed, people are more sensitive to the environment (Verma & Chandra, 2018), positively affecting sustainable tourism intentions. Considering the situations mentioned above, the following hypothesis was developed:

H5. Moral reflection positively affects sustainable tourism intention.

When making a travel plan to a touristic destination, the characteristics of the destination affect the attitude of tourists positively or negatively. This shows that there is a significant relationship between attitude and intention to visit the destination (Han et al., 2023). Ajzen (1991) states that there is a direct and positive connection between attitudes and intentions regarding a behaviour. Many studies have presented findings that confirm this relationship (Kim & Hall, 2019; Nowacki et al., 2021, Sujood et al., 2023). Consumers' attitudes towards a behaviour directly affect their intention to perform that behaviour (Cheng et al., 2006). When analysed in terms of intention to participate in sustainable tourism, it is seen that motivation (Chang et al., 2014) and moral reflection (Miao & Wei, 2016) are related to the formation of individuals' attitudes. It is seen that the concepts of motivation and moral reflection are related to attitudes towards sustainable tourism and intention to participate in sustainable tourism. Considering this relationship, the following hypothesis was developed:

H6. Travel motivation and moral reflection have a mediating role in the effect of attitude towards sustainable tourism on sustainable tourism intention.

Methods and Materials

This research was conducted within the framework of the relational survey model, which aims to reveal the degree and direction of the relationship between variables (Fraenkel et al., 2015). It is thought that research within the scope of relational surveys is important in terms of making predictions for the future (Sarikaya et al., 2018). In this study, the effect of domestic tourists' attitudes towards sustainable tourism on their sustainable tourism intentions was investigated. In addition, the mediating role of travel motivation (TM) and moral reflection (MR) in the effect of attitude towards sustainable tourism (AST) on sustainable tourism intention (STI) was examined. The model in Figure 1 was tested in the study.

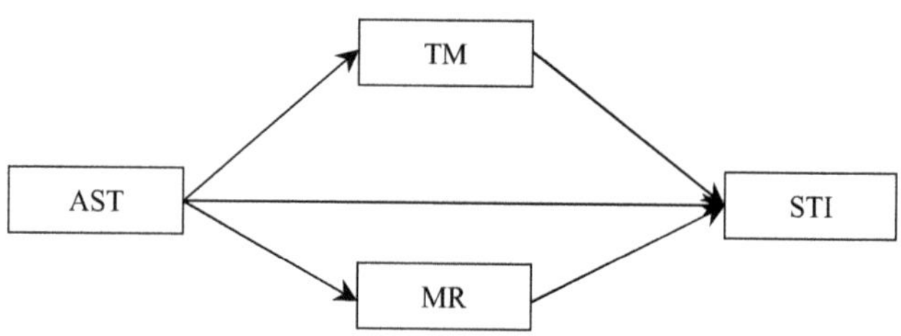

Figure 1. Research model

Within the scope of the implementation of the research, the survey technique was used to measure the sustainable tourism intentions of domestic tourists. A study by Nguyen et al. (2023) was utilised to determine attitude, moral reflection, and sustainable tourism intention. Motivation was taken from the study conducted by Palacios-Florencio et al. (2021). A questionnaire including the demographic characteristics of the participants was created and applied through face-to-face interviews with domestic tourists visiting the Eastern Black Sea region. Participants voluntarily completed 480 questionnaires, and the study was conducted with 472 questionnaire data after removing eight incorrectly completed questionnaires. The first part of the questionnaire includes questions about demographic characteristics. In the second part, 18 statements about the variables used in the study (attitude, motivation, moral reflection and intention) are made. A five-point Likert scale was used in order to reveal the opinions on the items in the second part. Likert rating is between strongly disagree (1) and strongly agree (5). After the collection of the data, the reliability and validity of the scale were determined with the AMOS statistical program. PROCESS macros were used to test the variables predicting sustainable tourism intention.

CFA was conducted to test the validity and reliability of the scale used in the study. The analysis is presented in Table 1.

Table 1. Factor analysis for variables

Dimensions	*Items*	*Factor Loading*	*CR*	*AVE*
Attitude towards Sustainable Tourism (AST)	AST1	0.906		
	AST2	0.869	0,934	0,781
	AST3	0.839		
	AST4	0.919		
Travel Motivation (TM)	TM1	0.889	0,885	0,794
	TM2	0.893		
Moral Reflection (MR)	MR1	0.920		
	MR2	0.909		
	MR3	0.848	0,948	0,786
	MR4	0.877		
	MR5	0.868		
Sustainable Tourism Intention (STI)	STI1	0.814		
	STI2	0.942	0,915	0,784
	STI3	0.895		

When Table 1 is analysed, it is seen that the lowest factor loading value is 0.76, and the highest factor loading value is 0.94. The reliability coefficient (Cronbach Alpha) of the scale was calculated as 0.96. However, within the scope of the measurement model, it was determined that the model fit values were at a good level [Chi-square(372.707)/df(120)= 3.10, RMSEA= .067, NFI= .96, TLI= .96, CFI= .97, IFI= .97, SRMR= .030, GFI= .93]. also the lowest CR value was calculated as 0.88 and the lowest AVE value was calculated as 0.72 (Hair et al., 2014).

Results

A positive and significant relationship was determined between the study variables (attitude towards sustainable tourism, travel motivation, moral reflection and sustainable tourism intention). It can be said that the attitude towards sustainable tourism (M=3.97, SD=.95), travel motivation (M=3.90, SD=1.13), moral reflection (M=3.71, SD=.99) and sustainable tourism intention (M=3.72, SD=.92) of the domestic tourists participating in the research are above average.

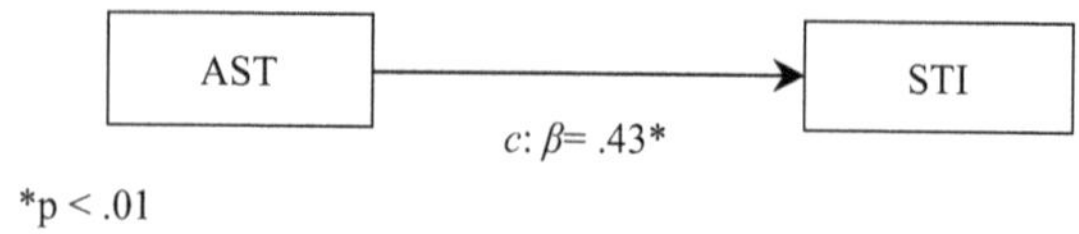

Figure 2. The total effect of AST on STI

The hypotheses were tested with PROCESS macros. The first hypothesis of the research is tested and shown in Figure 2. The total effect of domestic tourists' attitudes towards sustainable tourism on their sustainable tourism intentions is statistically significant (path c; $\beta = .43$, SE= .03 $p<0.01$). According to these results, H3 is accepted.

In addition, the results regarding the parallel mediating role of travel motivation and moral reflection in the effect of attitude towards sustainable tourism on sustainable tourism intention are presented in Table 2. Table 2 shows that attitude has a significant effect on travel motivation (path a1; $\beta = .51$, SE= .06, $p< 0.01$) and moral reflection (path a2; $\beta = .47$, SE= .06, $p< 0.01$). According to findings, H1 and H2 are accepted. In addition, the effects of travel motivation (path b1; $\beta = .60$, SE= .06, $p< 0.01$) and moral reflection (path b2; $\beta = .14$, SE= .06, $p< 0.01$) on intentions towards sustainable tourism are also significant. In this framework, H4 and H5 were accepted.

The indirect effect of domestic tourists' attitudes towards sustainable tourism on their sustainable tourism intentions through travel motivation is significant (path a1b1; $\beta = .31$; 95 % CI [.23, .38], SE=.03). Similarly, the indirect effect of participants' attitudes towards sustainable tourism on their sustainable tourism intentions through moral reflection is significant (path a2b2; $\beta = .06$; 95 % CI [.02, .12], SE=.02). The total indirect effect is also significant (path a1b1 + a2b2; $\beta = .37$; 95 % CI [.29, .45], SE=.02). When travel motivation and moral reflection were controlled as mediating variables, the effect of attitude on intention was not significant (path c'; $\beta = .06$, SE= .03, $p> 0.01$). In this context, travel motivation and moral reflection fully mediate the effect of domestic tourists' attitudes towards sustainable tourism on their sustainable tourism intentions. Within the framework of the findings, H6 was accepted, and the confirmed parallel mediation model is presented in Figure 3.

Table 2. Regression coefficients

Antecedent		Consequent										
		M_1 (TM)				M_2 (MR)				Y (STI)		
		β	SE	p		β	SE	p		β	SE	p
X (AST)	a_1	.51	.03	< .01	a_2	.47	.03	< .01	c'	.06	.03	.06
M_1 (TM)									b_1	.60	.04	< .01
M_2 (MR)									b_2	.14	.03	< .01
Constant	i_{M1}	1.99	.12	< .01	i_{M2}	1.85	.13	< .01	i_y	.58	.12	< .01
R^2		.36				.29				.54		
F (df)		274.65*** (1; 470)				197.44*** (1; 470)				238.76*** (3; 468)		

***p<.001

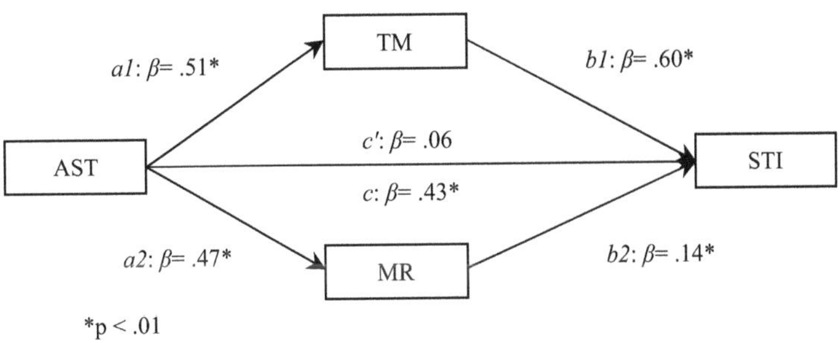

Figure 3. Results of the parallel mediation model

Conclusion and Discussion

In general, sustainable tourism helps protect the products of tourism types and transfers them to future generations. In this context, it is important to consult the views of tourists who consume tourism products. Because tourist behaviours and perceptions change day by day. Hosting more than 50 million tourists in Turkey, examining tourists' intentions towards sustainable tourism is important. Therefore, in this study, the intentions of domestic tourists towards sustainable tourism and the factors affecting them were revealed. In addition, the mediating role of travel motivation and moral reflection in the effect of attitude on intention was examined. Although many studies have been conducted on sustainable tourism, researching tourists' intentions will deepen the subject theoretically and practically.

The study results show that attitude towards sustainable tourism, travel motivation and moral reflection affect sustainable tourism intention. In addition, it was determined that travel motivation and moral reflection mediate the effect of attitude towards sustainable tourism on sustainable tourism intention. These results from Turkey, a developing country, support the results of previous studies (Nguyen et al., 2023; Luongo et al., 2024). Studies are revealing that attitudes and motivation are effective on behavioural intentions (Bayih & Singh, 2020; Wang et al., 2023). Kim et al. (2017) argue that individuals' ethical behaviour within the framework of moral reflection is related to their environmental behaviour. Meanwhile, Verma and Chandra (2018) state that moral reflection positively affects tourists' visits to green hotels.

Tourist behaviour is changing day by day. Based on the results of this article, various solutions can be developed to transfer tourism resources to future generations. Local governments can organise informative billboards or posters for tourists in tourism centres to improve tourists' intentions towards sustainable tourism. If tourism enterprises give messages about sustainability to tourists through social media, websites or various mass media, they can increase tourists' motivation to travel towards sustainable tourism. In addition, tourism policymakers can encourage tourists and tourism businesses to protect the natural environment.

Tourists' sustainable tourism intentions are relatively under-studied. Therefore, this article may shed light on possible future studies. This study investigated the relationships between attitude, motivation, moral reflection and sustainable tourism intention. The relationships between different variables and sustainable tourism intention can be analysed in the future. For example, environmental awareness, socio-economic status and various categorical variables (education, age, generation, gender, income status, etc.) can be added to the model. Meanwhile, the current study includes the views of tourists. Future studies on the views of other tourism stakeholders may differ from the results of this study. This may enrich the literature on sustainable tourism.

References

Afsar, B., & Umrani, W. A. (2020). Corporate social responsibility and pro-environmental behavior at workplace: the role of moral reflectiveness, coworker advocacy, and environmental commitment. *Corporate Social Responsibility and Environmental Management*, 27(1), 109–125.

Ajzen, I. (1991). The Theory of Planned Behavior. *Organizational Behavior and Human Decision Processes*, 50(2), 179–211.

Andereck, K. L. (2009). Tourists' perceptions of environmentally responsible innovations at tourism businesses. *Journal of Sustainable Tourism*, 17(4), 489e499.

Barber, N. A. (2014). Profiling the potential "green" hotel guest: Who are they and what do they want? Journal of Hospitality & Tourism Research., 38(3), 361e387.

Bayih, B. E., & Singh, A. (2020). Modeling domestic tourism: motivations, satisfaction and tourist behavioral intentions. Heliyon, 6, e04839. https://doi.org/10.1016/j.heliyon.2020.e04839.

Çeken, H. (2008). Turizmin bölgesel kalkinmaya etkisi üzerine teorik bir inceleme. *Afyon Kocatepe Üniversitesi, İ.İ.B.F. Dergisi*, 10(2), 293–306.

Chang, L. L., Backman, K. F., & Huang, Y. C. (2014). Creative tourism: a preliminary examination of creative tourists' motivation, experience, perceived value and revisit intention. *International Journal of Culture, Tourism and Hospitality Research*, 8(4), 401–419.

Cheng, S., Lam, T., & Hsu, C. H. (2006). Negative word of mouth communication intention: An application of the theory of planned behavior. *Journal of Hospitality & Tourism Research*, 30(1), 95–116. https://doi.org/10.1177/1096348005284269.

Dief, M. E., & Font, X. (2012). Determinants of environmental management in the Red Sea hotels: Personal and organizational values and contextual variables. *Journal of Hospitality & Tourism Research*, 36(1), 115–137.

Doran, R., & Larsen, S. (2016). The relative importance of social and personal norms in explaining intentions to choose eco-friendly travel options. *International Journal of TourismResearch*, 18(2), 159–166.

Fraenkel, J. R., Wallen, N. E., & Hyun, H. H. (2015). *How to Design and Evaluate Research in Education* (9th ed.). Mc Graw Hill Education.

Gürbüz, A., Ayaz, N., & Apak, Ö. C. (2019). Kış turizminin sürdürülebilirliği kapsamında seyahat acentası yöneticilerinin bakış açısı: Kayseri örneği. *Uluslararası Türk Dünyası Turizm Araştırmaları Dergisi*, 4(1), 35–46.

Hair, J. F., Sarstedt, M., Hopkins, L., & Kuppelwieser, V. G. (2014). Partial least squares structural equation modeling (PLS-SEM): An emerging tool in business research. European Business Review, 26(2), 106–121.

Han, H. (2015). Travelers' pro-environmental behavior in a green lodging context: converging valuebelief- normtheory and the theory of planned behavior. *Tourism Management*, 47, 164–177.

Han, L., Goetz, S. J., Eades, D., Entsminger, J., & Arbogast, D. (2023). An early assessment of COVID-19's impact on tourism in US countries. *Tourism Economics*, 29(5), 1355–1375. https://doi.org/10.1177/13548166221107814.

Hedlund, T., Marell, A., & Gearling, T. (2012). The mediating effect of value orientation on the relationship between socio-demographic factors and environmental concern in Swedish tourist' vacation choices. *Journal of Ecotourism*, 11(1), 16–33.

Hsu, C. H. C., Cai, L. A., & Li, M. (2010). Expectation, motivation, and attitude: a tourist behavioral model. *Journal of Travel Research*, 49(3), 282–296.

Huang, S., & Hsu, C. (2009). Travel motivation: linking theory to practice. *International Journal of Culture, Tourism and Hospitality Research*, 3(4), 287–295.

Imran, S., Alam, K., & Beaumont, N. (2014). Environmental orientations and environmental behavior: Perceptions of protected area tourism stakeholders. *Tourism Management*, 40, 290–299.

Juvan, E., & Dolnicar, S. (2014). The attitude behaviour gap in sustainable tourism. *Annals of Tourism Research*, 48, 76–95. https://doi.org/10.1016/j.annals.2014.05.012.

Kaygısız, A. D. (2023). Türkiye Cumhuriyeti'nin 100. Yılında turizm. *Bilge Uluslararası Sosyal Araştırmalar Dergisi*, 7(2), 124–133. https://doi.org/10.47257/busad.1389291.

Kim, A., Kim, Y., Han, K., Jackson, S. E., & Ployhart, R. E. (2017). Multilevel influences on voluntary workplace green behavior: individual differences, leader behavior, and coworker advocacy. *Journal of Management*, 43(5), 1335–1358.

Kim, M. J., & Hall, C. M. (2019). Can climate change awareness predict pro environmental practices in restaurants? Comparing high and low dining expenditure. *Sustainability*, 11(23), 6777. https://doi.org/10.3390/su11236777.

Luongo, S., Napolano, E., & Gul, K. (2024). Sustainable tourism intentions: Extending the theory of planned. *Turistica*, 32(3), 31–60.

Miao, L., & Wei, W. (2016). Consumers' pro-environmental behavior and its determinants in the lodging segment. *Journal of Hospitality & Tourism Research*, 40(3), 319–338.

Nguyen, V. H., Nguyen, T. H., Mai, L. H., Nguyen, T. T. P., Lan Nguyen, T. M., & Nguyen, T. P. L. (2023). The factors affecting Vietnamese people's sustainable tourism intention: an empirical study with extended the theory of planned behavior (TPB). *Foresight*, 25(6), 844–860. https://doi.org/10.1108/FS-06-2022-0065.

Nowacki, M., Chawla, Y., & Kowalczyk Anioł, J. (2021). What drives the eco friendly tourist destination choice? The Indian perspective. *Energies*, 14(19), 6237. https://doi.org/10.3390/en14196237.

Özkök, F., & Gümüş, F. (2009). Sürdürülebilir turizmde bilginin önemi. Yönetim Bilimleri Dergisi, 7(1), 51–72.

Palacios-Florencio, B., Santos-Roldán, L., Berbel-Pineda, J. M., & Castillo-Canalejo, A. M. (2021). Sustainable tourism as a driving force of the tourism industry in a Post-Covid-19 scenario. *Social Indicators Research*, 158, 991–1011. https://doi.org/10.1007/s11205-021-02735-2.

Puhakka, R. (2011). Environmental concern and responsibility among nature tourists in Oulanka PAN Park, Finland. *Scandinavian Journal of Hospitality and Tourism*, 11(1), 76–96.

Safshekan, S., Ozturen, A., & Ghaedi, A. (2020). Residents' environmentally responsible behavior: An insight into sustainable destination development. *Asia Pacific Journal of Tourism Research*, 25 (4), 409–423. https://doi.org/10.1080/10941665.2020.1737159.

Sarikaya, İ., Töman, U., & Öztürk, M. (2018). Examining the variables predicting attitudes towards teaching profession of preservice classroom teachers. *Acta Didactica Napocensia*, 11(1), 129–141. https://doi.org/10.24193/adn.11.1.10.

Scott, D. J., Lemieux, C. J., & Malone, L. (2011). Climate services to support sustainable tourism and adaptation to climate change. *Climate Research*, 47(1–2), 111–122. https://doi.org/10.3354/cr00952.

Scott, D. J., Lemieux, C. J., & Malone, L. (2011). Climate services to support sustainable tourism and adaptation to climate change. *Climate Research*, 47(1–2), 111–122. https://doi.org/10.3354/cr00952.

Shien, L. Y., Liu, C. H., & Li, Y. M. (2022). How positive and negative environmental behaviours influence sustainable tourism intentions. *Sustainability*, 14(11), 6922.

Süzer, Ö., & Doğdubay, M. (2022). Sürdürülebilir turizm hareketliliğinde yerel gıdaların önemi ve rekabet avantajı (kavramsal bir analiz). *Safran Kültür ve Turizm Araştırmaları Dergisi*, 5(2), 255–269.

Uçgun, G. Ö., & Narcı, M. T. (2023). Konaklama işletmelerindeki sürdürülebilir turizm uygulamalarının satın alma davranışına etkileri. *Afyon Kocatepe Üniversitesi Sosyal Bilimler Dergisi*, 25(4), 1595–1610.

UNWTO. (2023). Statistics. Available at: https://www.unwto.org/unwto-world-tourism-barometerdata. (Accessed: 15.05.2024).

Verma, V. K., & Chandra, B. (2018). An application of theory of planned behavior to predict young Indian consumers' green hotel visit intention. *Journal of Cleaner Production*, 172, 1152–1162.

Verma, V. K., & Chandra, B. (2018). An application of theory of planned behavior to predict young Indian consumers' green hotel visit intention. *Journal of Cleaner Production*, 172, 1152–1162.

Wang, C., Liu, T., Zhu, Y., Wang, H., Wang, X., & Zhao, S. (2023). The influence of consumer perception on purchase intention: Evidence from cross-border

E-commerce platforms. Heliyon, 9, e21617. https://doi.org/10.1016/j.heliyon.2023.e21617.

Weeden, C. (2011). Responsible tourist motivation: How valuable in the Schwartz value survey? *Journal of Ecotourism*, 10(3), 214–234.

Zengin, B., & Kiper, V. O. (2020). Destinasyon sürdürülebilirliği kapsamında de-marketing uygulamaları: Sapanca destinasyonuna dair önleyici bir çalışma. *Türk Turizm Araştırmaları Dergisi*, 4(3), 2102–2112.

Zhang, N., Ren, R., Zhang, Q., & Zhang, T. (2020). Air pollution and tourism development: An interplay. *Annals of Tourism Research*, 85, 103032. https://doi.org/10.1016/j.annals.2020.103032.

Metin Sürme* and Abdullah Akmaz†

Chapter 11 Examples of Successful Sister City Relations and Their Impact on Tourism

Introduction

The concept of Sister City refers to the relationship between two geographically distant settlements for cultural and commercial exchange. The establishment of special units in municipalities interested in Sister Cities, including the quality of the personnel working in these units and foreign language proficiency, is necessary for the success of these relations (Akman & Akman, 2017). Although the reasons that push municipalities to establish Sister City relations with other municipalities are diverse, the primary objectives are cultural exchange, economic development and promotion of tourism (Union of Municipalities of Turkey, 2017).

The concept of sister city can be defined as a set of relationships that municipalities facilitate in order to improve relations between local communities, including civil society, business world, education and municipalities; including many issues ranging from cultural exchanges to economic development. Since the 1950s, the sister city phenomenon has emerged in the USA for the development of peace and mutual understanding between its own cities and cities in the rest of the world. In Europe, at the same time and in a similar manner, twinning or fraternal relations called "Twinning" were started to be established in order to develop friendship especially between French and German cities. Looking at the emergence and development of the sister city concept in the world, it is seen that the first known modern sister city agreement was signed between Keighley in the United Kingdom and Poixdu Nord in France in 1920. Sister Cities International was established in 1956 at the White House citizen diplomacy summit of US President Eisenhower. This initiative envisaged a network that would be a campaign for peace and prosperity that would promote ties between people from different communities around the world (Pelit, et al., 2017).

* Associate Professor Dr. Gaziantep University Faculty of Tourism, surmemetin@gmail.com

† Assistant Professor Dr. Mersin University Faculty of Tourism, aakmaz@mersin.edu.tr

Sister city relations established by a municipality with a domestic or foreign municipality are relations established in the sense of exchanging information, technology and expertise between local governments (Bay & Çalışkan, 2016). The aims and objectives of sister city relations can be listed as promoting cultural exchange, promoting economic development and creating opportunities for international cooperation (Zeren & Aktulun, 2018).

Sister city relations play an important role in promoting tourism and generating tourism demand (Gezici and Kocaoğlu 2018). Similarly, Sister City cooperations have the potential to contribute positively to the development of tourism by promoting cultural exchange, encouraging economic development and creating opportunities for international co-operation (Zeren and Aktulun, 2018). Moreover, the additional economic power generated by a strong domestic tourism movement, mostly reflected in the statistics of visiting home and relatives, can contribute to the overall economic development of the municipality (Çulhaoğlu, 2019). Therefore, Sister City relations can be considered as an important tool for economic development and international cooperation as well as promoting tourism and cultural exchange (Bay & Çalışkan, 2016).

The impact of sister cities on the local economy and infrastructure cannot be overlooked. The development of tourism leads to job creation and the improvement of infrastructure such as transport and accommodation. One of the biggest advantages is the increased accessibility and convenience for tourists. The close proximity of sister cities makes it easier for tourists to visit both cities at once, which saves time and money. While sister cities offer many advantages for tourism, there are also limitations that need to be taken into account. In other words, there are limitations such as limited cultural diversity and over-tourism. Over-tourism can lead to overcrowding and strain local infrastructure. To overcome these challenges, sustainable tourism practices are crucial for the planned development of sister cities.

There are many successful sister cities that have developed their tourism industry. For example, Niagara Falls, located on the border of the United States of America and Canada, has become a popular tourist destination due to its natural beauty and unique attractions. Another example is Copenhagen and Malmö in Denmark and Sweden respectively. Many cities in Turkey have established sister city relations with different cities around the world for cultural and commercial exchange. For example, sister cities of Gaziantep include Duisburg (Germany), Braga (Portugal), Surabaya (Indonesia), Urumqi (China), Tartu (Estonia), Ludwigshafen (Germany), Minsk (Belarus) (Gaziantep Metropolitan Municipality, 2024). These cities have developed a successful tourism industry by emphasising their cultural differences and unique attractions.

Sister city relations have the potential to significantly impact local economies, including the tourism industry (Union of Municipalities of Turkey, 2017). When two cities establish a sister city relationship, they commit to promoting cultural, educational and economic exchange. This exchange can lead to increased tourism as visitors are attracted to experience the unique offerings of the sister city. For example, Ankara, Antalya, Bursa and Izmir Metropolitan Municipality have established the most sister city relationships, leading to increased tourism in these cities (Akman and Akman, 2017). Sister city relations can also lead to the transfer of policies and practices that are successful in one city to another, thus contributing more to economic growth and development (Güneş & Eren, 2021). In the light of all this information, successful examples of sister cities were examined in this study and the impact of sister cities on tourism was revealed.

Conceptual Framework

Sister City Definition and Development Process

In recent years, institutions and organisations such as the World Tourism Organisation (WTO), the World Bank and the European Union have been supporting regional and city projects more than countries. This situation has contributed to the establishment of more relations and cooperation between countries and cities. One of the concepts that emerged at this stage is the concept of "Sister City". The concept of "Sister City" can be defined as a whole that includes many issues from cultural interaction to economic development, covering civil society, business world, education and municipalities in order to improve the relations between local governments and societies (Pelit, Baytok and Gökçe, 2017: 177). "Sister cities" are associations established for cultural and commercial exchange. The concept of "Sister City", which emerged after the 1950s in order to ensure mutual interaction between the cities of the United States of America and the cities of countries around the world, started to be established for the same purpose under the name of "Twinning" in Europe, especially in France and Germany. After these years, as local governments have become more institutionalised, "Sister City" relations have become more prominent with their economic and cultural dimensions. Today, "Sister City" relations have an important place in the foreign relations of municipalities (Özdemir, 2011: 34). The concept, which is one of the effective cooperation arguments in today's globalising world, supports societies and local governments in many different fields both at national and international scales. The concept of "Sister City", which has expanded considerably today with bilateral relations and sharing of information, technology and experience, is a very comprehensive interaction area with collaborations such as agriculture, tourism, industrial investments, activities emphasising cultural, historical and religious loyalties, educational activities, etc. (Bay & Çalışkan, 2016: 75).

The first "Sister City" agreements in Turkey were signed between Konya Metropolitan Municipality and Torrance Municipality of the United States of America in 1958 after the Second World War. After the 1960s, as a result of the intensification of relations between Turkey and European countries, a rapprochement took place and a "Labour Recruitment Agreement" was signed between Turkey and Germany in 1961. In 1964, the "Sister City" agreement signed between İzmir Bergama Municipality and Boblingen Municipality of Germany was the first agreement signed with European cities in Turkey. Later, between 1964–1984, municipalities in Turkey signed 18 "Sister City" agreements, and with the "Metropolitan Law" enacted in 1984 and the collapse of the USSR in 1991, 29 "Sister City" agreements were signed between 1984–1991. The largest surplus occurred after the 2000s, with 315 "Sister City" agreements between 2000–2002 (Bay and Çalışkan, 2020: 75), and a total of 3042 "Sister City" agreements, both national and international, between 2002 and 2019 (Bay and Çalışkan, 2020: 70). The number of "Sister City" agreements with international cities is 1913 as of 2022. The continent where local governments in Turkey have the most "Sister City" agreements is the European continent, and Bulgaria, Republic of Cyprus, Bosnia-Herzegovina and Germany are located in the European continent respectively (Erdem, Ersavaş Kavanoz and Eker, 2024).

Sister City Practice and Tourism

In our globalising world, local governments cooperate not only in economic terms but also in cultural, religious and historical contexts and thus in the axis of tourism with the cooperation they have made with different regions in addition to their own opportunities. These relations have contributed to the interaction of people all over the world with different structures and cultures, the formation of close ties with each other, and the strengthening of ties of tolerance, brotherhood and love (Kocaalan, 2017: 301). There is a very close relationship between local governments and tourism. Considering that tourism is a very important factor in regional development in terms of local governments, tourism investments and practices to be implemented in the regions will have very positive effects on economic development and employment. For this reason, "Sister City" relations carried out by local governments can be considered as a very important marketing activity in the development of regional tourism (Saylan, 2019: 187).

Carruthers (2006), in his study analysing the impact of "Sister City" agreements between China and New Zealand on tourism, stated that the "Sister City" relations of the cities of the two countries are seen as an important factor in economic development and tourism growth, and that cultural interaction and tourism are the main elements that will contribute to economic development.

In his study, Cross (2010) stated that New Zealand's sister cities portfolio reflects its political and economic situation and that the increase in "Sister City" relations with China and Japan has shown positive results in terms of education, tourism and economy.

Liu and Hu (2018) examined the relationship between the number of foreign visitors and the number of international flights in the "Sister City" relations implemented in Chinese local governments, and found that there is a high correlation between the number of "Sister Cities" and international investments, the number of international flights and the number of foreign tourists, and that provinces that develop economically are more inclined to develop "Sister City" relations.

Özyurt (2020), in his study to determine the impact of "Sister City" relations in creating a value-oriented tourism alternative, stated that if the sister city collaborations established on the axis of different purposes are planned and directed correctly, they will provide very serious benefits to the development of tourism, give destinations the opportunity to be recognised in different parts of the world, will benefit promotion and tourism revenues as a result of the arrival of mutual visitors with the established relations, and promotion, advertising and marketing opportunities will be created.

Cohen (2021), in his study on Israel in order to determine how successful the sister cities project is, found that there was a dramatic increase in the number of tourists coming to Israel with bilateral cooperation between 2007–2017, and that culture, religion and common phenomena have a significant impact on tourism.

Ginanjar, Sinaga and Hidayat (2021), in their study investigating the effectiveness of the sister city relationship between West Java and Chongqing in the tourism sector, found that in this agreement between the two cities, both sides tried to take steps at this point by aiming to recognise tourism, prevent illegal travel, and improve each other's welfare and economic dimensions with tourism-oriented activities.

Tristofa and Tham (2022), in their study examining the effects of the "Sister City" cooperation between Yogyakarta and Kyoto in the tourism sector, found that there was a significant increase in the number of tourists coming to these two cities with the bilateral relations between 2015–2018, and that the bilateral agreement had significant contributions to the cities of Yogyakarta and Kyoto in terms of tourism.

Results

Sister cities, which have deep historical roots, are an initiative to produce solutions to potential problems in a much more effective and efficient way by strengthening mutual communication at the horizontal/local level in the 21st century, where individuals and societies have entered into multidirectional interaction with the globalisation process. In this context, studies on sister cities are becoming more and more important in Turkey as in the whole world (Gezici & Kocaoğlu, 2018). In this respect, the importance of the relationship between sister cities and the tourism sector is easily understood. In this study, successful examples of sister cities were analysed and their effects on the tourism sector were evaluated.

Modern tourism should be understood as a network system that needs to be integrated with other planning for social and economic development. At this point, the importance of sister cities can be understood more easily. When the relevant literature is analysed, it is seen that sister cities contribute to the creation of value-oriented strategic cooperation and are effective in creating tourism demand (Özyurt, 2020). Sister city programmes have positive contributions in many ways. In fact, the Sister Cities programme can help sister cities increase the chances of being productive in terms of the number of tourists visiting the country and the time they spend in the country. This kind of behaviour will ensure that the time and money spent in this process will be used more economically and efficiently. Tourism planners should also match the number of Sister Cities agreements signed with the resources available for the programme. They should bear in mind that success will require a much greater investment of resources and time if the compatibility requirement discussed in terms of size and culture is not met. Strategic tourism planners should also take into account other objectives, such as economic, political or cultural co-operation, as well as direct tourism growth, which will affect the participating parties in the long term (Gil, 2020).

As a result of the literature review, it has been observed that there is a limited number of studies examining the extent to which the sister cities programme affects tourism. In this context, it is recommended to examine different examples of sister cities on the subject. As another suggestion, guests coming from sister municipalities can be evaluated as a special status rather than tourists in relation to tourism, and a separate monitoring method can be used specifically for sisterhood.

This study can help tourism policy makers to design a more effective Sister Cities engagement policy for their countries and cities. A related scale can be developed and the impact of Sister Cities on tourism development at local and national level can be investigated.

References

Akman, Ç., & Akman, E. (2017). A qualıtatıve analysıs of town twınnıng ın turkey based on four metropolıtan munıcıpalıtıes. *Journal of Süleyman Demirel University Institute of Social Sciences* (27), 228–254.

Bay, A., & Çalışkan, V. (2019). National and international sister city relationships of İstanbul: determinative factors and characteristics of spatial distribution, 1st İstanbul İnternatıonal Geography Congress Proceedings Books, İstanbul, Türkiye.

Bay. A., & Çalışkan, V. (2020). Preference Factors and Spatial Distribution Characteristics of The International Sister City Agreement Of Turkey. *Eastern Geographical Review*, *25*(43), 73–92.

Carruthers, D. (2006). New Zealand and Chinese Sister Cities Their İmplications for Tourism. Unıversity of Otago, Department and Tourism Master Graduate Profile, Otago.

Cohen, G. (2021). How the Sister Cities Program Promotes Tourism, *Israel Affairs*, 27:2, 340–350.

Cross, B. (2010). Sister Cities and Economic Development: A New Zealand Perspective. Transylvanian Review of Administrative Sciences, 30, 104–117.

Çulhaoğlu (2019). İç turizmde kardeş şehirlerin önemi. Turizm Aktüel Gazetesi. https://www.turizmaktuel.com/haber/ic-turizmde-kardes-sehirlerin-onemi.

Erdem, N., Ersavaş Kavanoz, S. ve Eker, D. (2024). Analysis of Sister City Relations in Provincial and Municipal Level. Hacettepe University Journal of Economics and Administrative Sciences, 42(1), 47–65.

Gaziantep Büyükşehir Belediyesi (2024). https://www.gaziantep.bel.tr/tr/kardessehirler.

Gezici, H. ve Kocaoğlu, M. (2018). Dünyada ve Türkiye'de Kardeş Şehir Uygulamaları. Çağdaş Yerel Yönetimler, 27(2), 113–139.

Gil, C. (2020). The sister cities program and tourism. *Journal of Hospitality and Tourism Management*, 45, 182–191.

Ginanjar, Sinaga and Hidayat (2021). Effectiveness Of Sister Province Between West Java and Chongqıng in The Tourism Sector. Palarch's. *Journal of Archaeology of Egypt/Egyptology* 18(4), 4787–4799.

Güneş, I., & Eren, E. (2021). Town Twinning Implementation of Local Government Diplomacy Particular to Istanbul. *Strategic Public Management Journal*, 7(13), 35–59.

Kocaalan, M. L. (2017). Türkiye ve Suriye Arasındaki Kültürel Entegrasyonun Geliştirilmesinde Kardeş Şehir Uygulamasının Önemi. International West Asia Congress of Tourism (Iwact'17), Van.

Liu, H. and Hu, X. (2018). Are "Sister Cities" from 'Sister Provinces'? An Exploratory Study of Sister City Relations (SCRs) in China. Networks and Spatial Economics, 18, 473–491.

Pelit, E., Baytok, A., & Gökçe, Y. (2017). Reflections of Sister City Concept on Tourism: The Case of Afyonkarahisar. VI. Ulusal II. Uluslararası Doğu Akdeniz Turizm Sempozyumu 14–15 Nisan 2017/Gaziantep-TÜRKIYE.

Özdemir, A. T. (2011). Belediyelerin Uluslararası İlişkilerine İlişkin Temel Hususlar. İller ve Belediyeler Dergisi, Kasım-Aralık, 34–41.

Özyurt, P. M. (2020). An Alternative Tool for Creating Value-Oriented Tourism Demand: Sister City Cooperations. *Journal of Turkish Tourism Research*, *4*(3), 2378–2390.

Saylan, U. (2019). Studies on Interdisciplinary Economics and Business in: Sister Cities and Tourism: The Case of Turkey. 185 – 193, Berlin: Peterlang.

Tristofa, L. ve Tham, A. (2024). The Implementation of Sister City Agreement on Tourism between the Special Region of Yogyakarta Indonesia and Kyoto Japan. Journal of Paradiplomacy and City Networks, 1(1), 15–26.

Union of Municipalities of Turkey (2017). Neden Kardeş Şehir? https://www.tbb.gov.tr/Tr/icerik_neden-kardes-sehir_99.

Zeren, H. E., & Aktulun, E. (2018). Partnershıps and sıster-cıty relatıons of munıcıpalıtıes wıthın the framework of governance: the case of ankara metropolıtan cıty. *ASSAM International Refereed Journal*, 5(11), 1–14.

Yusuf Karakuş*

Chapter 12 Destination Management and Local Development: Economic and Social Impacts of Tourism

Introduction

The importance of Destination Management

Destination Management Organizations (DMOs) are crucial for local development as they coordinate various stakeholders and resources to plan, implement and market tourism development policies in a destination (Foriş et al., 2020). Sustainable refers to the use of resources subjected to economic, social and environmental variables arising from tourism activities and the concerns in ensuring its continuity and development for future uses (Costa et al., 2021). It is widely recognized that DMOs play a key role in coordinating local stakeholders and tourism resources to ensure sustainable tourism development (Chaniago, 2022). Effective destination management is becoming essential to prevent unsustainable tourism practices, such as over-tourism, which can negatively affect residents and the destination's quality of life (Mihalič and Kuščer, 2021). Through strategic leadership, effective governance and efficient implementation, DMOs can lead destinations towards sustainable tourism practices (Mihalič and Kuščer, 2021).

An important aspect of sustainable destination management is the segmentation of tourists into target markets based on their psychographic characteristics (Dolničar and Leisch, 2008). This strategy helps DMOs to tailor their marketing efforts to attract visitors who are more likely to engage in sustainable tourism practices. Moreover, DMOs are highly effective in developing plans for sustainable tourism activities, collaborating with service providers, identifying competitive advantages, promoting destination image, and creating tourism products to offer unique visitor experiences (FAUR & Ban, 2022). These activities can improve the well-being of local people and provide strategic options for sustainable tourism development and enhance the overall sustainability of tourism destinations (FAUR & Ban, 2022).

* PhD., Recep Tayyip Erdoğan University, Ardeşen Tourism Faculty, Department of Tourism Management, yusuf.karakus@erdogan.edu.tr

Sustainable destination management is based on three key parameters. These are knowledge management, collaboration and innovation (Martins, 2023). Knowledge management involves the sharing of knowledge, continuous education and training of individuals in the tourism sector to build a broad knowledge base. This accumulation of knowledge can lead to many improvements in the destination, such as standardization of processes, reduced costs, and increased quality. Collaboration involves engaging stakeholders, creating a shared vision and fostering strategic dialogues. Collaboration is vital as destinations are highly complex, with a wide variety of stakeholders and conflicting interests. Innovation is achieved through creative thinking and exploring new destination management approaches. These parameters collectively contribute to the sustainable development of tourism destinations (Dornier & Mauri, 2018).

Managing a tourist destination is a task involving several challenges and multifaceted variables (Kaya & Kaya, 2023). One of the most important reasons why destination management is complex is the competitive and multifaceted nature of the tourism industry, which requires the coordination of economic, social and geographical elements within a designated tourism area (Carlsen, 1999). DMOs face the challenge of balancing the interests of different stakeholders, including local communities, businesses, tourists and government agencies, to ensure the sustainable development and growth of the destination. However, it should be noted that the interests of these stakeholders are very different and sometimes conflict. In other words, an advantage for one stakeholder is a disadvantage for another. In such an environment, it is not easy to make the best decisions. Moreover, the management of tourist destinations is complicated by the need to understand and cater to the different preferences and behaviors of tourists. Factors such as tourist experience, inspiration, familiarity with the destination, age and family structure influence visitors' choice of activities and overall satisfaction (Xue et al., 2022; Tomić et al., 2019). DMOs should continuously adapt their marketing and management strategies to meet the changing demands of target markets. This is because in a world of ever-changing dynamics, consumers' needs, wants and expectations must always be predictable. In order to attract and retain tourists, it is crucial to establish a positioning in the minds of consumers and manage it continuously (Tomić et al., 2019).

Environmental factors also pose challenges for destination management, as destinations need to strike a balance between economic development and the preservation of the environment and cultural values. Regulations and policies implemented by destination management organizations play an important role in promoting sustainable tourism practices, economic benefits for local communities and environmental protection within the destination (Zulvianti et al.,

2022). Managing these various factors requires an approach and a systematic approach that can take into account the long-term impacts of tourism activities in the region. However, implementing such activities will not be that easy. This is because tourism activities, like other industries, are based on resource utilization. Therefore, scarce resources will make it difficult to make the right decisions about all these destination components. Because even if resources are scarce, there is no limit to what needs to be done for a sustainable tourism development. Therefore, establishing the above- mentioned balance can be realized by DMOs with authority, responsibility and competence in this sense.

The increasing use of social networks and digital marketing tools in destination management presents both opportunities and challenges for destination managers. While these tools can improve communication, engagement and marketing efforts, their effective use requires specialized knowledge and skills (Elizondo Saltos et al., 2021). At the destination level, the active use of information communication technologies by businesses alone may not have the expected impact. This is because the tourist product we call a destination has an integrated structure and a success that does not spread to the whole may not make sense for the destination in the long term. For this reason, systematic managerial and marketing actions should also be carried out in the digital world for tourist destinations. Furthermore, ensuring tourist safety and satisfaction is a critical issue that destination management must address. Tourism flows can be affected quickly and severely by many variables. Security breaches may also cause tourists not to prefer or even reject a destination later on (Karakuş & Kalay, 2017). In order to prevent such situations from occurring, a proactive approach should be adopted and the destination management mechanism should be operated. In doing so, challenges such as providing accurate information, addressing accommodation issues, analyzing risks, managing crises and disasters can affect tourists' perceptions and their overall satisfaction with the destination (Žabkar et al., 2010).

Moreover, the concept of destination image and brand personality plays an important role in shaping tourists' perceptions and influencing their travel intentions (Agustina, 2018). DMOs face the challenge of creating and maintaining a positive destination image that resonates with tourists and differentiates the destination from its competitors. This is because the image of the destination is one of the most important factors influencing the purchase of travel decisions for tourism activities. Although the destination image is formed in the minds of consumers, the DMO has tools to manage this image. However, by using these tools effectively, an image can be created, a bad image can be corrected or a good image can be strengthened.

The management of tourism destinations has important implications not only for the expansion and long-term viability of the economy, but also for the social and cultural interactions in these destinations. The strategies implemented by Destination Marketing Organizations (DMOs) have a wide range of social and economic outcomes that directly affect the well- being of local people and the satisfaction of tourists. Understanding the social and economic impacts of destination management and tourism is crucial to ensure sustainable and diverse tourism growth.

Impact of Tourism on Local Development

Local development is one of the very first and the most significant tool that tourism can be considered for and it will affect several aspects of societies. Tourism can play a significant role, especially in managing foreign trade balances for underdeveloped and developing countries. It can also help reduce differentiation in terms of development in a country. In other words, while some regions can develop through non-tourism industries, regions, which are not able to develop in this way, can also use tourism. Community-Based Tourism Community-based tourism has been an approach developed to attain and promote sustainable development in rural areas with the help of producing economic, social, and environmental benefits by the local community (Martínez-Salgado, 2021; Sarmiadi et al., 2021). All the advantages from the involvement of local communities in tourism activities and business, on the other hand, can provide income and improve the quality of life in rural areas by being a source of income for the local communities in that rural area (Juma & Khademi-Vidra, 2019). More generally, tourism empowerment programs have been supported to give direct influence on families in local communities, socio-economic development, and improvement of lifestyles (Sulaiman, Adi, & Runtiko, 2022). The benefits and many more that tourism flows can bring for the local development are therefore striking. However, further research indicated that tourism development can have both positive and negative impacts on local communities. These affect factors such as quality of life, economic growth, social well-being, and cultural understanding (Yu et al., 2018; Bajrami et al., 2020; Hwang et al., 2011). Residents' support on tourism development is often strongly influenced by their perceptions of the impacts of tourism flows on communities (Jansri, 2019). Positive perceptions lead to increased satisfaction with quality of life and willingness to support tourism initiatives (Bajrami et al., 2020; Jansri, 2019). Furthermore, tourism has been associated with economic development, natural resource conservation and socio-cultural enrichment in local communities (UMAM et al., 2022). It should be mentioned that long-term tourism activities in a region are not likely to find success if the local people in that region do not support the same. Therefore, local people will only embrace and support tourism activities if they are positively affected by tourism impacts. Therefore, the more local development is realized in regions where tourism activities are concentrated and the more negative impacts are minimized, the more the local community will embrace and support tourism and become a locomotive stakeholder for its development. Nevertheless, taking into consideration the plethora of stakeholders and their expectations, it cannot be deemed an easy task to manage the attitude of local people towards tourism. This is the reason why there is a need for a competent DMO that will assure the most effective use of the available resources. DMOs should be those entities that will be able to manage the impacts of tourism development on local communities; from economic growth and the creation of job opportunities to environmental degradation and cultural commodification (Thomas, 2015). The impacts of tourism on community wellbeing are closely related to factors such as residents' perception of tourism impacts, community participation, and support of tourism development (Li et al., 2019). Sustainable tourism practices and good governance are the approaches that can be effective in reducing negative impacts on local communities and biodiversity conservation (Din et al., 2014). As mentioned above, tourist destinations have a very complex structure, and the diversity of stakeholders is rather high. For this reason, a governance approach should be adopted in terms of destination management and a transparent management model should be established in which all the stakeholders are included in the decision-making mechanism and are liable to competencies and responsibilities.

The Role of Destination Management on Local Development

There are quite the different studies that can be found towards the examination of the impact of tourism on local development and the role of the DMO (Hristov, 2014; Permiakova, 2012; Tuohino & Konu, 2014). Sustainable tourism policies and destination management are very important in that they empower communities and revitalize the economy of the country through tourism activities (Khan et al., 2021). In particular, DMOs' efforts to promote sustainable tourism development could pay off in terms of economic benefits to the host community and cultural heritage preservation (Gato et al., 2022). Finally, as underlined by Buhalis and O'Connor, Destination management using ICT can be conducted differently since it can lead to interactive communication between all the stakeholders of tourism and tourists. Therefore, it is a potential way of rethinking and enhancing the procedure of developing and producing tourism products and services (Buhalis and O'Connor, 2005).

The tourism sector has been extensively affected by the COVID-19 pandemic. This has indicated that destination management has become a strategic issue in guaranteeing sustainable and competitive tourism development, especially in rural areas (Guerreiro, 2021). At the same time, it should be noted that destination branding and image can play an important role in attracting tourists and affecting the destination choice. It is also a proven fact that effective destination management is crucial in shaping tourist perceptions (Tamitiadini et al., 2021). The 'collaborative approach to knowledge sharing in rural tourism destinations' by Chin et al. (2021) has made it clear that the destination marketing is a necessity so that the sustainable development of tourism in each region will be achieved. DMOs, through strategic planning, through their interactions with tourism service providers, and through the design of unique tourism products, can effectively participate in the promotion of tourism development and destination competitiveness, as mentioned above (FAUR and Ban, 2022). In fact, as Šerić and Marušić (2019) have highlighted, the participation of industry actors in the management of a tourist destination can have a key role in encouraging innovation and sustainable growth, especially in tourism destinations (Štumpf et al., 2021). Destination marketing as promoted by DMOs is an activity that is considered to be of high importance for the presentation improvement of destinations and the facilitation of collaboration among stakeholders for the creation of a united strategy for increasing tourism attractiveness and competitiveness (Wang, 2008).

The direction taken for tourism development is important and can produce an effect on communities with respect to economic development, cultural preservation, and ecological preservation. DMOs, through the adoption of sustainable practices of tourism and through effective collaboration with local stakeholders, can provide benefits to host communities and minimize the negative effects of tourism development. As Virág argues, the development of a DMO system is essential to managing sustainable and competitive tourism throughout the globe (Virág, 2013).

Conclusions

Managing destination tourism is complex and requires detailed consideration of many factors. Destination Management Organizations (DMOs) play a pivotal role in the mobilization of diverse stakeholders and resources and consequently the successful development and implementation of tourism development policies. Destination management pegs on sustainability to ensure continued viability of the tourism activities with minimal adverse impacts on the environment as well as the local community.

Proper destination management requires strategic leadership, appropriate governance, and timely implementation. Destination management organizations (DMOs) must manage and prioritize the interests and needs of the stakeholders, including the local community, businesses, tourists, and government agencies. This requires continuous improvement of the shifting market trends and consumer needs. Social media platforms and digital marketing tools can enhance the destination marketing initiatives, but it also introduces challenges in information management and promoting brand discipline.

Tourist experience is critical in destination management, given that it has a great influence on the satisfaction and ultimately the loyalty of the tourists. Destination management organizations (DMOs) must harness unique and unforgettable tourist experiences which resonate well with the destination brand. This entails the crafting of exciting activities, quality service delivery, and the establishment of a social bond among the tourists.

Sustainable destination management means properly and effectively managing the environmental and social impacts that are associated with tourism. DMOs need to develop policies and strategies that create a harmonious balance between economic growth and the protection of the environment and society. This requires careful planning and coordination with the surrounding communities, businesses, and government bodies. In other words, the sustainable management of tourist destinations in the long term depends to a great extent on destination management. DMOs must adopt an integrated approach to the management of economic, social, and environmental issues in a balanced manner. They can create positive and memorable tourist experiences in a long-lasting way that brings benefits to the tourists and local people. The management of tourist destinations is a complex process that requires the careful consideration of multiple factors. DMOs must follow an integrated approach to the management of economic, social, and environmental issues in a balanced manner. They can create positive and memorable tourist experiences in a long-lasting way that brings benefits to the tourists and local people.

Destination management is a complex process that requires careful consideration of many factors. Destination Management Organizations (DMOs) are key in arranging different stakeholders and resources to develop and implement tourism development policies. Destination management involves sustainability as a key element, seeking to ensure the long-term feasibility of the tourism activities with a reduced adverse impact on the environment and local communities.

Destination management can only succeed if there is strategic leadership, effective governance, and implementation. Destination management organizations must be able to balance the needs and interests of various stakeholders, including the local community, businesses, tourists, and government agencies. This needs to take into consideration the continuous changes in the market trends and consumer behaviors. Destination marketing can be supported using digital marketing tools and social media platforms, though it can also be challenging in terms of information management and managing a consistent brand image.

The tourist experience is an essential element in destination management. The tourist experience will significantly affect tourist satisfaction and loyalty. DMOs should, therefore, focus on offering unique, memorable tourist experiences consistent with the destination's existing brand. This involves creating attractive activities, delivering high-quality services and fostering socializing among tourists. Sustainable destination management must involve the effective management of the environmental and social impacts of tourism. DMOs must develop policies and strategies to strike a harmonious balance between economic growth and the protection of the environment and social well-being. This has to be achieved through careful planning and coordination with local communities, businesses, and government agencies.

References

Agustina, N. K. W. (2018). The Influence of Destination Images on Revisit Intention in Mount Batur. *Journal of Business on Hospitality and Tourism*, *4*(2), 157. https://doi.org/10.22334/jbhost.v4i2.125

Bajrami, D. D., Radosavac, A., Cimbaljević, M., Tretiakova, T. N., & Syromiatnikova, Y. A. (2020). Determinants of Residents' Support for Sustainable Tourism Development: Implications for Rural Communities. *Sustainability*, *12*(22), 9438. https://doi.org/10.3390/su12229438

Buhalis, D., & O'Connor, P. (2005). Information Communication Technology Revolutionizing Tourism. *Tourism Recreation Research*, *30*(3), 7–16. https://doi.org/10.1080/02508281.2005.11081482

Carlsen, J. (1999). A Systems Approach to Island Tourism Destination Management. *Systems Research and Behavioral Science*, *16*(4), 321–327. https://doi.org/10.1002/(sici)1099-1743(199907/08)16:4<321::aid-sres255>3.0.co;2-5

Chaniago, A. (2022). *Digital Marketing and Destination Management Models in Shaping Tourist Behaviour*. https://doi.org/10.46254/eu05.20220609

Chin, C. H., Wong, W. W. P. M., Chin, C. L., & Franklin, G. (2021). Discovering the Intangible Innovation of Knowledge Sharing for Improving Rural

Tourism Destinations' Competitiveness: A Collaborative Approach. *International Journal of Academic Research in Business and Social Sciences, 11*(8). https://doi.org/10.6007/ijarbss/v11-i8/10655

Costa, J., Varum, C., Montenegro, M., & Gomes, J. P. (2021). Measuring Tourism Success: Are We on the Brink of a New Paradigm? *Worldwide Hospitality and Tourism Themes, 14*(1), 3–7. https://doi.org/10.1108/whatt-10-2021-0130

Din, B. H., Habibullah, M. S., & Choo, W. C. (2014). The Impact of Sustainable Tourism and Good Governance on Biodiversity Loss in Malaysia. *SHS Web of Conferences, 12,* 1063. https://doi.org/10.1051/shsconf/20141201063

Dolničar, S., & Leisch, F. (2008). Selective Marketing for Environmentally Sustainable Tourism. *Tourism Management, 29*(4), 672–680. https://doi.org/10.1016/j.tourman.2007.07.010

Dornier, R., & Mauri, C. (2018). Overview: Tourism sustainability in the Alpine region: The major trends and challenges. *Worldwide Hospitality and Tourism Themes, 10*(2), 136–139.

Elizondo Saltos, A., Flores-Ruiz, D., & Barroso González, M. D. L. O. (2021). Applying Social Networks in the Management of Sustainable Tourist Destinations: An Analysis of Spanish Tourist Destinations. Land, 10(11), 1142.

Faur, M., & Ban, O. (2022). The Role of Tourism Destination Management Organizations in the Development of Tourism Activities. *The Annals of the University of Oradea Economic Sciences.* https://doi.org/10.47535/1991auoes31(1)004

Foriş, D., Florescu, A., Foriş, T., & Barabas, S. (2020). Improving the Management of Tourist Destinations: A New Approach to Strategic Management at the DMO Level by Integrating Lean Techniques. *Sustainability.* https://doi.org/10.3390/su122310201

Gato, M., Dias, Á., Pereira, L., Costa, R. L. D., & Gonçalves, R. (2022). Marketing Communication and Creative Tourism: An Analysis of the Local Destination Management Organization. *Journal of Open Innovation Technology Market and Complexity, 8*(1), 40. https://doi.org/10.3390/joitmc8010040

Guerreiro, S. (2021). Destination Management in a Post-Covid Environment. *Worldwide Hospitality and Tourism Themes, 14*(1), 48–55. https://doi.org/10.1108/whatt-10-2021-0137

Hristov, D. (2014). Co-creating destination development in the post-austerity era: destination management organisations and local enterprise partnerships. *Revista Turismo & Desenvolvimento, 3*(21/22), 25–34.

Hwang, D., Stewart, W. P., & Ko, D. W. (2011). Community Behavior and Sustainable Rural Tourism Development. *Journal of Travel Research, 51*(3), 328–341. https://doi.org/10.1177/0047287511410350

Jansri, W. (2019). Perception of Residents, Community Participation and Support for Tourism Development in the Old Town Muang Songkhla, Thailand. *SSRN Electronic Journal*. https://doi.org/10.2139/ssrn.3433965

Juma, L. O., & Khademi-Vidra, A. (2019). Community-based tourism and sustainable development of rural regions in Kenya; Perceptions of the citizenry. *Sustainability*, *11*(17), 4733.

Karakuş, Y., & Kalay, N. (2017). A Study on the Concept and Causes of Destination Rejection. *International Journal of Management Economics and Business*. https://doi.org/10.17130/ijmeb.2017331320

Kaya, E., & Kaya, M. C. (2023). The role of contingency theory for tourist destinations management. *Tourist Destination*, *1*(1 SE-Theoretical paper), 25–32. https://doi.org/10.5281/zenodo.10251165

Khan, M. R., Khan, H. U. R., Lim, C. K., Tan, K. L., & Ahmed, M. F. (2021). Sustainable Tourism Policy, Destination Management and Sustainable Tourism Development: A Moderated-Mediation Model. *Sustainability*, *13*(21), 12156. https://doi.org/10.3390/su132112156

Li, Q., Xiao, C., Liu, C., & Yang, Q. (2019). *Research on the Effect of Tourism Impact Perception on the Subjective Well-Being of Community Residents*. https://doi.org/10.2991/ichssd-19.2019.66

Martínez-Salgado, O. (2021). Study of Tourism Empowerment in the Region El Chagüe With a Gender Perspective. *Apuntes De Economía Y Sociedad*, *2*(2), 37–46. https://doi.org/10.5377/aes.v2i2.13088

Martins, L. P. (2023). The Enduring Political Viability of Governance Models in Tourist Destinations and Their Implications for People and Planet. *Worldwide Hospitality and Tourism Themes*, *15*(6), 669–676. https://doi.org/10.1108/whatt-09-2023-0115

Mihalič, T., & Kuščer, K. (2021). Can Overtourism Be Managed? Destination Management Factors Affecting Residents' Irritation and Quality of Life. *Tourism Review*, *77*(1), 16–34. https://doi.org/10.1108/tr-04-2020-0186

Permiakova, A. (2012). *The engagement of local residents in Destination Management Organizations*.

Šerić, N., & Marušić, F. (2019). Tourism promotion of destination for Swedish Emissive market. *Advances in Economics and Business*, *7*(1), 1–8.

Štumpf, P., Business, J., Hradec, C. R. J., Mattyašovská, J., & Šťastná, A. (2021). *The Role of Entrepreneurs in Tourism Destination Management System: A Conceptual Model*. https://doi.org/10.31410/eman.s.p.2021.123

Sulaiman, A. I., Adi, T. N., & Runtiko, A. G. (2022). Community empowerment as socio-economic education of tourism villages based on local wisdom. *Education Quarterly Reviews*, *5*(3).

Tamitiadini, D., Hapsari, R., & Hussein, A. S. (2021). Destination Branding: Perceived Sacredness and Spiritual Tourism in Indonesia Amid the COVID-19 Pandemic. *Kne Social Sciences*. https://doi.org/10.18502/kss.v5i8.9396

Thomas, T. K. (2015). Measuring Community Impact Assessment for Internal Destination Performance Evaluation in an Exploring Tourist Destination. *Atna Journal of Tourism Studies, 10*(1), 53–71. https://doi.org/10.12727/ajts.13.5

Tomić, S., Leković, K., & Tadić, J. (2019). Consumer Behaviour: The Influence of Age and Family Structure on the Choice of Activities in a Tourist Destination. *Economic Research-Ekonomska Istraživanja, 32*(1), 755–771. https://doi.org/10.1080/1331677x.2019.1579663

Tuohino, A., & Konu, H. (2014). Local stakeholders' views about destination management: who are leading tourism development? *Tourism Review, 69*(3), 202–215.

Umam, K., Kurniawati, E., & Widianto, A. A. (2022). The Dynamics of "Pokdarwis Capung Alas" in the Development of Community - Based Tourism in Pujon Kidul Village During the Covid-19 Pandemic. *Geojournal of Tourism and Geosites, 43*(3), 850–857. https://doi.org/10.30892/gtg.43302-896

Virág, Á. (2013). Cooperation to Achieve Successful Tourism. *Visegrad Journal on Bioeconomy and Sustainable Development, 2*(2), 68–71. https://doi.org/10.2478/vjbsd-2013-0014

Wang, Y. (2008). Collaborative Destination Marketing: Roles and Strategies of Convention and Visitors Bureaus. *Journal of Vacation Marketing, 14*(3), 191–209. https://doi.org/10.1177/1356766708090582

Xue, J., Zhou, Z., Majeed, S., Chen, R., & Zhou, N. (2022). Stimulating Tourist Inspiration by Tourist Experience: The Moderating Role of Destination Familiarity. *Frontiers in Psychology, 13*. https://doi.org/10.3389/fpsyg.2022.895136

Yu, C., Cole, S., & Chancellor, C. (2018). Resident Support for Tourism Development in Rural Midwestern (USA) Communities: Perceived Tourism Impacts and Community Quality of Life Perspective. *Sustainability, 10*(3), 802. https://doi.org/10.3390/su10030802

Žabkar, V., Brenčič, M. M., & Dmitrović, T. (2010). Modelling perceived quality, visitor satisfaction and behavioural intentions at the destination level. Tourism management, 31(4), 537–546.

Zulvianti, N., Aimon, H., & Abror, A. (2022). The Influence of Environmental and Non-Environmental Factors on Tourist Satisfaction in Halal Tourism Destinations in West Sumatra, Indonesia. *Sustainability, 14*(15), 9185. https://doi.org/10.3390/su14159185

Ilkay Kanik*

Chapter 13 The Presentation of Different Forms of Danger in Turkish Cinema Through Food and Beverages

Introduction

The concept of danger is described by the Turkish Language Association (n.d.) as "a situation with a great capacity for harm that can even lead to annihilation" and is defined as "a potentially hazardous situation." The Merriam-Webster (n.d.) dictionary also defines danger as "an unsafe situation with a high chance of loss and injury, leading to harm and loss." Where there is danger, harmful situations and violence are very close.

Danger inherently contains potential risks and harms. Turkish cinema defines dangerous situations and features danger in its films. This article argues that food relationships and food metaphors are used in very different ways in places where danger is imminent. Through this, society recognizes danger and learns to cope with it. Following the diversity of food metaphors within areas and situations of danger also allows for the development of different perspectives in the analysis of social structures. The situations that cinema depicts as dangerous demonstrate the functioning of social values, social relationships, and social dynamics. Additionally, this article, primarily focusing on Yeşilçam era cinema, illustrates how period-specific societal dangers are portrayed in different styles in films.

Paul Ekman (1992, p. 170) discusses a series of distinct, different emotional states such as fear, anger, and enjoyment. Danger is a condition that evokes many emotions. A person's recognition, confrontation, and relationship with danger vary from society to society. Therefore, danger is also a sociological phenomenon. Each society defines its own dangerous situations. When danger emerges as a situation that must be dealt with, emotions such as humor, fear, and anger may sometimes arise. Following food and nutrition relationships reveals the social conditions of people coping with danger. Danger indicates a tense situation. This dangerous situation is defined in Yeşilçam Cinema, is nourished by the dynamics of society, and gains meaning through social relationships.

---

* Associate Professor, Visiting Researcher, Center for the History of Emotions, Max Planck Institute for Human Development, Berlin, Germany

In this study, representations of danger in films are classified and distinguished by following food relationships and food metaphors. The humor, characters, situations, and emotional depictions associated with danger have been analyzed.

Danger in Humor, Comedic, and Ludicrous Situations

When the language of humor is created, the presence of danger within this language also changes the style of its transmission. Henri Bergson writes about how comic situations arise. He argues that the natural habitat of the comic is indifference and claims that the greatest enemy of the comic is emotions. Indifference emerges with a detached view of situations. One of the consequences that can arise where uncontrollable events occur is indifference. This situation leads to a detached view. As Bergson suggests, through a detached view, it is possible to see the comic situation by distancing oneself from the accompanying emotions (Bergson, 2011, p. 5–8). The theories that laughter and the comic prevail go back to Plato. John Morreall, who examines the relationship between laughter and the theory of superiority of the comic, argues that laughter changes the situation. Because the comic situation empowers the laugher (Morreall 1997, p. 8–24).

In times where people do not feel safe, where economic inequality and violence have become a threat, a detached perspective establishes a relationship between danger and humor. A humor and laughter that prevails over evil provides a way to cope with the emotional situations brought about by danger. Selected films from the Yeşilçam era can serve as examples of the relationship between danger situations and humor, comedy, and the comic.

Danger: The gangster Malik tries to take away people's living spaces as part of a racketeering scheme.

Comic Situation: In the film *Yedi Bela Hüsnü* (Baytan 1982), Kabadayı Malik attempts to evict residents of Hisarüstü Neighborhood in Istanbul to turn it into a racketeering area. He initiates legal proceedings for this purpose and sends warnings to all the shops in the neighborhood. Malik begins to pressure the residents to move out by threatening them. The residents claim that Hüsnü, a poor and naive character like themselves, is their protector against Malik's intimidation. Hüsnü, believing himself to be a tough guy, adopts the name *Yedi Bela Hüsnü* (Seven Troubles Hüsnü). Despite all of Malik's efforts, he cannot defeat Yedi Bela Hüsnü, so he decides to try another method: feeding him a bomb. He calls his men and orders them to find the bomb maker Niyazi and prepare a pear bomb. He also sends his regards to the bomb maker, saying, "*If the bomb doesn't throw Hüsnü ten meters forward, I'll throw him myself.*" One of Malik's men argues that the bomb maker had made an apple bomb last time and claims that the apple bomb is better. However, Malik insists on a pear bomb. Then, Malik invites Hüsnü to dinner to talk and settle their issues. Before resolving the matter with Hüsnü, Malik insists on having dessert. He then orders his men to bring pears for dessert. Despite Hüsnü's suggestion to eat pilav first, Malik insists on having fruit before the meal. Malik explains to Hüsnü what tough guys eat: "*The best part of the pear is reserved for the tough guys, here you go*". However, the person making the bombs, Niyazi, does not listen to Malik; he places the bombs inside the apples again. The pears that Hüsnü eats do not explode, but the apples that Malik eats do. In the incident of apple and pear bombs, we see that the situations become intertwined and quite complicated. The expected series of events is disrupted, and an unexpected and surprising situation arises. Malik and his men, who are expected to be strong, find themselves in a comic situation. A naive person turns the tide in his favor through coincidences. The apple bombs serve this purpose. Situation comedy is created by the interweaving of lines. Repetitions such as apple or pear bombs, repetitions, mixing apples and pears, confusion of the question of who will eat the pear, the chaos created by the good and bad eating the pear reveal the situation comedy.

The cruel Malik, who is powerful and wants to evict people from their homes, is brought into a comic situation and defeated by the naive, poor, and good-hearted Hüsnü, establishing a dynamic of superiority. Laughter directed at the person who is prevailed over signifies belittling them in some way, to overcome and assume a superior position. Incongruities regarding the object and situation disrupt the superiority, and the powerful one is thus defeated in this manner.

Danger: A threat to male identity.

Comic Situation: In the film *İyi Aile Çocuğu* (Seden, 1978), Cemal is a famous tough guy and has a twin brother. According to the script of the movie, whatever one of the twins feels, the other feels the same. Comedy arises when they feel similar things even in inappropriate situations. While Kemal is having dinner with the girl he likes, Cemal is having dinner with another tough guy. Seeing his tough guy friend tearing off a big piece of meat and looking at it lovingly, Cemal says, "*This is the first time in my life I'm eating with a beauty like you* " Upon hearing this, his friend leaves the table in anger, stating that he will never sit at the same table again. Cemal's men, hesitantly, say, "*Brother, we get it, you're a disaster flirt, but it's shameful to say this. Where did these perverted ways come from? Is he a soloist, brother?*" The comedic situation is achieved through incongruity. John Morreall (1997, p. 29) discusses how incongruity about an object or person can ridicule the object or person. This incongruity brings about a comedic situation. In this film, physical defects, i.e., inability to control behaviors, acting like the twin, reverse of what is expected in conversations, emergence of the opposite of societal expectations from a man, Cemal's interest in a man by saying "*This is the first time in my life I'm eating with a beauty like you*" which highlights masculine traits, creates the comedy. The mismatch that deviates from social order creates the ludicrous. The reason for this incongruity is the twin's mildness and Kemal, who is in harmony with society, eating with a beautiful woman at the same time, causing Cemal to give similar reactions due to being twins. The compliments Kemal gives to the beautiful woman while eating are not funny. But it is comical for Cemal to unintentionally say the same words to a man, in contrast to expectations. Cemal, who resorts to brute force, being ridiculed and defeated by his twin Kemal, who complies with societal rules and does not resort to brute force, also brings about superiority and relief.

Danger: Being caught in the midst of a war between gangs.

Comic Situation: In the film *Bitirim Kardeşler* (Ökten, 1973), Ali and Veli, who want to sabotage the collaboration and friendship dinner between the biggest smuggling gang in the Middle East and the fastest gangster gang in the Near East, have spaghetti with a bomb hidden in it served at the dinner. After the bomb explodes, both sides accuse each other of sabotage and start to fight. The winners are Ali and Veli. While Ali and Veli are two characters that the gangs could easily defeat, they find a way to defeat these two gangs by turning them against each other. The means to this end is the spaghetti with a bomb. Here, these two people, who are poor and good-hearted, turn the situation around with situational comedy, both establishing superiority and turning the situation into one where the weak prevail.

Portrayal of Dangerous Characters

Danger: The desire for revenge of a woman whose husband has been killed. The emotion that arises is anger. In order to eliminate this anger, a murderer is sought and found. The dangerous situation is hidden in the portrayal of a killer's character.

In the film *Dila Hanım* (Aksoy, 1977), Dila Hanım, whose husband has been killed, wants revenge. She sends her steward to find Kara Haydar (Erol Taş), who will do this. The steward finds Kara Haydar in a cave, eating a large piece of meat savagely and voraciously. Kara Haydar's way of eating the meat is depicted in a manner resembling that of a wild animal silently and mercilessly devouring its prey. Without looking at the steward as if he doesn't care, Kara Haydar asks, "*What do you want?*" This shows that he doesn't acknowledge any hierarchy. The steward responds, "*I am the steward of Barazoğlu İbrahim Bey.*" Kara Haydar leaves the half-eaten leg of meat and licks his fingers one by one. He has been invited to speak for Dila Hanım to take revenge. Kara Haydar accepts this invitation.

The scene of Erol Taş eating meat is a highly iconic scene in Turkish Cinema, showcasing the actor's powerful portrayal of his character. The close-up shots of Erol Taş eating meat, showing his teeth tearing into the meat, provide insight into the potential for evil deeds he is capable of. Another notable feature here is the appetite of the character Kara Haydar, portrayed by actor Erol Taş. Appetite is an indicator of danger. The proper management of appetite is a sign of civilization. Societies demand the control of common appetites. Excessive appetite is a sign of unvirtuous eating and indicates danger (Tigner & Carruth, 2017, p. 48). At the same time, the meal eaten by Kara Haydar is represented distinctly by a piece of meat that symbolizes a portion of the whole animal. We see the rawness and juiciness of the meat in this scene, and Kara Haydar's behavior of licking his fingers disregards societal norms, indicating the lack of virtue in eating, where many social and moral rules can also be violated. The danger now arises not only from the individual but also from the way of eating and the food consumed, evoking a sense of fear.

Depiction of The Loss of Societal Values and Danger

Danger: A situation where values are compromised and innocence is lost.

In Turkish Cinema, characters portrayed by actor Nuri Alço mix drugs into the drinks of innocent girls to lead them astray. These malicious characters take advantage of innocent and naive girls using drugs that cause loss of consciousness mixed into soda or alcohol. Girls who venture beyond the boundaries of traditional life and are curious about modern lifestyles are faced with great danger. The means to evil is the drinks. Women are important viewers in Turkish Cinema. Presenting different forms of danger to these women through examples in films also aims to draw attention to situations they may encounter in real life.

In the film *Vazife Uğruna* (Elmas, 1986), Gamze, the daughter of a police officer, meets Akrep(Scorpion) Naci (Nuri Alço), falls for him, and believes his promises of marriage. Akrep Naci takes her to a house, saying "*I'll introduce you to my mother*" and introduces her to a woman claiming to be his mother. At that moment, he suggests, "*Let's have a drink while waiting for my father*" Gamze declines, saying, "*Can't we do without drinks?*" but Naci insists, saying, "*No, don't embarrass me in front of my family, I told them you're a modern girl*". The woman introduced as Naci's mother is not his real mother. Her intention is to gain the trust of the girls by acting like a mother and convincing them to drink drugged beverages. The fake mother also insists that Gamze drinks, saying, "*We all like to drink as a family, my dear, I hope you'll adapt to us in the future*". Gamze, as a prospective bride, agrees, saying, "*Of course, ma'am, I'll try to adapt*" Naci mixes drugs into Gamze's drink and abuses her. Gamze and her family are deeply affected by this situation. A similar incident of drugging drinks and abuse is the theme of the film *Adını Anmayacağım* (Elmas, 1971). Once a famous singer, Gül, believes her soldier husband has been killed in Korea. This loss greatly affects her, causing her to lose control of her life. Cemil, Gül's violinist and the composer of her songs, is malicious. Cemil drugs Gül's drink and attempts to assault her. In both films, it is seen that women's lives change completely after consuming drugged drinks and they are ostracized by society. Drinking drugged beverages and being abused are not only personal emotional wounds in women's lives but also dangerous situations of societal contamination and exclusion.

In both of these films, danger arises from stepping outside the boundaries of the family. There is a close relationship between eating and drinking patterns and the sense of family. The sharing of meals draws the boundaries of the family's symbolic and emotional existence. Involvement in family boundaries, entering or leaving the family, also affects eating relationships (Beardsworth & Keil, 2002, p. 73–74). In the film *Vazife Uğruna*(Elmas, 1986), although Gamze does not want to drink alcohol, she feels safe because she thinks she is with Naci's family. In *Adını Anmayacağım* (Elmas, 1971), it is Cemil, a trusted colleague, who abuses Gül with drugged drinks, in a family that has fallen apart. Cemil is the person closest to the family boundaries.

Sara Ahmed (2019, p. 92), who examines the relationship between home and fear, expresses this situation as follows: "What is threatening and what is not, who is eligible to be the object of fear, is circumscribed by statements about circulation. Such statements call for action, implying that women must be vigilant at all times outside their homes. Statements not only construct the "outside" as inherently dangerous but also claim that the home is safe". Turkish Cinema has a significant family audience. Family values, discourse targeting women such as honor, are conveyed to the audience in many films. Therefore, drugged drinks are a harbinger of dangerous situations. Emotions about why one should be afraid are taught to women. While these two films draw attention to the danger of women drinking alcohol, they also emphasize that the home and relationships within it are a space and relationships that protect women from dangers.

Danger in Locations Where İndividuals Subject to Crime and Violence, Such as Prisons, Reside: Sense of Justice, Power Struggles, and Anger

Danger: Prisons, where individuals convicted of crimes reside, are places where danger and violence can easily arise due to the nature of the crimes committed and the resulting punishments. In places where societal justice condemns individuals, injustice creates dangerous situations.

In prisons, inmates are often housed together in the same cells, necessitating the need to learn to live together. Bunk bed arrangements are implemented to accommodate more prisoners in these cells. In addition to the cells, there is a common area where all inmates gather. In many Turkish films, this area is used as a kitchen and a living room. In *Tatar Ramazan Sürgünde* (Melih Gülgen, 1992), there are several meal scenes that highlight power dynamics. Tatar Ramazan is known for his opposition to injustice. He is sent to a new prison. In this prison, a hierarchical system similar to feudalism has been created, and power balances are established accordingly. The leaders are those who apply violence and have strong financial relationships. Whoever holds power determines the order in the prison. Abdurrahman Çavuş and his right-hand man, Akseli Ali Ağa, are at the top of this system. When Tatar Ramazan is transferred to this prison, Abdurrahman Çavuş invites him to dinner to demonstrate his power. Inequalities in the prison are depicted through what and how people eat. After inmates line up to receive their meals, they sit down wherever they can find a place and eat their soup with helva, while at Abdurrahman Çavuş's table, there is roasted chicken and fruit. Abdurrahman Çavuş, who has established a feudal system in the prison, demonstrates his wealth and power with a flashy menu. Abdurrahman Çavuş confiscates meals from poor inmates in exchange for money in the gambling system he has established within the prison. He encourages them to gamble among themselves. Inmates who lose at gambling go hungry and feel even more vulnerable and powerless. Tatar Ramazan is physically strong, fearless, and acts with a sense of justice that supports the oppressed. He opposes Abdurrahman Çavuş, who thrives on power and self-interest. The oppression and exploitation of the weak bring forth both the desire for justice and danger. If the violence perpetrated by the powerful is not controlled, their power increases even more. Abdurrahman Çavuş starves, subjugates, and humiliates inmates, thus feeling even more powerful. As Byung-Chul Han (2017, p. 25) expresses, "applying violence increases the sense of power. More violence means more power" Accessing or not accessing quality food is an indicator of being powerful or powerless. Danger thrives on relationship networks where justice is absent.

Conclusion

Mary Douglas (2007, p. 20) asserts that danger is "varied and everywhere." Through Cinema, danger draws attention to situations where there are many risks for both individuals and society. Films, as cultural products representing everyday life, serve to both educate and reinforce understanding of the risks inherent in danger. Turkish cinema fulfills this function. Through nourishment relationships depicted in films, viewers learn various messages about danger:

- ✓ By using the metaphor of food, relationships are established through humor to gain superiority over dangerous situations and individuals. This superiority provides relief.
- ✓ The future of danger and risky situations is portrayed through what characters eat and drink.
- ✓ The dimensions of danger and risky situations are depicted. The extent of danger becomes visible through what is consumed.
- ✓ The presence of justice is easily understood by what the powerful and the powerless eat. When justice disappears, there is an anticipation that violence will ensue.

Douglas (2007; 25) notes that socially defined dangers serve as sanctions for maintaining society. Beliefs about danger produce expected, convinced behaviors within society without any problems. In Yeşilçam Cinema and subsequent films, nourishment relationships and food symbols are portrayed in different ways to define danger for the continuity of society.

Notes

İ- The period spanning from the 1950s to the late 1980s, which is also known as the Yeşilçam Cinema era, is a significant phase within Turkish Cinema. While this study primarily focuses on the Yeşilçam era, it also includes films produced beyond this period. Therefore, a general definition of Turkish Cinema is used in this study.

The handling of dangerous situations in Turkish Cinema is diverse and unique. The food and drink culture frames the dangerous situation and the dangerous person in a way that is suitable for family audiences. By defining the danger, it draws attention to the dangerous situation and person. Cinema can convey danger to the audience in different ways. Danger can become a humorous element. Comedy and humor are developed as a method of coping with danger. In Yeşilçam cinema, the eating and drinking habits of the expected villains are different and memorable. Attention is drawn to the values of society through this difference. Recognizing danger helps to avoid exposure to it. The dangerous person is recognized by what they eat and drink. The eating and drinking habits of the villains are highlighted to protect the pure and naive characters. The films draw attention to drinks spiked with drugs that lead to the violation of women's honor. The importance of the family institution is emphasized. In films made after the Yeşilçam era, danger is redefined. Scenes depicting the feeding of prisoners in prisons, where the sense of danger is heightened, allow the audience to form a human connection with the inmates. The sense of justice is reexamined in places like prisons, where justice is expected.

Films

Baytan, N. (Director). (1982). Yedi Bela Hüsnü [Film]. Turkey: Cumhur Film.

Seden, O. (Director). (1978). İyi Aile Çocuğu [Film]. Turkey: Gülşah Film.

Gülgen, M. (Director). (1992). Tatar Ramazan Sürgünde [Film]. Turkey: Gülgen Film.

Ökten, Z. (Director). (1973). Bitirim Kardeşler [Film]. Turkey: Erler Film.

Elmas, O. (Director). (1986). Vazife Uğruna [Film]. Turkey: Erler Film.

Elmas, O. (Director). (1971). Adını Anmayacağım [Film]. Turkey: Erler Film.

References

Ahmed, S. (2019). *Duyguların kültürel politikası* (S. Komut, Trans.). Sel Yayıncılık.

Aksoy, O. (Director). (1977). *Dila Hanım* [Film]. Turkey: Arzu Film.

Beardsworth, A., & Keil, T. (2002). *Sociology on the menu: An invitation to the study of food and society*. Routledge.

Bergson, H. (2011). *Gülme: Gülüncün anlamı üstüne deneme*. Türkiye İş Bankası Kültür Yayınları.

Douglas, M. (2007). *Saflık ve tehlike* (E. Ayhan, Trans.). Metis Yayınları.

Ekman, P. (1992). An argument for basic emotions. *Cognition and Emotion, 6*(3–4), 169–200. https://doi.org/10.1080/02699939208411068

Han, B.-C. (2017). *Şiddetin topolojisi* (D. Zaptçıoğlu, Trans.). Metis Yayınları.

Merriam-Webster. (n.d.). *Dangerous*. In *Merriam-Webster Thesaurus*. Retrieved May 29, 2024, from https://www.merriam-webster.com/thesaurus/dangerous.

Morreall, J. (1997). *Gülmeyi ciddiye almak* (K. Aysevener & Ş. Soyer, Trans.). İstanbul: İris Yayınları.

Tigner, A., & Carruth, A. (2017). *Literature and food studies*. Routledge.

Türk Dil Kurumu. (n.d.). Tehlike. In *TDK Türkçe Sözlük*. Retrieved May 29, 2024, from https://sozluk.gov.tr/tehlike.

Sena Bakir[*] and Gulsun Yildirim[†]

Chapter 14 Vertical Farming and Its Applications in Gastronomy

Introduction

A food crisis will likely occur in the future due to the increasing world population, climate change and degradation of agricultural lands (Pakirdasi, 2023). The world's population is expected to increase by 34 % from 6.8 billion in 2018 to 9.1 billion in 2050, according to the most recent version of the UN Population Prospects (medium variation). The rate of population increase will be significantly lower than fifty years ago. Nevertheless, the absolute increase—2.3 billion extra humans—will still be noteworthy given the much larger base. Most of this population growth will occur in the region of the planet currently home to emerging nations. The least developed countries of today are predicted to have the biggest relative increase—120 percent. Growing populations, greater urbanization, and higher wages are the main socioeconomic factors in rising food consumption. Due to increased food consumption, sustainable and efficient use of natural resources and lands is crucial. There are concerning indications of degradation in a large portion of the global natural resource base now under use. Out of the 24 ecosystem services that were looked at, 15 are already being used unsustainably or are being degraded, according to the Millennium Ecosystem Assessment. Water supply and catch fisheries are two examples of this. Degradation of one ecological service is frequently the result of efforts to enhance another, such as the ecosystem service "food production." Clear indications include depletion of freshwater reserves, erosion, desertification, loss of tropical forests, and biodiversity. If land use practices are not made more sustainable and maintained and rehabilitation efforts are not increased, the productive potential of land, water, and genetic resources might continue to diminish at alarming rates (FAO, 2009).

---

* Recep Tayyip Erdogan University, Tourism Faculty, Gastronomy and Culinary Arts, and Recep Tayyip Erdogan University Blueberry Application and Research Center, sena.bakir@erdogan.edu.tr

† Recep Tayyip Erdogan University, Tourism Faculty, Gastronomy and Culinary Arts, ecem.incekaraceper@kent.edu.tr

With the intense urbanization and population movement to cities, novel approaches in bio-architecture and agro-urban planning became necessary. Global demand for food, feed, and fiber is predicted to increase by 70 % in the first half of this century while more and more crops are being used for bioenergy and other industrial uses. Thus, the existing limited agricultural resources would come under increasing strain from both traditional and new demand for agricultural produce. Agriculture will also need to play a significant role in mitigating the effects of climate change, preserving natural habitats, and sustaining biodiversity, in addition to competing with expanding urban settlements for land and water. Higher temperatures, increased carbon dioxide (CO_2) concentration, altered precipitation patterns, and increased pressure of weeds, pests, and diseases are all effects of climate change that will impact agriculture and forestry systems. According to all available quantitative evaluations, food security will be negatively impacted by climate change. Climate change is predicted to result in an average increase in food prices (FAO, 2009).

Agriculture can assist to lessen the effects of climate change, but it will also need to adapt to it. It is imperative to discover novel approaches in rural regions, particularly in developing nations, to improve climate change adaptation and mitigation. Farmers will require innovative technologies to produce more with fewer laborers on less land in order to meet those demands. Vertical farming practices have a vital role at this stage. In this study, Vertical Farming and Its Applications in Gastronomy were investigated.

Development of Vertical Farming

Vertical farming, which was first developed only for urban agriculture and later enabled the more efficient use of natural resources and the fast and reliable way to meet the nutritional needs of the rapidly growing population, was developed in 1999 by Dickson Despommier, Professor of Public Health Sciences at Columbia University. Despommier put forward urban agriculture and vertical farms designed in cities to solve the food deficiency caused by the rapid increase in the world population and the resulting increase in the rate of urbanization. Agricultural practices such as roof gardens, greenhouses, wall gardens and soilless agriculture have effectively emerged in vertical farming (Uzun, 2023).

As phrased at Pakirdasi (2023)'Vertical farming is a vegetable production system in which, instead of conventional farming methods, soilless farming practices such as hydroponic, aeroponic and aquaponic are used'. These techniques use greenhouses or other indoor controlled environments in place of arable land. Shipping containers, home basements, and warehouses can all be utilized indoors. Because all the environmental factors that a plant needs—lighting, temperature, humidity, airflow, and nutrients—can be regulated indoors. Plant growth is accelerated and sustained consistently throughout the year, irrespective of external weather conditions (Pakirdasi, 2023).

Although modern vertical farming is a newly developed concept by Despommier (2010), vertical farming practices have a long history. The Hanging Gardens of Babylon, with their gardens consisting of a series of vaulted terraces stacked one after another in 600 BC, can be considered the first example of a 'vertical farm'. In 1150, the Aztecs used a form of hydroponic farming by growing plants in swampy areas. The theory of hydroponic gardening and farming methods was first mentioned in Sir Francis Bacon's book "Sylva Sylvarum" in 1627.

The idea of soilless gardening was developed by the English scientist John Woodward in 1699 through a series of water culture experiments with mint. The concept of 'Vertical Farming' was first described by American geologist Gilbert Ellis Bailey in his book of the same name in 1915. This book focuses on lower agriculture rather than upstream. In 1929, William F. Gericke, an agronomist at the University of California, Berkeley, outlined his work on modern soilless plant cultivation. The term "hydroponics" was first used for soilless agriculture by Botany Professor Dr. William Albert Setchell in 1937 AD. In World War II, more than 8,000 tons of fresh vegetables were produced to feed allied forces deployed on South Pacific islands using soilless growing systems for the first time in modern history. At the Vienna International Horticultural Exhibition in 1964, a vertical farm in the form of a tall glass tower was exhibited. The contemporary vertical farm concept was developed in 1999 with his students under the leadership of Dr. Dickson Despommier, professor of Public Health Sciences at Columbia University, USA. The first modern vertical farm was built in 2009. Sky Green Farms' of Singapore consists of more than 100 towers, each 9 meters high, that grow green vegetables using sunlight and rainwater (Crumpacker, 2018).

Vertical farming, which is a farming method that provides maximum water saving compared to traditional farming methods and produces ornamental plants or plants used in the food and pharmaceutical industry without soil using daylight or artificial light, is called Controlled Environment Agriculture or Building Integrated Agriculture (Celebi, 2019).

Vertical farming has several advantages (Despommier, 2010; Uzun, 2023). These can be listed as follows:

- The ability to obtain a variety of products in urban areas with limited land,
- Since no soil is used, it does not require large land areas. Thus, the ecosystem is not harmed,

- Resource management based on sustainability,
- Maximum savings in areas such as energy resources (e.g. wind, solar, biogas) and especially water conservation. It consumes 90 % less water than traditional farming methods.
- Since fast production can be made, reliable and healthy products are obtained,
- Reducing the use of agricultural machinery, including transport vehicles, tractors and combine harvesters, and reduces machinery, an essential agricultural input, and the cost of diesel fuel,
- Since weeds and pests can be controlled quickly, chemical pesticides are not used in crop cultivation,
- Vertical farms, where each production stage is meticulously controlled through computer and robot systems, are resistant to natural and human-made disasters, thus ensuring a continuous supply chain,
- Since controlled agriculture can be done, there is no product loss,
- Local product diversity can be increased. Thus, it can bring economic vitality to its local market,
- Provides significant savings in chemical fertilizer consumption,
- By using idle buildings in city centers for vertical farming, long transportation distances between the product and the buyer are eliminated,
- The ability to achieve year-round production thanks to the use of advanced LED lighting specifically designed for plant growth eliminates concerns about acreage, rotation and seasonal limitations,
- While traditional agriculture continues to be affected by climatic conditions, indoor vertical farming conditions are kept constant by adjusting as desired with a closed circuit system. Thus, the product can be harvested with increased efficiency all year round, regardless of climatic conditions,
- Since access to sufficient and reliable products is quick and effortless, the eating habits of the urban population will change, healthy generations will grow, and diseases that develop due to nutrition will decrease.

Some Plants and Herbs Produced with Vertical Farming

Gastronomy and culinary arts are developing areas focusing on quality, taste, and traditional know-how. The general trend in kitchen goal is to generate the awareness required to maximize the manufacturing, processing, and consuming agricultural food products.

Even in the times, when chemistry and molecules were unknown to humanity, humans were able to distinguish and categorize the various "aroma" that the plant created (Bianchi, 2015). The word "spices," which is still widely used today, has a long history and refers to the group of plants used in cooking because of their potent aromatic qualities. Still, there is a significant pharmacognostic distinction between spices and aromatic plants that is based on the drying procedure (Bianchi, 2015). From ancient times to dates, aromatic herbs are primarily used in fresh form in cooking to enhance flavor and fragrance. However, because many herbs degrade quickly and lose some organoleptic qualities when cooking, they have a limited shelf life (Monsalve, Gaviria, & Correa, 2021). On the other hand, dried herbs keep well for months at room temperature when kept in an airtight container in a dark place (Purba & Silalahi, 2018). Chefs use many dried or fresh herbs varieties in baked goods, salads, and other sweet and savory recipes (Arnott, 2011).

The use of exotic spices, particularly those from the Americas and the Far East, gradually enhanced the use of wild, aromatic plants that are still common in many Mediterranean dishes. Indeed, a combination of antibacterial, antiperoxidative, and antioxidant properties found in rosemary, pepper, savory, cinnamon, thyme, cloves, and other spices can help conceal the off-flavours caused by deteriorating meat and fish thanks to their potent aroma (A & L., 1998). Many herbs, such as thyme, sage, lavender, fennel, mint, dill, coriander, and basil, have become staples in kitchens with their aromatic qualities (Mohammed, Uysal, & Sevindik, 2023).

Gastronomy and Vertical Farming Applications

The developed the idea of "vertical farming," which has had a significant influence on scientists, engineers, government decision-makers, and architects, in order to cultivate healthy food and benefit individuals and communities, Allen (2012) advocated the "good food revolution" and launched a social enterprise with the goal of producing agricultural goods locally for consumption locally.

A plant production facility with a nearly airtight, thermally insulated warehouse-like building is referred to as a "plant factory with artificial lighting" (PFAL) (Kozai, 2013) and the market for PFAL-grown vegetables would establish significant development if they were to be utilized commercially for both fresh salads and as raw materials, which would result in the creation of "a new branch of urban agriculture" or "a new field of food industries." The range of applications for plants cultivated in PFAL in the upcoming decades includes vegetables that have been sliced and prepared fresh, vegetables that are dried and frozen, paste, sauce, juice, soft drinks, and drink additions, salted vegetables, pickles, vegetables for fried and soup dishes, fire-pot meals, colors, aromatics, and spices (Kozai, 2020). One type of vertical agriculture, hydroponic farming, uses water to supply the nutrients and minerals required for plant growth, making it a creative and ecologically friendly farming method, in contrast to conventional agriculture (Pandey, Jain, & Singh, 2009). The hydroponic system mainly cultivates tomatoes, cucumbers, beans, lettuce, spinach, and broccoli (Sönmez, Mamay, & Söylemez, 2019). Water culture is a viable method for growing red curly lettuce, as Okudur (2018) demonstrated. The agri-gastronomy concept, an integrated idea that covers the production, processing, and consumption of agriculturally produced food, the idea is sound analytically because it integrates multidisciplinary knowledge, techniques, and technologies in a synergistic way; it produces a wide range of agricultural products from environmentally conscious farms; it comprehends food products through minimal or technologically optimized processing; it maintains respectable traceability, transparency, and ethics regarding food quality, cost, and catering, retail, or institutional food (Romulus, 2023).

In recent decades, the function of the chef has considerably developed. Chefs are becoming increasingly acknowledged for their ability to improve environmental sustainability and public health. Due to changes in dietary trends favoring a greater consumption of vegetables with high nutritional content, Europe has boosted its production, processing, and export of vegetables in recent years. There are calls for sustainable vegetable production and consumption due to the increased interest in environmental issues (Pedretti, Duca, Ballarini, Boakye-Yiadom, & Ilari, 2023). There is no denying the connections between agriculture, health, and nutrition; nonetheless, traditional agricultural and healthcare systems sometimes function in isolation, which contributes to these urgent issues. In the US, there is plenty of food. Still households cannot afford or achieve a healthy diet, and because of that, as a solution, launching teaching kitchens in tandem with urban, vertical, and community gardens has the potential to spark a movement that highlights the importance of the food system in fostering resilience in communities, advancing human and planetary health, and fostering interdisciplinary collaboration (Cole, Pethan, & Evans, 2023). According to interviews with chefs and system users in Turkiye, there is a lot of potential for hydroponic agriculture in Turkey, and chefs want to use these goods (Ozay, 2021). In a study conducted by Jaeger, Chheang, Roigard, and Frøst (2023) with the participation of 190 people in Denmark, vertical farming-grown products, including salad greens, herbs, and fruits, did not consistently receive less favorable reviews than their organic counterparts. Furthermore, on average, there were no reports of consumers disliking produce farmed by vertical farming. The significance of various characteristic flavors was discovered by examining the sensory drivers of love for herbs and salad greens.

Conclusions and Future Aspects

Around the world, most of the agricultural lands are currently under use. At the same time the remaining is a wasteland, whose potential for agriculture has all but vanished in recent decades as a result of inadequate land management. To provide customers with great and genuine culinary experiences and support the social and economic advancement of the surrounding area and local community, vertical farming is a good alternative to traditional agriculture. Conceptualized group food initiatives are feasible in theory, particularly through the restaurant sector, with plans to ensure that visitors and locals alike have equal access to nutritious, high-quality food with significant potential for national and world-wide marketing. More awareness is yet required, which highlights how crucial it is to conceptualize different food systems, like the agri-gastronomic idea.

References

A, B., & L., M. (1998). *CD-Rom Le piante officinali del Giardino delle Erbe di Casola Valsenio (The medicinal andaromatic plants of the del "Giardino delle Erbe di Casola Valsenio")*. Bologna.

Allen, W. (2012). *Good Food Revolution: Growing Healthy Food, People and Communities*. USA: Gotham Books.

Arnott, M. L. E. (2011). *Gastronomy: The anthropology of food and food habits*: Walter de Gruyter.

Bianchi, A. (2015). The Mediterranean aromatic plants and their culinary use. *Natural Product Research, 29*(3), 201–206.

Celebi, E. S. (2019). *Mevcut Yapilarin Sürdürulebilirlik Açısından Yeniden Kullanimlarinda Dikey Tarim Uygulamalari Üzerine Bir Araştirma: İstanbul Porselen Fabrikasi Örneği* (Master's Thesis), Gebze Teknik University.

Cole, A., Pethan, J., & Evans, J. (2023). The Role of Agricultural Systems in Teaching Kitchens: An Integrative Review and Thoughts for the Future. *Nutrients, 15*(18), 4045.

Crumpacker, M. (2018). A Look at the History of Vertical Farming.

Despommier, D. (2010). *The Vertical Farm: Feeding the World in the 21st Century*. New York: St. Martin's Press.

FAO. (2009). Global Agriculture towards 2050,How to Feed the World in 2050.

İpar, M. S., Babaç, E., & Kök, A. (2020). Evaluatıon Of Customer Revıews For Green Generatıon Restaurants By Content Analysıs. *Journal of Gastronomy, Hospitality, and Travel*, 3(2), 260–269.

Jaeger, S. R., Chheang, S. L., Roigard, C. M., & Frøst, M. B. (2023). Consumers' expectations and experiences of salad greens, herbs, and fruits from vertical farming: Comparison with organic produce. *Food Quality and Preference, 112*(105020).

Kozai, T. (2013). Plant factory in Japan: current situation and perspectives. *Chron. Hortic., 53*(2), 8–11.

Jo, J., Joung, H.-W., & Taylor, J. J. (2023). Impact of Sustainability Attributes of Menu Labels on Restaurant Green Image and Customer Attitudes. *Journal of Marketing Development and Competitiveness*, 17(4). https://doi.org/10.33423/jmdc.v17i4.6646.

Kozai, T., & Niu, G. (2020). Role of the plant factory with artificial lighting (PFAL) in urban areas *Plant Factory* (pp. 7–34). Academic Press.

Mohammed, F. S., Uysal, I., & Sevindik, M. (2023). A review on antiviral plants effective against different virus types. *Prospects in Pharmaceutical Sciences, 21*(2), 1–21.

Monsalve, J. R., Gaviria, L. J., & Correa, A. A. (2021). Use of freeze-dried aromatic herbs with quality organoleptic characteristics in gastronomic products. *International Journal of Gastronomy and Food Science, 24*(100341).

Okudur, E. (2018). The Effect of Fertilization on the Yield and Quality of Three Different Lettuce. *Süleyman Demirel Üniversitesi Ziraat Fakültesi Dergisi*, 394–399.

Ozay, C. (2021). *Hydroponic agriculture.* (Graduation Project), Yeditepe University.

Pakirdasi, O. (2023). *Economic Evaluation of Vertical Farming in İstanbul, Turkey, by Black Scholes Real Option Model.* (The Degree of Master of Science), Galatasaray University.

Pandey, R., Jain, V., & Singh, K. (2009). Hydroponics Agriculture: Its Status,. Division of Plant Physiology. *Division of Plant Physiology, Indian Agricultural Research Institute*, 20–29.

Pedretti, E. F., Duca, D., Ballarini, M., Boakye-Yiadom, K. A., & Ilari, A. (2023). Environmental impact assessment of producing frozen spinach in central Italy. *Resources, Environment and Sustainability 12*(100110).

Purba, E. C., & Silalahi, M. (2018). Gastronomic ethnobiology of "terites"—a traditional Batak Karo medicinal food: A ruminant's stomach content as a human food resource. *Journal of Ethnic Foods, 5*(2), 114–120.

Romulus, G. (2023). The concept of agri-gastronomy. *Annals of the Academy of Romanian Scientists*, 5–14.

Şimşek, N. & Akdağ, G. (2017). Investigation of Green Genetic Restaurants within Content of Sustainable Gastronomy Tourism. Journal of Academic Social Science Studies. (60):351–368.

Sönmez, C., Mamay, M., & Söylemez, S. (2019). *Determination of The Effect of Different Hydroponic Culture and Different NH4:NO3 Ratio on The Density of Aphid Aphis Spp.(Hemiptera:Aphididae) Population In Greenhouse Lettuce Cultivation.* . Paper presented at the 1st International Göbeklitepe Agriculture Congress, Şanlıurfa: Harran University.

Uzun, D. (2023). *Dikey Tarim Sistemlerinde Farkli Işik Yoğunluklari Ve Kompozisyonlarinin Yemlik Arpa Çiminin Verim, Kalite Ve Fizyolojik Parametreleri Üzerindeki Etkileri.* (Master Science Thesis), Ege University.

Tuba Aldemir*

Chapter 15 Gastronomy-Themed Recreation Activities

Introduction

Gastronomy tourism aims to provide an experience of local food and explore different culinary cultures. In contrast, recreational tourism allows people to relieve stress and relax through nature, music, sports, and other activities. Combining gastronomy with recreational activities will enable visitors to enjoy delicious food, natural beauty, and fun experiences, promoting physical and mental relaxation.

Activities such as dining in restaurants surrounded by natural beauty and trying local snacks to recharge while mountain climbing or cycling after a wine-tasting tour can combine gastronomy and recreation to offer a rich and varied vacation experience. Thus, providing gastronomy tourism and recreation offers tourists a comprehensive holiday experience encompassing delicious food, enjoyable activities, cultural immersion, and natural beauty. This study aims to examine these increasingly popular activities, including gastronomy-themed recreational events, activities carried out in this context and their contribution to tourism development.

Leisure Time and Recreation

The engagement of individuals in recreational activities during their leisure time has steadily increased with urbanisation and industrialisation. People participate in recreational activities to make the most of their free time and escape the monotonous routine of daily life and the fast pace of urban living (Sevil, 2012; Doldur & Nesipoğlu, 2022). Before exploring the relationship between recreation and gastronomy, it is important to clarify the definitions of leisure time and recreation, as these two concepts are closely interconnected and share similar meanings (Karaküçük & Gürbüz, 2007; Şenel & Kılıç, 2022).

* PhD., İstanbul Esenyurt University, Vocational School, Department of Hotel, Restaurant and Catering Services, tubaaldemir@yahoo.com

Leisure is a concept used in the meaning of "permission" and "freedom", which is the equivalent of the Latin word "Lyceum, Licere" and the English word "Leisure" (Goodale & Witt, 1980; Kandemir Altunel, 2021). It is the time left over from compulsory time such as personal needs, sleep and work, and essential activities to be done by individuals separated from physical, social, and economic constraints (Goodale & Witt, 1980; Wichasin, 2007). In essence, it represents the discretionary time individuals can freely allocate according to their preferences within their free time (Uğurlu, 2005).

The term "recreation" originates from the Latin word "recreare", which means "re-formation, renewal" (Jensen & Naylor, 2000). Torkildsen (2005) states that engaging in leisure activities and meeting individual needs is voluntary. In other words, recreational activities offer rest, revitalisation, renewal, personal satisfaction, and are voluntary after obligatory work (Kraus & Washington, 1985; Jensen & Guthrie, 2006). The key criteria for recreational activities are that they occur during leisure time, are chosen and participated in by the individual entirely of their own free will, bring happiness and pleasure, and provide personal and social benefits (Bucher, 1972; Karaküçük, 2014). As can be seen, leisure time encompasses a specific period, while recreation involves engaging in activities during that leisure time (Kandemir Altunel, 2021). In summary, recreational activities are the pursuits individuals select and partake in during their leisure time to experience enjoyment and entertainment, whether indoor or outdoor areas, with or without vehicles, within or outside urban areas, within or outside an organisational settings (Karaküçük, 2014).

Relationship Between Recreation and Gastronomy Tourism

The term "tourism" refers to the travel and accommodation activities that individuals engage in during their leisure time, either alone or with others, outside of their usual place of residence, temporarily and without financial gain (Öztürk, 2019). On the other hand, "recreation" encompasses activities that individuals choose to participate in during their free time (Kandemir Altunel, 2021). While tourism and recreation differ in various aspects and are considered complementary, they are often used interchangeably. The connection between recreation and tourism becomes particularly evident when the concept of leisure is applied in both areas. Additionally, there is a close relationship between tourism and recreation concerning the activities undertaken (Brey & Lehto, 2007; Öztürk, 2019).

The United Nations World Tourism Organization (UNWTO) (2012) defines gastronomy tourism as travelling to destinations with gastronomic values to participate in various recreational activities and to experience the food and beverages of a particular place. The popularity of these activities highlights the increasing significance of gastronomy in tourism, drawing visitors eager to explore and take part in these culinary experiences. As a result, people's participation in gastronomy-themed activities to spend their leisure time has become an important aspect of recreational tourism (Karakan, 2021).

Gastronomy tourism is based on the concept of knowing and discovering the gastronomic culture of a region or country (Nesterchuk et al., 2022). Therefore, tourists travel to different places to discover new culinary traditions, taste foods and drinks they haven't tried before, and learn about them. This demonstrates the strong connection between gastronomy and tourism. This connection has attracted the attention of tourism authorities and stakeholders, leading destinations to be promoted not only for their cultural and natural resources but also for their gastronomic values (Şenel & Kılıç, 2022). The relationship between recreation and gastronomy tourism is based on people's interest in food experiences and exploring local culinary cultures during their vacations and leisure travel. Recreation activities involve travelling to relieve stress, have fun, and seek new experiences, while gastronomy tourism focuses on discovering food and culinary cultures. When these two types of tourism converge, people can participate in cooking classes during their holidays, visit local markets, savour local cuisine, and engage with local chefs to learn regional recipes.

The main motivation behind gastronomy tourism is to visit specific areas, such as producers, restaurants, and food festivals, to sample a particular type of food or observe its production. People are willing to travel long distances to pursue this culinary experience, sometimes even travelling kilometres to enjoy a meal prepared by a famous chef (Hall et al., 2003; Zengin et al, 2014). Engaging in activities such as visiting olive and olive oil producers, local cheese makers, vineyards and wineries, organic farms, beekeeping facilities, and participating in mushroom picking not only enhances travellers' holiday experiences but also contributes to the preservation and promotion of local culinary traditions (Sormaz et al., 2015; Düzgün, 2022). When recreation and gastronomy tourism converge, individuals can enjoy relaxing holidays and exceptional culinary adventures.

Gastronomy-Themed Recreation Activities

The involvement in gastronomy-themed activities holds a crucial place in recreational pursuits, offering individuals the opportunity to utilise their leisure time in an informative, enjoyable, and impactful manner. These activities, an integral part of cultural tourism, play a significant role in fostering intercultural interaction and contributing to social and economic progress (Onat, 2021). By incorporating such activities, destinations can enhance their attractiveness and desirability, increasing the number of visitors, as well as the frequency and duration of accommodations (Williams, 2003; Hazar, 2014; Şenel & Kılıç, 2022). Gastronomy-themed recreational activities unite tourists and locals through a shared interest in cooking, eating, and discovering diverse culinary traditions. These activities provide food enthusiasts an enjoyable and educational experience, offering socialising and entertainment opportunities.

Gastronomy-themed recreational activities typically include cooking classes, culinary workshops, food tours, festivals, gastronomy museums, and restaurant visits (Karakan, 2021).

Gastro-Tours

According to the World Food Travel Association (WFTA), 80 % of tourists travelling to a new destination research food and beverage offerings, and 53 % identify themselves as culinary travellers (University of Central Florida, 2024). The food tour is one of the most common gastronomy-themed recreational activities. These guided excursions take participants through a city or region, stopping at local eateries, markets, and speciality food producers to sample the region's signature dishes, ingredients, and culinary traditions. Food tours provide an opportunity to learn about the local culture and history while indulging in delicious local cuisine.

These gastro tourists are individuals whose travel motivations are mainly driven by the desire to experience local food (Hall et al., 2003) and to taste local dishes produced using raw materials and cooking techniques (Smith & Xiao, 2008). Therefore, combining local flavours with authentic artistic elements such as music, decoration, and ambience in gastronomy tourism is important in creating a recreational activity (Santich, 2004; Barakazı & Önçel, 2017).

A culinary tour, commonly referred to as a food tour, is a thoughtfully planned excursion that offers a unique perspective on a region's heritage and traditions through its cuisine. It includes visits to different food and drink establishments to taste dishes and beverages that are culturally significant. Gastronomy tours provide an immersive way to experience a destination through intimate and off-the-beaten-path adventures, elevating dining into an exciting journey beyond dining luxury. (Tasty Tours NYC, 2024). These tours make a significant contribution to the local economy by supporting food businesses and providing a deeper understanding of the cultural significance behind each dish (Tasty Tours NYC, 2024). As gastronomy tours become an integral part of modern tourism, food enthusiasts are encouraged to explore local hidden gems and family-owned establishments, enhancing their travel experience while also supporting the hospitality industry (Le Cordon Bleu, 2024a).

During a food tour, which typically lasts two to four hours, an expert leads you through a city's culinary landscape. You'll learn about the local history and visit various eateries and culinary hubs to sample authentic local dishes. In addition to appealing to your palate and visual senses, the region's unique street flavors offer a fun and engaging experience by allowing you to participate in the kitchen actively. Therefore, participants' preference for and experience of street food in daily city tours for leisure purposes is an excellent example of gastronomy-themed recreational activities (Solunoğlu, 2021).

Cooking Courses and Cooking Workshops

Another popular option for visitors is to participate in cooking classes and workshops. These hands-on experiences provide the opportunity to learn from skilled chefs and local experts, gaining valuable insights into traditional cooking techniques, recipes, and indigenous ingredients. Participants may have the chance to prepare and enjoy a full meal or even acquire new culinary skills to incorporate into their cooking repertoire. These cooking courses can be organised by hotels and restaurants in the region for interested tourists or held as daily workshops. These events cater to tourists interested in the region's culinary culture, professionals, and individuals seeking to participate in gastronomy-themed recreational activities (Ahiler Kalkınma Ajansı [AKA], 2023).

"The Pizza Village World Masterclasses" stand out as a prime example of international success, having gained substantial recognition since its inception. Organised live from Naples, the birthplace of pizza, the Pizza Village World Masterclasses are broadcast in Italian and English in 180 countries worldwide (Pizza Village, 2024b).

"Le Cordon Bleu" is a prestigious culinary school, annually training over 20,000 students in more than 35 international schools across 20 countries. Established in 1895 in Paris, the institution imparts knowledge of classical French cuisine and international cookery techniques through the guidance of master chefs hailing from Michelin-starred restaurants and renowned culinary experts with impressive competition accolades (Le Cordon Bleu, 2024b).

Gastronomy Festivals

Certain travel destinations host gastronomy-themed festivals to provide a more immersive experience. These events often feature foraging expeditions, wine or cheese tastings, and multi-day culinary retreats that offer a deeper exploration of local food culture and traditions. These activities not only cater to the palate but also provide opportunities for cultural exchange and personal enrichment. These events are effective marketing tools, significantly enhancing a destination's reputation and drawing visitors from all over the world. They boost local businesses like restaurants and hotels, leading to job creation in the hospitality and tourism industries (Yang et al., 2020). Positive gastronomic experiences at these events leave a lasting impression, strengthening visitors' connection with the destination and influencing their future travel decisions, behaviours, and intentions to revisit or recommend the destination (Sandybayev, 2018).

Gastronomy festivals have become increasingly popular as food-themed events in recent times. The opportunity to encounter world cuisines and explore new and exciting flavours has become a compelling reason to travel (Skyscanner, 2019). In line with this trend, local administrations and tourism stakeholders have begun organising festivals that showcase local gastronomic elements, often named after these elements themselves (Kargiglioğlu & Kabacık, 2017).

Thousands of people from many countries participate in the "La Tomatina" Tomato Festival, organised annually on the last Wednesday of August in Bunol in Valencia, Spain. This renowned event is famous for being the world's biggest food fight, with attendees engaging in a massive tomato-throwing battle. In addition to the main event, the festival features Paella competitions, traditional dances, musical performances, and a spectacular fireworks display (Festival Turu, 2024).

"Sydney International Food Festival" is another world-famous gastronomy-themed festival that takes place every October. It features world-famous dishes prepared by star chefs (Skyscanner, 2019).

Every year in October, a vegetarian festival is organised in Bangkok, the capital of Thailand. The nine-day festival honours nine emperors and showcases a variety of vegetarian dishes such as hummus, noodles, cakes, and other vegetable-based delicacies. One of the festival's highlights is the skilful preparation of vegetable-based dishes that closely resemble meat. (Skyscanner, 2019).

"The Pizza Village Festival", held in Naples, Italy, is renowned for its wide variety of pizzas. The festival showcases pizzas of different sizes and shapes, prepared with an assortment of ingredients. Drawing over 1 million attendees and pizza makers from around the world, the event includes competitions, amateur and professional pizza-making classes, entertainment and educational workshops, as well as conferences and seminars (Pizza Village, 2024a).

The "Fete Le Vin" is a significant wine festival that takes place every two years in Bordeaux, a renowned wine city. The festival spans four days and features wine tastings, captivating drone shows, and stunning light and fireworks displays (All Exciting, 2024a).

"Alaçatı Herb Festival" has been a popular event, attracting thousands of local and international visitors each year since 2010. This four-day festival showcases hundreds of unique herbs from the Aegean Region of Türkiye and offers visitors a taste of local flavours. The event features workshops, tastings, entertainment, and various other activities. Renowned figures in Turkish gastronomy come together at the festival to provide visitors with valuable insights and an enjoyable experience (Alaçatı Ot Festivali, 2024).

Gastronomy Museums

There are successful examples of gastronomy museums in many cities around the world. Gastronomy museums, which provide information about the culinary culture as well as the history of the region, are considered an important attraction element under the roof of gastronomy tourism and contribute to the travel motivation of gastro tourists (Çekal et al., 2022; Lokman et al., 2022; AKA, 2023). However, gastronomy museums are seen as an essential tool in terms of the preservation of culinary culture, its written and visual transfer to future generations, and its sustainability (Çetin & Küçükkömürler, 2019; Sezgin & Akbıyuk, 2021; Şeyhanlıoğlu et al., 2023).

Gastronomy museums can focus on a specific culinary product or seek to celebrate the culinary heritage of a region. They showcase local food and beverages, traditional cooking and storage techniques, as well as the kitchen tools used for these purposes (Çekal et al., 2022). These gastronomy museums, where many different elements of gastro culture can be seen, from culinary culture to wine, bread to chocolate, and aromatic herbs to olive oil, attract many local and foreign tourists throughout the year (Food in Life, 2020).

"The Escoffier Museum of Culinary Art" is an authentic house from the 18th century, nestled in the village of Villeneuve-Loubet, between Cannes and Nice. The Escoffier Culinary Arts Museum was converted into a museum after the death of the famous French chef Auguste Escoffier and opened in 1966 in France. In addition to the Escoffier's memoirs, the museum contains objects and utensils of the period, a wealth of documents on culinary art, dressed tables, a Provencal kitchen, a collection of over 3000 menus, sugar and chocolate sculptures, an extensive library and a research centre on gastronomy (Musee-Escoffier, 2024).

"The Alimentarium Food Museum" in Switzerland provides visitors to explore the historical evolution of eating habits and their transformations over time. This gastronomy museum is the world's first food-themed museum, founded by Nestle in 1985 (Alimentarium, 2024). The museum hosts a wide array of daily workshops for visitors of all ages, offering them the opportunity to create and savour local and international dishes. Furthermore, Alimentarium grants access to its wealth of knowledge through its digital platform. The museum aims to be a reference on food and nutrition, offering exclusive digital content to both amateurs and professionals, as well as children and adults (Vaud, 2024).

"The Bread Culture Museum" was established as an association in 1955 by Willy Eiselen and his son Hermann Eiselen in Ulm, Germany. The museum was transferred to the Eiselen family foundation in 1991 and hosted its one-millionth visitor in 2004. The museum showcases historical objects, paintings, and sculptures depicting cereal cultivation. Additionally, it features films about dough making and presentations highlighting the historical development of bread-baking ovens (Otay, 2017).

One of the world's most famous ice cream museums is "Gelato Museum Caprigiani" in Bologna, Italy. The museum provides information about the history of ice cream, a guided tour of the museum, ice cream tastings, ice cream production classes, hands-on gelato classes, and workshops suitable for all age groups (Gelato Museum, 2024). The museum features multimedia stations, over 20 artisanal machines, 10,000 photographs and historical documents, valuable period tools and accessories, and unpublished video interviews (Bologna Welcome, 2024).

Turkey's first culinary gastronomy museum, "Emine Göğüş Culinary" Museum, opened in Gaziantep in 2008. The museum aims to introduce Gaziantep's culinary culture by exhibiting kitchen utensils from the past to the present. Mannequins dressed in local clothes also animated details about dishes cooked on special occasions (Food in Life, 2020; TC. Kültür ve Turizm Bakanlığı [TCKTB], 2024).

"Choco-Story", the Chocolate Museum in Bruges, Belgium, which aims to tell and promote the history of the world-famous Belgian chocolate, introduces the 5500-year history of chocolate, the techniques, materials, and raw materials used in chocolate making throughout history (Choco-Story Brugge, 2024). Activities such as film screenings about the history of chocolate, chocolate-making workshops, and chocolate tastings are also organised in the museum (All Exciting, 2024b).

"The Köstem Olive Oil Museum" is the world's largest museum in its field, and it opened in 2017 in Urla, Türkiye. It exhibits artworks explaining the history of olives from ancient times to the present day and displays a variety of olive oil machines and tools (TCKTB, 2017).

Results

Gastronomy-themed recreation activities have gained popularity in recent years as a unique and immersive way for individuals to engage with food and culinary experiences. These activities allow participants to learn about different cuisines, ingredients, and cooking techniques while enjoying the social aspect of sharing a meal. Individuals can enhance their culinary skills and knowledge by participating in these activities and fostering creativity and self-expression. Furthermore, gastronomy-themed recreation activities can positively impact mental and emotional well-being.

Food tours, cooking events, cooking courses, tasting days, gastronomy museums, and festivals are the most common activities among these events. Regardless of the specific activity, gastronomy-themed recreational experiences provide a unique and memorable way for travellers to connect with a destination's culinary heritage. They also support local businesses and communities while allowing visitors to indulge in the flavours that make each destination unique.

Integrating gastronomy into recreational tourism enhances the travel experience and plays a pivotal role in tourist destinations' economic and cultural development. In this context, it is essential to increase gastronomy-themed events, especially for destinations that stand out with their gastronomic values and culinary culture. Utilising their unique culinary heritage, local destinations can create and promote gastronomy-themed events that attract tourists and promote local cuisine, thereby enriching the overall tourism experience (Razpotnik Visković & Komac, 2021). Highlighting, promoting, and sustaining local gastronomic elements should contribute to the development of tourism in the region and the economic well-being of the region's people. In this context, the public, private sector, and non-governmental organisations should support the activities to be carried out. In addition, the participation of the region's people in the organised activities can also increase attractiveness.

Effective promotion and publicity of these events through social media and broadcasting organs will attract more people who want to spend their free time with gastronomy-themed events. In this way, attracting the attention of more gastro tourists and potential candidates living in remote locations worldwide is also possible. Facilitating access and accessibility and organising package tours with an affordable price policy and package tours can also encourage the participation of those with limited budgets. To ensure all these developments, supporting the involvement of individuals interested in gastronomy-themed recreation activities, planning to be prepared by regional and national tourism authorities, and developing strategies are of great importance.

References

Ahiler Kalkınma Ajansı (2023). *Dünyada ve Türkiye'de İyi Uygulama Örnekleri ile Gastronomi Turizmi ve Nevşehir İli Uygulanabilirlik Analizi.* Ahiler Kalkınma Ajansı, Nevşehir.

Alaçatı Ot Festivali (2024). Festival hakkında. https://alacatiotfestivali.com.tr/hakkimizda/

Alimentarium (2024). Our history. https://www.alimentarium.org/en/basic-page/our-history

All Exciting (2024a). Bordeaux şarap festivali – Féte le vin. https://tr.allexciting.com/bordeaux-wine-festival/

All Exciting (2024b). Choco-story and chocolate museum in Bruges. https://allexciting.com/choco-story/

Barakazı, M., & Önçel, S. (2017). Rekreasyonel turizm faaliyeti olarak Urfa sıra geceleri. *Disiplinlerarası Akademik Turizm Dergisi, 2*(1), 87–97. https://doi.org/10.31822/jomat.349580

Bologna Welcome (2024). Museums and collections. https://www.bolognawelcome.com/en/places/museums-and-art-galleries/gelato-museum-carpigiani-eng

Brey, E. T., & Lehto, X. Y. (2007). The relationship between daily and vacation activities. *Annals of Tourism Research, 34*(1), 160–180. https://doi.org/10.1016/j.annals.2006.08.001

Bucher, A. C. (1972). *Foundation of physical education.* The C.U., Mosby Company, Saint Louis.

Choco Story Brugge (2024). Discover the world of chocolate. https://choco-story-brugge.be/en/

Çekal, N., Doğan, E., & Aktürk, H. (2022). Türkiye'de gastronomi müzeleri ve özellikleri. *Sosyal, Beşerî ve İdari Bilimler Dergisi, 5*(11), 1655–1673. https://doi.org/10.26677/TR1010.2022.1122

Çetin, M., & Küçükkömürler, S. (2019). Destinasyon çekicilik unsuru olarak gastronomi müzeleri: Oleatrium zeytin ve zeytinyağı tarihi müzesi örneği. *Ankara Hacı Bayram Veli Üniversitesi Turizm Fakültesi Dergisi, 22*(1), 1–17.

Doldur, H., & Nesipoğlu, Ö. (2022). Antakya şehri ve yakın çevresinde rekreasyon faaliyetlerinin özellikleri ve dağılışı. *Coğrafya Dergisi* (44), 31–47. https://doi.org/10.26650/JGEOG2021-896150

Düzgün, E. (2022). Kültür turlarında gastronomik ürünlerin tercih edilmesinde turist rehberlerinin önemi. *Hitit Sosyal Bilimler Dergisi, 15*(1), 120–141. https://doi.org/10.17218/hititsbd.1111524

Festival Turu (2024). La Tomatina – domates festivali. http://www.festivalturu.com/la-tomatina-domates-festivali-turu

Food In Life (2020, 3 June). Gastronominin tarih ile buluştuğu noktalar: Gastronomi müzeleri. https://foodinlife.com/gastronominin-tarih-ile-bulustugu-noktalar-gastronomi-muzeleri/

Gelato Museum (2024). History. https://www.gelatomuseum.com/en/history

Goodale, T.L., & Witt, P. A. (1980). *Recreation and leisure: Issues in an era of change.* Venture Publishing, USA.

Hall. M. C., Sharples, L., Mitchell R., Macionis, N., & Cambourne B. (2003). *Food tourism around the world: Development, management and markets.* Butterworth-Heinemann, Elsevier, Oxford.

Hazar, A. (2014). *Rekreasyon ve animasyon.* Detay Yayıncılık, Ankara.

Jensen, C. R., & Guthrie, S. (2006). *Outdoor recreation in America* (6th ed.). Human Kinetics, Champaign, Illinois.

Jensen, C. R., & Naylor, J. (2000). *Opportunities in recreation and leisure careers,* VGM Career Books, USA.

Kandemir Altunel, G. (2021). *Elements of a memorable visitor experience in cultural recreation areas: the case of historic areas of İstanbul.* (Unpublished doctoral thesis), İstanbul Üniversitesi Sosyal Bilimleri Enstitüsü, Turizm İşletmeciliği Anabilim Dalı, İstanbul.

Karakan, H. İ. (2021). *Gastronomi temalı rekreasyon etkinlikleri.* In: S. Kül Avan, E. K. Şimşek, Ö. Yayla, M. C. Birinci (Eds.), Rekreasyon ve Turizm Araştırmaları, pp. 41–52, Çizgi Kitabevi Yayınları, Konya.

Karaküçük, S., & Gürbüz, B. (2007). *Rekreasyon ve kent(li)leşme.* Gazi Kitabevi, Ankara.

Karaküçük, S. (2014). Rekreasyon: Boş Zamanları Değerlendirme. Google Books.

Kargiglioğlu, Ş., & Kabacık, M. (2017). Gastronomi turizmi kapsamında Urla enginar festivaline gelen turistlerin festival hakkındaki görüşleri. *Journal of Tourism and Gastronomy Studies, 5*(3), 409–421. https://doi.org/10.21325/jotags.2017.102

Kraus, R., & Washington, N. (1985). Needs assessment in urban recreation. A university's role. *Parks & Recreation, 20*(7), 40–44.

Le Cordon Bleu (2024a). Culinary tourism, now trending. https://www.cordonbleu.edu/news/culinary-tourism/en

Le Cordon Bleu (2024b). Home. https://www.cordonbleu.edu/home/en

Lokman, U., Yarmacı, N., & Güdül, H. (2022). Turizmde çekicilik unsuru olarak gastronomi müzeleri: Adatepe zeytinyağı müzesi örneği. *Journal of Tourism & Gastronomy Studies, 10*(1), 380–399. http://doi.org/10.21325/jotags.2022.995

Musee Escoffier (2024). The Escoffier museum of culinary art. https://en.musee-escoffier.com

Nesterchuk I., Matsuka V., Balabanyts A., Skarha O., Pivnova L., & Kondratenko I. (2022). Tools and development drivers for the gastronomic tourism. *Economic Affairs, 67*(4), 579–587. https://doi.org/10.46852/0424-2513.4.2022.24

Onat, G. (2021). *Rekreasyon Faaliyetleri Bağlamında Gastronomik Etkinliklerin Değerlendirilmesi.* In: M. C. Birinci, Ö. Yayla (Eds.), Rekreasyon ve Turizm Araştırmaları, pp. 281–299, Çizgi Kitabevi Yayınları, Konya.

Otay, O. (24 December 2017). Dünyanın en tuhaf 8 müzesi. https://gezimanya.com/Yazilar/dunyanin-en-tuhaf-8-muzesi

Öztürk, Y. (2019). *Relationship Between Tourism and Recreation From a Current Perspective.* In: İ. Yazıcıoğlu, Ö. Yayla, A. Solunoğlu (Eds.), Current Issues in Tourism and Hospitality Management, pp. 271–281, SRA Academic Publishing, Lithuanian.

Pizza Village (2024a). Eventi. https://www.pizzavillage.it/eventi

Pizza Village (2024b). World masterclass. https://www.pizzavillage.it/worldmasterclass

Razpotnik Visković, N., & Komac, B. (2021). Gastronomy tourism: A brief introduction. *Acta Geographica Slovenica, 61*(1), 95–105. https://doi.org/10.3986/AGS.10258

Sandybayev, A. (2018). The impact of street and food festivals in gastronomic tourism through visitor's emotions and satisfaction. A case of Abu Dhabi food festival. *International Journal of Research in Tourism and Hospitality, 4*(1), 27–32. http://dx.doi.org/10.20431/2455-0043.0401004

Santich, B. (2004). The study of gastronomy and its relevance to hospitality education and training. *International Journal of Hospitality Management, 23*(1), 15–24. https://doi.org/10.1016/S0278-4319(03)00069-0

Sevil, T. (2012). *Boş zaman ve rekreasyon: Kavramlar ve özellikler.* In: S. Kocaekşi (Ed), Boş Zaman ve Rekreasyon Yönetimi, pp. 2–25, Anadolu Üniversitesi Açıköğretim Yayınları, Eskişehir.

Skyscanner (2019, 6 June). Dünyanın en ilginç yemek festivalleri. https://www.skyscanner.com.tr/haberler/ilham-verici-oneriler/dunyanin-en-ilginc-yemek-festivalleri

Smith, S. L., & Xiao, H. (2008). Culinary tourism supply chains: A preliminary examination. *Journal of Travel Research, 46*(3), 289–299. https://doi.org/10.1177/0047287506303981

Solunoğlu, A. (2021). *Rekreatif Bir Etkinlik Olarak Sokakta Yemek.* In: Ö. Yayla, H. Çetiner (Eds.), Turizm Alanında Güncel Konular ve Yaklaşımlar, pp. 212, 219, Çizgi Kitabevi Yayınları, Konya.

Sormaz, Ü., Özata, E., & Güneş, E. (2015). Turizmde gastronomi. *Sosyal Bilimler Dergisi, 1*(2), 67–73. https://sobider.com/?mod=sayi_detay&sayi_id=616

Şenel, E., & Kılıç, İ. (2022). *Rekreasyon Endüstrisi İçerisinde Gastronomi Faaliyetlerinin Yeri.* In: İ. Yıldız, S. Batal (Eds.), Sosyal ve Beşerî Bilimlerde Uluslararası Araştırmalar-1., pp. 59–76, Serüven Yayınevi, İzmir.

Sezgin, A. C., & Akbıyuk, T. (2021). Dünyadaki gastronomi müzelerinin tematik analizi. *Journal of Tourism & Gastronomy Studies, 9*(1), 153–184. https://doi.org/10.21325/jotags.2021.783

Şeyhanlıoğlu, H. Ö., Duman, H., & Arıca, R. (2023). Türkiye'de yer alan gastronomi müzelerinin tematik analizi. *Journal of Society, Economics and Management, 4 (Special)*, 121–142. https://doi.org/10.58702/teyd.1352183

Tasty Tours NYC (2024). What is a food tour? https://www.tastytoursnyc.com/what-is-a-food-tour

TC. Kültür ve Turizm Bakanlığı (2017). Urla ilçemizde Köstem zeytinyağı müzesi açıldı. https://izmir.ktb.gov.tr/TR-194266/urla-ilcemizde-kostem-zeytinyagi-muzesi-acildi.html

TC. Kültür ve Turizm Bakanlığı (2024). Emine Göğüş mutfak müzesi. https://gaziantep.ktb.gov.tr/TR-174064/emine-gogus-mutfak-muzesi.html

Torkildsen, G. (2005). Leisure and recreation management (5th ed.). Routledge Taylor and Francis Group, New York.

Uğurlu, A. (2005). Rekreasyonel amaçlı doğa sporlarının turizmde kullanılması Antalya Köprülü kanyon rafting uygulaması. (Yayımlanmamış Yüksek Lisans Tezi), Akdeniz Üniversitesi Sosyal Bilimler Enstitüsü, Spor Yöneticiliği Anabilim Dalı, Antalya.

United Nations World Tourism Organization (2012). *Global Report on Food Tourism.* In P. Jordan (Ed.), (Vol.4). World Tourism Organisation, Madrid.

University of Central Florida. (2024). Food tourism: the impact of food tv shows on local industries. https://www.ucf.edu/online/hospitality/news/food-tourism/

Vaud (2024). Alimentarium, Food Museum. Founded by Nestlé and featuring a delightful selection of exhibitions, interactive presentations, and cooking workshops, the Alimentarium invites young and old alike to take another look at an essential activity: eating! https://www.myvaud.ch/en/P380/alimentarium-food-museum

Wichasin, P. (2007). *A study of Thai women as health tour participants in relation to lifestyle and leisure practice.* (Unpublished doctoral dissertation), Bournemouth University, Philosophy, Bournemouth.

Williams, S. (2003). *Tourism and recreation.* Pearson Education Limited, Edinburgh.

Yang, F. X., Wong, I. A., Tan, X. S., & Wu, D. C. W. (2020). The role of food festivals in branding culinary destinations. *Tourism Management Perspectives, 34*, 100671. https://doi.org/10.1016/j.tmp.2020.100671

Zengin, B., Uyar, H., & Erkol, G. (2014). Gastronomi turizmi üzerine kavramsal bir inceleme, *15.Ulusal Turizm Kongresi*, 13–16 Kasım 2014, 511–524, Detay Yayıncılık, Ankara.

Sercan Kadam* and Gökhan Şalli†

Chapter 16 Street Foods from Historical Perspective

Introduction

People's eating out habits have increased worldwide due to reasons such as population growth, people's greater involvement in business life, limited time for eating, and meeting the need for socialization. In this context, street flavors have gained importance in the process. (Şeker, 2018, s. 392).

People's street food consumption habits and snacking behaviors; It has become a part of the food and drink culture. According to research, it has been determined that approximately 2.5 billion people consume street food. Preparation and consumption of street food is carried out within the framework of food culture all over the world. (Ballı, 2016, 4).

The Concept of Street Tastes

The concept of street delicacies was first included in a regional workshop on Asian street foods held in Indonesia by the United Nations Food and Agriculture Organization in 1986 (Solunoğlu, 2018).

Street food by FAO (World Food and Agriculture Organization); İt is explained as food and beverages prepared for sale on streets and avenues or in shops. It is also defined as mobile food services by some countries. In addition to the presentation and sale of street food with motorized and non-motorized vehicles, service quality is increasing due to the development of the gastronomy sector. Legal regulations are being made in order to increase service quality. Hygiene and sales permit certificates have become mandatory to provide better quality products and services. Sanctions on this issue will increase further in the future. (Choi vd, 2013, 223; Haryani vd., 2007).

A classification process has been made by researchers regarding street food. It is classified as meals, beverages, snacks, fruits and vegetables by Draper (1996).

* Lecturer, Bülent Ecevit University, Devrek Vocational School, Department of Culinary, sercan.kadam@beun.edu.tr

† Lecturer, Anadolu University, Eskisehir Vocational School, Department of Culinary, gsalli@anadolu.edu.tr

Groupings were made by Ballı (2016) as meals, drinks, snacks, desserts and fruits. Classification was made by Demir, Akdağ, Sormaz and Özata (2018) as seafood, animal-derived products, pastries, beverages, desserts, fruits and other gastronomic values. WHO (World Health Organization) (2010) grouped meals according to where they are prepared as meals prepared in small-scale businesses or traditional workshops, at home, in the marketplace and on the street.

Although street delicacies do not have the feature of creating a specific menu, they have a great variety within the framework of consumption habits (Tinker and Cohen, 1985). People's food culture is effective in creating this diversity (Ohiokpehai, 2003).

History of Street Tastes

When the history of street food is examined, it is known that it dates back to Ancient Greece. The ancient Greeks were the first people to introduce fried fish to their street delicacies. These flavors were carried to Ancient Rome and their taste was improved (StreetFoody, 2014).

During the gladiator wars during this period, the Romans consumed street delicacies such as sausages or fried fish for betting purposes (Kargiglioğlu, 2019). During the excavations in the ancient city of Pomei, it was seen that street vendors and food sales areas around the amphitheater were drawn in the paintings dating back to 59 AD (Holleran, 2016).

Street delicacies such as shish kebab and grilled fish are consumed in the coastal areas of the Eastern Mediterranean, which extends from Egypt to Lebanon. In the book 'Viaggia' written by the Florentine traveler Leonardo di Niccolo in 1384, it is reported that tourists visiting Cairo bought rice, kebabs and fried vegetables from street vendors (Snodgrass, 2004).

Global migrations and wars; It enabled the spread of street delicacies to a wide geography (Aydın, 2020).

As a result of the increase in street vendors due to urbanization in North America, administrators began to take precautions with legal regulations. With the regulation issued in 1691, food vendors were restricted. In 1707, the sale of food was completely banned on the streets of New York (Solunoğlu, 2018). Consumption of street delicacies increased as a result of the increase in the number of workers as a result of the industrial revolution in the 19th century. (Calloni, 2013).

Street delicacies in the Ottoman Empire were first served by Sultan II. It was discussed during the Beyazıt period. The Ottoman Empire was the first state to regulate street delicacies in 1502. Foods such as fish, liver, sherbet, rice, pudding, sugar pastes, bread types, boza, bagels and pastries were widely consumed. Hot meats such as roasted chicken and lamb were among the most consumed foods during the Renaissance (İrigüler vd., 2016; Bozkurt, 2021).

For Brazilian street flavors; 'Pao de queijo', known as cheese bread, is one of the delicacies that can be tasted. 'Sardinha frita', which is deep-fried fish, is one of its important products. 'Feijoada', in which plav is served with sautéed garlic cabbage and farofa, is usually served on Saturday. 'Churrasco', the name of barbecue in Brazil, is prepared from various meats (https://www.brezilyakultur.com/brezilya-sokak-yiyecekleri/, 2024).

Street Taste Varieties

Etiyopya'da 'injera' adlı krep şeklinde ince bir katman üzerine 'selata', 'shiro', 'alecha key wot', 'tibs', 'injera' ve 'doro wot' adlı yemekleri, sokak lezzetleri açısından önemli yer tutmaktadır (https://bugusto.blogspot.com/2015/04/tadm-etiyopya-mutfag-habesha-restaurant.html, 2024).

When street foods in America are examined, fast food flavors such as hamburgers and cheeseburgers come to the fore. Due to Spanish immigration, flavors such as 'Burrito' and 'Taco', which belong to the Spanish food culture, are also included in American street food. The hot dog called 'hot dog' also has an important place. For dessert, 'churro', known as 'pate a choux', is consumed by frying the dough and then sprinkling it with a mixture of sugar and cinnamon (Gönülgül ve Durlu-Özkaya, 2017).

'Jianbing', which resembles pancakes, is among the street breakfast dishes among Chinese street delicacies. Dough balls called 'Jiaozi', prepared by stuffing vegetables and meat into manti, are very popular. 'Tea Egg', known as boiled eggs coated with special sauce, is very popular. 'cifantuan' prepared with 'aha cai (pickled vegetables), 'rousing' (pork floss) and youtiao (fried brown dough); It is among the most popular breakfast snacks. Desserts called 'tanghulu', prepared by pouring hardened sugar syrup over fruit grains, are widely consumed. 'Baozi', which is prepared by boiling dough in water and stuffed with boiled pork or vegetables, has taken its place among Chinese street delicacies. 'Donkey burger', also known as a donkey meat sandwich, is served with various spices. (www.neoldu.com/cin-sokak-yemekleri-37844h.htm, 2024).

In Italy, 'arancini', which is prepared by filling it with meat, peas and cheese and frying it, has spread all over the world. 'Trapizzino', which is a combination of pizza and sandwich, is consumed with pleasure. 'Olive ascolana', which is prepared by removing the pits of the olives, filling them with beef and chicken meat, and then rolling them in breadcrumbs, is prepared as a breakfast dish.

'Lampredotto', which is prepared by boiling beef tripe with tomatoes, onions, celery stalks, parsley and spices and then chopping it, can be served with various sauces. 'Piadina Romagnola', which is prepared by stuffing cheese, tomatoes and arugula into pancakes, is consumed with pleasure. (https://flyista.com/blog/italyanin-10-populer-sokak-lezzeti, 2024).

Argentinian street delicacies include; 'Empanada', a type of pastry prepared by filling it with different ingredients, is among the indispensable items of Argentinian street delicacies. Choripan, a type of hot dog, is served with tomatoes and parsley. 'Milanesa', which is prepared by coating veal meat in flour and eggs, can also be served with chicken meat. 'Tortilla' prepared from wheat flour can be served with a variety of cheese and vegetables. 'Gebrannte mandeln', which refers to roasted sugar-coated almond kernels, is consumed with pleasure. 'Pancho', which means a type of hot dog, is served with its own sauce (https://www.neoldu.com/arjantin-sokak-yemekleri-37821h.htm, 2024).

In Morocco, 'harira', a soup enriched with tomatoes, lentils, chickpeas and noodles, is consumed throughout the year. Sheep head is a popular delicacy on the streets of Marrakech. It is also served as a offal plate, such as tongue and head meat. 'Ma'qooda', which are deep-fried potato balls, are consumed as an appetizing snack (https://fabetravel.com/marakes-sokak-yemekleri/, 2024).

Street Taste Examples From Turkey

Various street flavors are consumed throughout Türkiye. Some of these are as follows:

Simit: Simit, derived from the word 'samid' meaning pure white flour; It is defined as a ring-shaped bun sprinkled with sesame seeds. Simit prepared with flour, water, salt, yeast and additives; It is flavored with various foods such as molasses, sesame and black cumin. Bagel varieties are prepared according to regional differences such as Ankara, Manisa, Rize, Izmit, Kastamonu, Izmir and Devrek bagels (Kazkondu, 2023, 205).

Nohutlu pilav: Chickpea rice, which is a common consumption habit throughout Türkiye; It is sold in unique special mobile vehicles. It can be prepared by diversifying it with products such as chickpeas, chicken breast and liver, on top of a delicious rice (Şeker, 2018, s. 396).

Kokoreç: The intestines of large and small cattle are used in making kokoreç. The intestines must first be cleaned. The inner part of kokorec is prepared from the drum, and the outer part is prepared from the small intestine. Bumbars are wrapped around a long skewer. Then the small intestine is wrapped around the skewer. Thin layers of fat are placed between them. Cooked on a charcoal grill and served with spices if desired (Kara, Aslan, Yaman ve Akkaya, 2013).

Döner Kebap: It is reported that it was first prepared in Bursa in 1855 by skewering lamb meat and cooking it on a wood stove. After red and white meat is shaped into leaves or minced meat, it is marinated for around 3–12 hours. It is prepared by positioning the doner kebab tightly and vertically. In recent years, it has been consumed in many countries such as the USA, Germany, France and England (Cebirbay ve Aktaş, 2008).

Et köfte: Et köfte, which are widely known in world cuisines; It is prepared by grinding the meat into minced meat and mixing it with onion, parsley, bread and spices. It varies from region to region, depending on the materials used and the variation in preparation processes. It includes variety such as Tekirdağ köftesi, Akçaabat köftesi, İnegöl köftesi, ıslama köfte, kasap köftesi. (Oğan, 2022).

Çiğ köfte: Minced meat that is lean and raw; It is a dish unique to Turkey, prepared by mixing it with onion, bulgur and spices. It is prepared by adding pepper, cumin, allspice and tomato paste. It is not subjected to any heating process. In recent years, it has been prepared without minced meat as a precaution against the negative effects of raw minced meat on health. Accordingly, the main ingredient consists of bulgur (Şireli, Göncüoğlu ve Pehlivanlar, 2008).

Şerbetli Tatlılar: Sherbet desserts, prepared by mixing foods such as sugar, honey and molasses with other ingredients, have also taken their place among street delicacies. Tulumba dessert, which is prepared with flour, water, salt, oil, sugar, oil and eggs and then shaped, is consumed with sherbet. In addition, lokma dessert prepared with flour, starch, water, sugar, yeast and salt is also served with sherbet. (Bayram, 2012).

Street Taste Festivals

As street delicacies began to come to the fore, the number of events held in this area began to increase. As a result of these activities, the value given to street flavors is increasing. Association of British Street Food Associations, Street Food Square Association, Italian Street Food Association, National Association of Street Vendors of India, Atlanta Street Food Coalition; They were established to develop street delicacies (Solunoğlu, 2018). In order to promote street delicacies within the Vendy awards, the best street vendor was determined (Hendersen vd., 2012).

Street food festivals are held in various parts of the world. Countries such as Italy, Mexico, America, Italy and Türkiye can be given as examples in this context. As a result of these studies, the ability of countries to attract local and foreign tourists has been increased. (Solunoğlu, 2018).

Kaynakça

Aydın, B. (2020). Sokak Lezzetlerinin Gıda Güvenliği Açısından Değerlendirilmesi. (Yüksek Lisans Tezi), İstanbul Gedik Üniversitesi, İstanbul.

Ballı, E. (2016). Gastronomi turizmi açısından Adana sokak lezzetleri. *Journal of Tourism & Gastronomy Studies, 4*(Special Issue 1), 3–17.

Bayram, F. (2021). Şerbetli Tatlılar (Ed: A. Kurnaz, S. Şen), *Tatlı, Pasta ve Hamur İşleri Kavramlar, İlkeler ve Örnek Uygulamalar,* Detay Yayıncılık, Ankara.

Bozkurt, İ. (2021). Gastronomi Turizmine Sokak Lezzetlerinin Etkileri: İstanbul Avrupa Yakası Örneği. (Yüksek Lisans Tezi), İstanbul Ticaret Üniversitesi, İstanbul.

Calloni, M. (2013). Street Food on the Move: A Socio-Philosophical Approach. Journal of the Science of Food and Agriculture, 93, 3406–3413.

Cebirbay, M. A., & Aktaş, N. (2008). Türk Mutfağı'nın Geleneksel Yiyeceği: Döner Kebap. *Uluslararası Asya ve Kuzey Afrika Çalışmaları Kongresi, 38*, 329–339.

Choi, J., Lee, A., & Ok, C. (2013). The effects of consumers' perceived risk and benefit on attitude and behavioral intention: A study of street food. *Journal of Travel & Tourism Marketing, 30*(3), 222–237.

Demir, Ş., Akdağ, G., Sormaz, Ü. & Özata, E. (2018). Sokak lezzetlerinin gastronomik değeri: İstanbul sokak lezzetleri. Güncel Turizm Araştırmaları Dergisi, 2(Ek.1), 589–601.

Draper, A. (1996). Street foods in developingcountries: The potential for micronutrient fortification. London: London School of Hygiene and Tropical Medicine.

Gönülgül, S. & Durlu-Özkaya, F. (2017). Los Angeles ve Toronto Sokak Lezzetlerinin Menü İçeriği ve Üretim Bakımından. *Journal of Tourism and Gastronomy Studies, 454*, 465.

Haryani, Y., Noorzaleha, A. S., Fatimah, A. B., Noorjahan, B. A., Patrick, G. B., Shamsinar, A. T., … & Son, R. (2007). Incidence of Klebsiella pneumonia in street foods sold in Malaysia and their characterization by antibiotic resistance, plasmid profiling, and RAPD–PCR analysis. *Food control, 18*(7), 847–853.

https://bugusto.blogspot.com/2015/04/tadm-etiyopya-mutfag-habesha-restaurant.html, Erişim Tarihi: (25.05.2024).

https://www.neoldu.com/cin-sokak-yemekleri-37844h.htm, Erişim Tarihi: (21.05.2024).

https://flyista.com/blog/italyanin-10-populer-sokak-lezzeti, Erişim Tarihi: (25.05.2024).

https://www.neoldu.com/arjantin-sokak-yemekleri-37821h.htm, Erişim Tarihi: (20.05.2024).

https://www.brezilyakultur.com/brezilya-sokak-yiyecekleri/, Erişim Tarihi: (20.05.2024).

https://fabetravel.com/marakes-sokak-yemekleri/, Erişim Tarihi: (18.05.2024).

Henderson, J. C., Yun, O. S., Poon, P., & Biwei, X. (2012). Hawker Centres as Tourist Attractions: The Case of Singapore. International Journal of Hospitality Management, 31(3), 849–855.

Holleran, C. (2016). Representations of Food Hawkers in Ancient Rome. Calaresu, M. & Van den Heuvel D. (Ed.). Food Hawkers: Selling in the Streets from Antiquity to the Present, London: Routledge.

İrigüler, F., Öztürk B. & Güven, S. (2016). Sokak Yemekleri. Özdoğan, O. (Ed.). Yiyecek İçecek Endüstrisinde Trendler II. Ankara: Detay Yayıncılık.

Kara, R., Aslan, S., Yaman, H., & Akkaya, L. (2013). Afyonkarahisar'da tüketime sunulan kokoreçlerin mikrobiyolojik kalitesinin belirlenmesi. *Kocatepe Veterinary Journal, 6*(1), 7–10.

Kargiglioğlu, Ş. (2019). Planlı Davranış Teorisi Bağlamında Sokak Lezzetlerinin Destinasyon Seçimi ve Tekrar Ziyaret Etme Niyetine Etkisi: İstanbul İli Örneği. (Doktora Tezi), Gazi Üniversitesi, Ankara.

Kazkondu, İ. (2023). Simit (Ed: A. Arman, C. Çetinkaya, A. Kurnaz), *Türkiye'nin Sokak Lezzetleri*, Nobel, Ankara.

Oğan, Y. (2022). Kirşehir yöre mutfağına özgü bir lezzet: Besmeç. Journal of gastronomy, hospitality and travel (Online), 5(1), 16–23. doi.org/10.33083/joghat.2022.109.

Ohiokpehai, O. (2003). Nutritional aspects of street foods in Botswana. Pakistan Journal of Nutrition, 2(2), 76–81.

Snodgrass, M. E. (2004). The Encyclopedia of Kitchen History. New York: An Imprint of the Taylor & Francis Group.

Solunoğlu, A. (2018). Sokak Lezzetlerine İlişkin Tüketici Tutumları: Gaziantep ve İstanbul Örnekleri. Doktora Tezi. Gazi Üniversitesi / Sosyal Bilimler Enstitüsü, Ankara.

StreetFoody (2014). Street Food: The Story. https://www.streetfoody.it/en/street-food-the-story/, Erişim Tarihi: 24.05.2024.

Şeker, İ. T. (2018). Gastronomi ve Yiyecek Tarihi. Akbaba, A. ve Çetinkaya, N (Ed.), Sokak Yemekleri (s. 392–400). Ankara: Detay.

Şireli, U. T., Göncüoğlu, M., & Pehlivanlar, S. (2008). Growth of Listeria monocytogenes in çiğ köfte (raw meat ball). *Ankara Üniversitesi Veteriner Fakültesi Dergisi*, *55*(2), 83–87.

Tinker, I., Cohen, M. (1985). Street foods as a source of income for women. Ekistics (52)310, 83–89.

WHO. (2010). Essential Safety Requirements for Street-Vended Foods. Geneva: World Health Organization.

Nazik Çelikkanat Pasli*

Chapter 17 Recreation and Leadership

Introduction

Recreation is a broad concept that includes activities that individuals perform voluntarily in order to meet their physical and mental needs in their leisure time. It can be said that commonly preferred activities include areas such as sport, travel and the arts.

The various benefits (physical, mental, social, etc.) provided by recreation activities to individuals have made the need for recreation more apparent and this has led to an increase in the number of recreation areas and the diversity of recreation activities. This increase has given rise to the concepts of recreation leader and recreation leadership. Recreation leaders are the people who ensure that the activities to be offered to the participants are planned, organised and carried out smoothly and in accordance with the determined objectives. Recreation leaders work in places where recreational activities are offered such as sports centres, accommodation establishments, parks. Recreation leadership involves the process of managing the activities briefly.

In this section, firstly, the concept of recreation, leader and leadership concepts are explained. Then, under the title of recreation and leadership, recreation leader and recreation leadership issues are examined and presented in depth.

Concept of Recreation

From historical times to the 21st century, there are three elements that have not changed in human life and affect life; the first of these is work, the second is basic needs (sleeping, etc.), and the third is leisure time, which refers to the time left over from these two elements (Bayer, 1992: 205). The concept of leisure time utilisation has emerged as people voluntarily participate in satisfying activities in their leisure time and this concept has been widely used as 'recreation' (Hazar, 2014: 12).

Recreation consists of activities and experiences voluntarily chosen in leisure time, often providing satisfaction, pleasure, creativity, personal or social benefits

* Asst. Prof., Giresun University, Tourism Faculty, Department of Recreation Management, nazik.celikkanat@giresun.edu.tr

(Torkildsen, 1994: 121). Various explanations and definitions have been made by different authors for the concept of recreation. Williams (2003: 4) defines recreation *"is most commonly connected with the idea of activity- with purposeful and constructive engagement with pursuit or event"*. Jensen and Naylor (1999: 2) define recreation as *"the participant is recreated physically, psychologically, spiritually, or mentally; that he or she becomes refreshed and enriched; that he or she is revitalized and more ready to cope with the routines and trials of life"*. Gulam (2016: 157) summaries recreation *"is life itself, without recreation life is meaningless"*.

Concept of Leader and Leadership

The leader is the pioneer of a creativity and change-oriented management approach based on business ethics (Tekin, 2016: 9). A leader is a dedicated personality who paves the way for people, creates an indescribable sense of loyalty among them, makes it a struggle among his/her followers to tend towards the same goal and to achieve it, has a vibrance, depth, mental and social skills, communication and persuasion skills different from ordinary people, can see the interests of people above his/her own interests, is sensitive to local and universal morality and basic human values, draws his/her source from objective knowledge and logic rules, but at the same time has a deep and busy world of heart and soul (Fındıkçı, 2012: 62–63). Good leaders create a vision and define a strategy to get there. They provide guidance (showing the way, setting goals and eliminating uncertainty); driving force (providing motivation, giving confidence and creating team harmony); communication within the organization and representation to the outside world and to the team from outside (Torkildsen, 1994: 73–74). In short, a leader is defined as a person or persons who influence people and change their behaviours in order to achieve certain goals, and this process is defined as leadership (Vurgun, 2020: 7).

Northouse (2019) states that the following components are at the centre of the leadership phenomenon: (a) leadership is a process, (b) leadership involves influence, (c) leadership takes place within groups, and (d) leadership involves shared goals. In line with these components, Northouse (2019: 43) defines leadership *"is a process whereby an individual influences a group of individuals to achieve a common goal."*. Leadership is the ability to mobilise people in line with determined goals (Güney, 2012: 26). In other words, leadership is the sum of the knowledge and abilities to gather a group of people around certain goals and to mobilise and influence them to achieve these goals (Zel, 2006: 110).

Various approaches have been developed in order to define the leader and leadership process more closely. These are trait, behavioural, situational and modern approaches (Zel, 2006: 114–191; Fındıkçı, 2012: 73–90; Güney, 2012: 364–413; Özkalp & Kırel, 2013: 311–331):

Trait Approaches: It is the most basic and oldest approach to define the leader and the leadership process. At the core of this approach is the fact that leaders have personality traits and qualities that are different from other people. According to this approach, the most important thing that distinguishes leaders from other people is their individual characteristics. In other words, the characteristics of the leader are the main factors that determine the effectiveness of the leadership process. These characteristics can be summarised as being knowledgeable, intelligent, self-confident and energetic, as well as having a natural ability to lead, manage and direct.

Behavioural Approaches: Due to the inadequacy of trait approaches in the leadership process, the focus has been on the behaviours of leaders. These approaches argue that a leader's behaviours are more easily observable and identifiable than his/her personality traits. According to behavioural approaches, behavioural characteristics of an effective leader differ from those of a less effective leader. Therefore, behavioural approaches focus on the main behavioural differences that distinguish leaders from others.

Situational Approaches: It is argued that in order for leadership traits and behaviours to emerge, appropriate conditions and environments should support them. In other words, these approaches state that the main power that enables the emergence of leaders is the environment in which they take place rather than their personality traits, qualities and behaviours. The environment in question emerges as a result of conditions and needs. In summary, situational approaches reveal that individual characteristics and behaviours of leaders can only be successful in certain environments and situations.

Modern Approaches: It has developed as a result of the need to see the leader more deeply, comprehensively and holistically. Charismatic leadership and transformational leadership approaches are examples of these approaches. In charismatic leadership, the leader displays his/her different personality and behaviour patterns in such a balanced and systematic way that this display places him/her in a privileged position in the eyes of the followers. Transformational leadership is based on the fact that leaders and followers can transform each other. In other words, the leader can transform his/her subordinates and the subordinates can transform the leader towards the level of motivation, desire and values that enable the achievement of higher goals determined in common.

Recreation and Leadership

Recreation enterprises are institutions managed by the public or private sector that provide recreational activities and produce recreational products and services (Paslı & Çelikkanat Paslı, 2019: 73). The continuous growth trends and increasing complexity of recreation enterprises increase the dimensions of the need for leaders (Karaküçük, Yenel & Akgül, 2017: 213).

The need for leaders in the presentation and realisation of recreation activities has led to the emergence of the concept of recreation leader. Various explanations and definitions have been made by different authors for the concept of recreation leader. Hazar (2014: 82–83) defined recreation leaders as individuals who plan recreation activities, prepare activity areas to minimise the risks that may occur by evaluating weather conditions and geographical features, inform participants and ensure that the activities are carried out in the most efficient way. Yayla (2014: 158–159) defined a recreation leader as a person who, in line with the wishes, needs and aims of the community in which he/she is involved, includes them in certain activities by taking into account the characteristics of the group such as age, gender, ethnicity, directs the group in accordance with the purpose, communicates closely with the group members, helps individuals to socialise and ensures that group members get the highest level of pleasure from the activity. Ardahan (2016: 139) defined a recreation leader as a person who has the knowledge, ability and power to mobilise himself and/or a group of people around certain aims by using communication processes and tools and to mobilise himself and/or others to achieve these aims.

It is possible to see recreation leaders in outdoor and indoor sports organisations (coaches, trainers, etc.), entertainment departments of hotels, tours (tourist guide, tour leader, etc.), luxury cruise ships, camps (summer camps, science camps, etc.), health and wellness centres, non-governmental organisations. In short, a recreation leader can be found wherever there are people who want to engage in purposeful activities or learn more about the world around them (13; Erkol Bayram & Bayram, 2019: 94; Özdemir Uçgun, 2021: 243). Akoğlan Kozak (2017: 87) summarised leader types as 'outdoor activity leader' and 'indoor activity leader' depending on recreation activities. The author defines an outdoor leader as a person who manages activities in outdoor areas such as mountains, countryside, parks, seas, streams and pools and gives tourist guides and city park managers as examples of outdoor recreation leaders. Akoğlan Kozak defines an indoor recreation leader as a person who manages activities in an indoor setting such as sports halls and stages (theatre etc.), and exemplifies stage animators as indoor recreation leaders. Ardahan (2016: 137–138) analysed leadership in recreation in two ways as 'leadership of institutions' and 'leadership of individuals'. The author defines institutional leadership as the leadership of organisations, whether private sector, public sector or non-governmental organisations, whose main purpose is to change and/or transform society, by organising various activities or supporting the activities of other organisations. Ardahan also defined the leadership of individuals as directing the people participating in recreational activities. Başarangil & Akyıldız Munusturlar (2018: 674–675) summarised recreational leader typologies, leadership roles and tasks in the table below:

Table. 1. Recreational Leader Typologies, Leadership Roles, Task.

Recreational Leader Typologies	Leadership Roles	Tasks
Face-to-Face Leader / Direct Recreation Leader	– Contact with group members is face-to-face. – This leader has a role that is advisory, authoritative, mediatory and persuasive. – It is this leader who gives the members of the group the opportunity to be satisfied and to develop their skills. – This type of leader builds strong relationships with people. – These leaders don't like to manage from where they live. – This leader is the leader of the lower level of the hierarchy.	Activity organiser in a holiday village or club, camp adviser, dance instructor, tennis coach, tour guide, etc.
Supervisor / Consultant Leader	– This leader supports team members' daily activities. – These leaders can be acting individually or part of a team whilst working towards the achievement of their goals. – This leader takes her/his strength from the organisation. – Mid-level managers play this role.	Animation supervisors, club officers, sports managers, they take on positions such as recreation officers.

(*continued*)

Table. 1. Continued

Recreational Leader Typologies	Leadership Roles	Tasks
Executive / Managerial Leader	– This type of leadership is responsible for the development and implementation of recreation programs in the desired manner. – Finance management and control, long-term strategic planning, resource development and policy-making are also the responsibility of this leader. – These leaders are senior executives.	The manager of the club, recreation leaders in top positions such as the Chef de Village.
Other Typologies	– These leaders lead in outdoor areas such as the mountain, the countryside, the sea, the river, the park and the swimming pool. – These leaders are in charge of recreational activities which take place in such enclosed venues as fitness centres, sports halls or stages.	These leaders are "indoor and outdoor activity leader". For instance, they are outdoor leaders like facility managers in city parks or tourist guides. These leaders are in charge of indoor activities such as group shows and stage entertainers, they can also be leaders in sports events as trainers and coaches.

Source: Başarangil & Akyıldız Munusturlar, 2018: 674–675.

The characteristics/qualities that a recreation leader should have can be summarised as follows (Karaküçük, Yenel & Akgül, 2017: 211; Kılbaş, 2010: 204–205; Özdemir Uçgun, 2021: 236, 243; Genç, 2022: 108, 113):

- To have physical and mental health, to be dynamic,
- To be attractive, to be effective with their clothes,
- To have a variety of knowledge and skills (e.g. knowledge of first aid, organisational skills, music and drama skills), to be well-equipped and constantly update oneself, to be creative and open to innovations,
- To be hardworking, to be enthusiastic about work and to work with love and pleasure,
- To be smart, intelligent, quick-thinking and capable of making the right decisions,
- To be honest, to be a trusted person, to treat people sincerely,

- To be able to communicate well, to express oneself well, to have a clear voice, good diction and body language, to attract people's attention with one's opinions and to be persuasive,
- To love people, to recognise and understand them, to empathise with them,
- To be impartial, to act equally, to be fair, to balance conflicting expectations, to manage cultural differences,
- To be stable,
- To be successful in crisis management, to be able to analyse possible conflicts and act proactively,
- To be able to control their emotions, to be patient and cautious,
- To be sensitive to social issues, to encourage people in this issue, to be an role model to his/her environment.

The general tasks and responsibilities of leaders in recreation activities can be summarised as follows (Karaküçük, 2008: 257; Kılbaş, 2010: 207; Yayla, 2014: 160):

1. To prepare a recreation programme according to the layout of the activity areas and the needs of the people,
2. In addition to the prepared programme, diversify the programme with special activities to attract more interest of the group,
3. To try to emphasise the talents of individuals by showing the necessary techniques in creative and interesting games, to help their personality development,
4. To encourage participation in events,
5. Directing, guiding and influencing participants to certain activities,
6. To provide the participants with the pleasure of membership, to help them gain satisfaction from being in that group,
7. Keeping records of information such as the number of participants, materials used in activities,
8. To check the areas where recreational activities will be carried out on a daily basis and to take precautions against accidents that may occur,
9. To provide first aid to people injured in accidents that occur,
10. Ensuring clean use of recreation areas and protecting these areas,
11. Developing collaborative relationships with other organisations in the neighbourhood.

To best fulfil their roles and responsibilities, recreation leaders must be aware of the scale and pace of change on a global scale and be prepared to respond. For this, it is very important to identify the factors affecting recreation leaders. In the

21st century, social trends, political environment, economic situation, technological developments and globalisation are the factors affecting recreation leaders (Figure 1.) (O'Connell, 2014: 4).

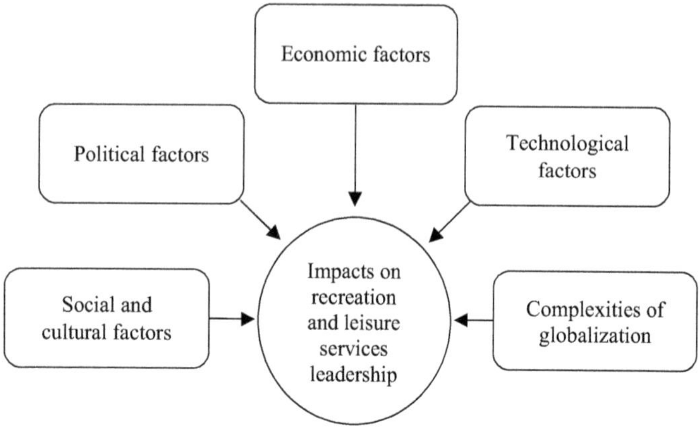

Figure. 1: Factors Affecting Recreation and Leisure Services Leadership.
Source: O'Connell, 2014: 5.

In line with all this information, the leader should keep him/herself up to date with current trends. It is important for the relevant institutions and organisations to develop measures to enable the leader to keep up to date. For this purpose, it is of great importance that leaders follow the social, political, economic developments and changes in the country and the world, people's leisure time tendencies, new activities, changing types of activities and innovations in application areas (Karaküçük, Yenel & Akgül, 2017: 227).

Results

Recreation activities offer positive contributions to the participants at both individual and social level from entertainment, health, social interaction to personal development. With the increase in the number and variety of recreation activities, there is a need for recreation workers who will accompany the participants, guide them and ensure that the activity is successful. Recreation workers who will undertake this mission are called recreation leaders.

It is a qualified, expert recreation leader who is highly effective in making recreational activities successful and sustained (Erkol Bayram ve Bayram, 2019: 95).

The recreation leader is responsible for ensuring that activities are carried out according to the demands of the individual(s) and for the successful completion of these activities. Having a person leading the individual or group in the activities is very important for the individual or group to be satisfied with the activities they participate in at the maximum level (Yayla, 2014: 158).

It is very important for recreation leaders to continuously improve themselves both to support their personal development and to increase their professional success. The continuous renewal of leaders' knowledge, skills and abilities will facilitate their adaptation to innovations in their profession. Recreation leaders who continuously improve themselves and show the right leadership approach will be able to bring their careers and the organisations they serve to the highest level. For this reason, the education they have received in the past, as well as the education they will continue to receive in the future, stands out in professional success. The primary task here is the educational institutions. Developing recreation education institutions, providing quality education services, training students as successful and competent individuals will meet the need for recreation leaders in the future to a great extent.

References

Akoğlan Kozak, M. (2017). Liderlik. In M. Akoğlan Kozak (Ed.), *Rekreasyonel Liderlik ve Turist Rehberliği* (pp. 61–93). Ankara: Detay Yayıncılık.

Ardahan, F. (2016). Rekreasyon ve Liderlik. In F. Ardahan (Ed.), *Her Yönüyle Rekreasyon* (pp. 137–172). Ankara: Detay Yayıncılık.

Başarangil, İ. & Akyıldız Munusturlar, M. (2018). Assessment of type a and type B personality according to typologies of recreational leader in the tourism industry. In R. Efe (Managing Ed.), *Social Sciences Researches in the Globalizing World* (pp. 668–678). Sofia: St. Kliment Ohridski University Press.

Bayer, M. Z. (1992). *Turizme Giriş*. İstanbul: İşletme Fakültesi Yayın No 253.

Erkol Bayram, G. & Bayram, A. T. (2019). Rekreasyon ve Liderlik. In Ö. Yayla & E. Karaçar (Eds.), *Rekreasyon Yönetimi (Güncel Animasyon Uygulamaları)* (pp. 87–108). Ankara: Detay Yayıncılık.

Fındıkçı, İ. (2012). *Hizmetkâr Liderlik* (3. Baskı/Edition). İstanbul: Alfa.

Genç, K. (2022). Rekreasyon Liderliği. In B. Eryılmaz (Ed.), *Rekreasyon: Kavram ve Uygulamalar* (pp. 100–117). Ankara: Detay Yayıncılık.

Gulam, A. (2016). Recreation- Need and importance in modern society. *International Journal of Physiology, Nutrition and Physical Education*, *1*(2), 157–160.

Güney, S. (2012). *Liderlik*. Ankara: Nobel.

Hazar, A. (2014). *Rekreasyon ve Animasyon* (4. Baskı/Edition). Ankara: Detay Yayıncılık.

Jensen, C. & Naylor, J. (1999). *Opportunities in Recreation and Leisure Careers.* United States of America: McGraw-Hill.

Karaküçük, S. (2008). *Rekreasyon: Boş Zamanları Değerlendirme.* Ankara: Gazi Kitabevi.

Karaküçük, S., Yenel, İ. F. & Akgül, B. M. (2017). Rekreasyon Liderliği. In S. Karaküçük, S. Kaya & B. M. Akgül (Eds.), *Rekreasyon Bilimi 2* (pp. 165–233). Ankara: Gazi Kitabevi.

Kılbaş, Ş. (2010). *Rekreasyon: Boş Zamanı Değerlendirme* (4. Baskı/Edition). Ankara: Gazi Kitabevi.

Northouse, P. G. (2019). *Leadership: Theory and Practice* (Eighth Edition). United States of America: SAGE Publications.

O'Connell, T. S. (2014). Introduction to 21st-Century Leadership in Leisure Services. In T. S. O'Connell, B. Cuthbertson, T. J. Goins (Eds.), *Leadership in Recreation and Leisure Services* (pp. 2–17). United States of America: Human Kinetics.

Özdemir Uçgun, G. (2021). Rekreasyonel Liderlik. In G. Güneş & F. Varol (Eds.), *Rekreasyon: Disiplinlerarası Yaklaşım ve Örnek Olaylar* (pp. 229–249). Ankara: Nobel.

Özkalp, E. & Kırel, Ç. (2013). *Örgütsel Davranış* (6. Baskı/Edition). Bursa: Ekin.

Paslı, M. M. & Çelikkanat Paslı, N. (2019). Rekreasyon İşletmeleri ve Yönetimi. In Ö. Yayla & E. Karaçar (Eds.), *Rekreasyon Yönetimi (Güncel Animasyon Uygulamaları)* (pp. 73–86). Ankara: Detay Yayıncılık.

Tekin, S. (2016). *Sihirli Liderler.* İstanbul: Olympia Yayınları.

Torkildsen, G. (1994). *Leisure Management A-Z: A Dictionary of Terms.* Great Britain: Longman.

Vurgun, L. (2020). *İş Dünyasında Örneklerle Liderlik: Gelenekten Geleceğe.* Ankara: Detay Yayıncılık.

Williams, S. (2003). *Tourism and Recreation.* England: Prentice Hall.

Yayla, Ö. (2014). Rekreasyon Faaliyetlerinde Liderlik. In A. Yaylı (Ed.), *Rekreasyona Giriş* (pp. 155–162). Ankara: Detay Yayıncılık.

Zel, U. (2006). *Kişilik ve Liderlik* (2. Baskı/Edition). Ankara: Nobel.

Nevzat Emrah Özçelik[*] and Sena Ekinci[†]

Chapter 18 Evaluation of Geographical Indication Products in the Scope of Gastronomy Tourism

Introduction

Worldwide, the tourism sector is rapidly evolving, expanding, and becoming an increasingly influential economic force. Factors such as globalization, socio-economic changes, and technological advancements that enable tourists to access real-time information have led to shifting needs and expectations among travelers. This change particularly results in a growing demand for destinations offering niche tourism experiences (Amira, 2009; Richards, 2002). Alongside this growth in the tourism sector, competition among countries and destinations is intensifying day by day. Therefore, many countries recognizing the significant potential of tourism are seeking new and alternative forms of tourism to increase their share of the global tourism market and attract more tourists to their destinations (Sanchez-Canizares & López-Guzmán, 2011: 230).

When we look at the history of food, its close relationship with tourism becomes evident. The convergence of local cuisines, beverages, and traditional hospitality has led to the emergence of the concept of gastronomic tourism. Travelers have started considering the gastronomic values of a region when choosing destinations for their vacations. Recently, there has been an increased desire among tourists not only to explore new places and cultures but also to experience the authentic regional foods and beverages of those places. The desire to discover new foods and beverages naturally increases demand (Özkaya and Gündeğer, 2014: 188). Gastronomic tourism stems from tourists' desire to experience the foods and beverages existing in different cultures (Long, 2004). This type of tourism includes visiting food producers, attending food festivals, and dining at specific restaurants to experience a special meal (Yarış, 2014). Alongside changes in tourists' consumption habits, the importance of cultural

* Bolu Abant İzzet Baysal University, Mengen Vocational School, nevzatemrah.ozcelik@ibu.edu.tr

† Bolu Abant İzzet Baysal University, Mengen Vocational School, sena.ekinci@ibu.edu.tr

and traditional products in tourism is increasing. Legal regulations are being implemented in different countries to protect and prevent unfair competition against these local products and their imitations (Richards, 2001; Deveci et al., 2013). Geographical indication systems are widely used legal regulations worldwide. The registration and protection of geographically indicated local products play a significant role in their promotion and evaluation for tourism (Yenipınar et al., 2014). Destinations are crucial for local development and promotion. These destinations are closely associated with the region, possessing a specific reputation and characteristics. To gain an advantage over competitors, they opt to register these characteristics with geographical indications (Bayram and Işın, 2022:65).

Gastronomy Tourism

There are different views regarding the existence of various motivations that influence participation in travel (Albayrak and Özkul, 2013: 19). The tourism industry utilizes gastronomic values to sustain its development and enhance various forms of tourism. Gastronomy is considered a highly effective marketing strategy in tourism for attracting both foreign and domestic tourists (Küçükaltan, 2009). Gastronomic tourism encompasses participation in food festivals, visits to food producers, and visits to specific restaurants to experience a special meal (Yarış, 2014). The foundations of gastronomic tourism are based on culture, agriculture, and traditions. These factors are essential for marketing a region as a gastronomic tourism destination. While the products are typically derived from the agricultural sector, the services are provided by the tourism industry. The convergence of these sub-elements forms the framework of gastronomic tourism (Du Rand and Heath, 2006: 209). To increase its economic contribution, gastronomic tourism needs to be integrated with other forms of tourism. This integration positively impacts the income levels of the local population. Furthermore, it is believed that increasing the number and quality of food and beverage establishments in the region can further enhance this contribution. In this way, developments in gastronomic tourism can have positive effects on the regional economy (Aksoy and Sezgi, 2015).

It's evident that in destination marketing, the advantages and characteristics of a region play a crucial role in competitiveness. Local products and culture can enhance the value of a region and increase the attractiveness of the destination. Gastronomic tourism strengthens a region's cultural identity and economy, contributing to the preservation of the local identity and the growth of the regional economy (Barkat and Vermignon, 2006). Similarly, establishing a positive relationship between the local population and tourists visiting the destination can help enhance the destination's image. In this way, gastronomic tourism can enrich the experiences of both the local community and tourists, thereby increasing interest in the region (Kalkstein, 2007).

Gastronomic tourism refers to travel undertaken to learn and experience any kind of information within the broad scope of gastronomy. The primary objective of such travel is to focus on the journey rather than a specific destination, with the aim of exploring, experiencing, and enjoying various gastronomic activities (Sevimli et al., 2021). Gastronomic tourism triggers the desire to obtain new food and beverage experiences, significantly motivating these travel behaviors. It is also considered a form of gastronomic mobility (Çalışkan, 2013; Gökdeniz et al., 2015).

According to another definition, gastronomic tourism refers to travel undertaken by tourists to experience local flavors instead of ordinary meals in restaurants and hotels (Yüncü, 2010: 29). Stakeholders of gastronomic tourism include hotels, tour operators, individual producers, restaurants, tourists participating in package tours, and tourists determining their own travel routes (and YiLiu, 2011). Local foods and beverages, which are significant elements of gastronomic tourism, can attract the interest of tourists traveling for gastronomic tourism purposes as well as tourists traveling under other forms of tourism, influencing their preferences. In this regard, local flavors and culinary culture can enhance a region's potential for gastronomic tourism and lead tourists to visit the region, positively impacting the region's tourism revenues and economy (Şengül & Türkay, 2015).

In today's world, all sectors undergo continuous change and development. Particularly, the tourism sector, which holds a significant share in the service sector, is significantly affected by these changes and developments. Changes in visitor expectations bring about significant developments that will influence marketing strategies (Öztürk et al., 2021: 367). Gastronomy tourism, which has long been in demand in Europe, plays a significant role in fostering local development. Furthermore, it holds great importance in terms of sustainability. Achieving sustainability depends on the accurate promotion of the region and attracting potential visitors (Bayram et. al., 2018:95). Gastronomic tourism is of great importance for our country and can highlight various destinations. Turkish cuisine boasts the most diversity among world cuisines. With the right policies and a creative image, Turkish cuisine can be effective in branding as a gastronomic tourism destination.

Geographical İndication

The concept of geographical indication and registration expresses the relationship between a product and its production area. In this concept, it is emphasized that the geographical area of the product and the product definition are clearly defined (Lukose, 2007). Such indications are given to products that have distinguishing features from similar products and are associated with the region they originate from. While these indications express the connection with the origin of the product, they generally reflect the regional, local, or national identity of the product (Dimara et al., 2004).

The concept of geographical indication is primarily a mechanism developed to preserve the authenticity and quality of products. This mechanism aims to prevent consumers from being deceived by counterfeit and imitation products. Additionally, it ensures that the production methods of products and their reputation are legally protected. Moreover, it helps to prevent the emergence of potential counterfeit products marketed under similar names (Doğan, 2015).

Geographical indications are signs representing products that are associated with a specific region, area, or country in terms of distinct qualities, reputation, or other characteristics. These signs can be registered as a designation of origin or an appellation of origin. Products produced in various fields such as food, agriculture, mining, handicrafts, and industrial goods can be subject to geographical indication registration (Industrial Property Law No. 6769 - Article 34).

Name of Origin

The designation of origin is a term used when all or essential qualities of a product originate from natural and human factors in a specific geographical area. For a product to be registered as a designation of origin, all its production, processing, and other procedures must take place within the boundaries of the designated geographical area. Therefore, the connections of geographical indications registered as designations of origin with their places of origin are extremely strong.

Appellation of Origin

The appellation of origin signifies products that are associated with a specific geographical area in terms of distinct qualities, reputation, or other characteristics. This sign is used for products where at least one of the production, processing, or other procedures takes place within the designated geographical area. When a product, its raw material, or at least one of its production or processing stages occurs in a specific geographical area and this situation is registered as an appellation of origin, other production and processing stages can also be carried out outside the region of origin.

Traditional Product Name

For products that do not fall within the scope of designation of origin or appellation of origin, names proven to have been used in the relevant market for at least thirty years are accepted as traditional product names. These names are defined under the following conditions:

a) Originating from traditional production or processing methods or traditional composition.
b) Produced from traditional raw materials or ingredients.
(Industrial Property Law No. 6769 - Article 34).

The Decree Law No. 555, which entered into force in 1995 and concerns the Protection of Geographical Indications, is the first regulation in Turkey related to geographical indications. However, with the enforcement of Industrial Property Law No. 6769 in January 2017, interest in geographical indications has started to increase across the country. Geographical indications are granted for various products, ranging from natural to industrial products, and this grant allows the relevant products to be associated with specific definitions and conditions. The provider of this authority is the Turkish Patent and Trademark Office (Boyraz, 2019).

Turkey, with its rich cultural heritage and accumulation, has the potential to obtain geographical indications for many products with its unique soil structure and distinctive production methods. Registering this diversity of products with geographical indications contributes positively to rural development, supports producers, and protects consumer rights. Therefore, registering Turkey's geographical indication products contributes to preserving national values and generating tourism revenues for the economy (Polat, 2017). Geographical indication products not only contribute to the economy of their region but also significantly contribute to its development. In regions with intensive tourism activities, geographical indication products make significant contributions to the local economy through tourism (Mercan and Üzülmez, 2014). Taking Gaziantep's baklava, which has a geographical indication, as an example, tourists who travel to taste it often visit other cultural elements in the city as well. In summary, products with geographical indications have the potential to initiate a movement that can affect all elements in tourism (Orhan, 2010). When aiming to increase tourism revenues and develop tourism, it is important for regions to obtain geographical indications for their local products and support them with advertising and promotion. This makes it easier for tourists to access these products (Yıldız, 2021).

The Statistics of Geographical İndication Products in Turkey

Since 2014, the Turkish Patent and Trademark Office (TPTO) has been handling application and registration procedures. In the first year, they received 72 applications and registered 6 products. In 2021, it became the year with the highest number of applications and registrations, with 577 applications and 352 registrations. As of 2023, there have been 386 applications and 209 products registered.

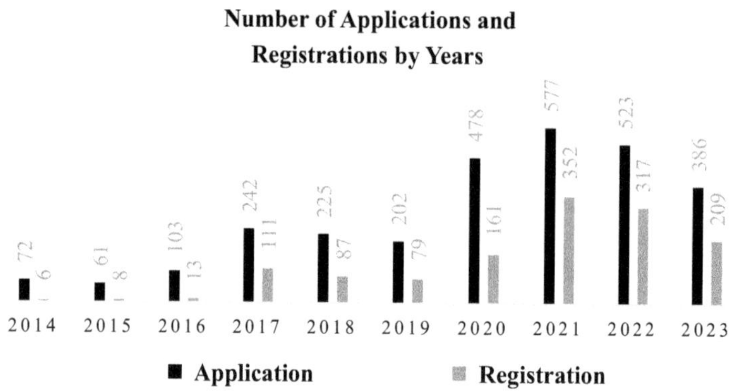

As of 2023, the number of products receiving Appellation of Origin is 1186, the number of products receiving Designation of Origin is 390, and the number of products receiving Traditional Product Name is 6.

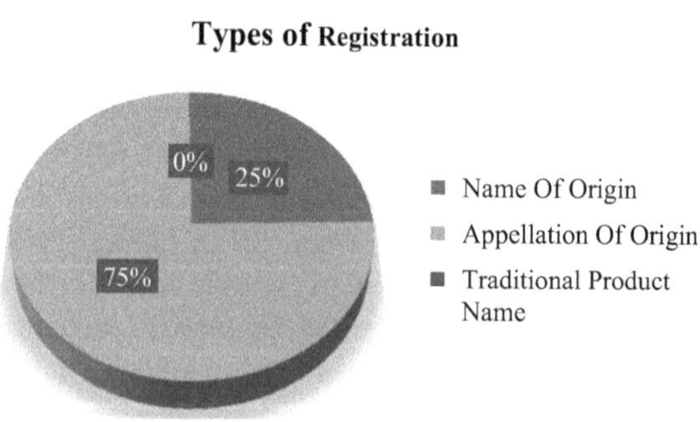

The product groups that have received geographical indications are classified as follows:

- Non-alcoholic beverages
- Honey
- Beers and other alcoholic beverages
- Chocolate, confectionery, and related products
- Other products
- Textiles
- Ice creams and edible ices
- Bakery and pastry products, dough products, desserts
- Carpets and rugs
- Handicraft products other than carpets, rugs, and textiles
- Processed and unprocessed meat products
- Processed and unprocessed fruits, vegetables, and mushrooms
- Cheeses
- Dairy products other than cheeses and butter
- Butter and both solid and liquid fats
- Tobacco
- Meals and soups
- Seasonings/flavorings, sauces, and salt for food

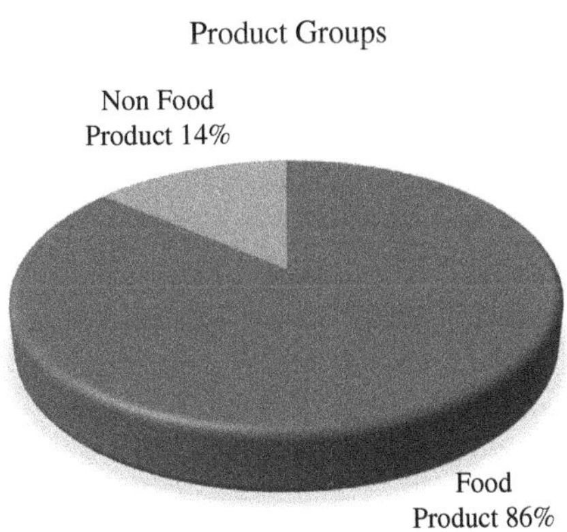

As of 2023, a total of 1370 products in the food and beverage category and 218 products in the non-food category have been registered with a Geographical Indication.

Conclusion

With the increasing popularity of gastronomy and geographical indication concepts, people are showing interest in exploring not only the cultural and historical heritage of the regions they travel to but also the region-specific food and beverages. This development has highlighted the need to discover new places as well as experience various local foods and drinks. Additionally, the existence of region-specific and geographically indicated products has positively impacted the promotion and image of these destinations (Şengül and Çılgınoğlu, 2023).

Gastronomic products registered with geographical indication demonstrate a significant difference from other products in terms of quality, naturalness, characteristics, and taste. These products provide consumers with a sense of confidence due to their unique features, thus offering advantages in marketing. This sense of confidence plays a crucial role both nationally and internationally. The presence of geographical indications contributes to the development of gastronomic tourism at both national and international levels. These developments have a significant impact on the economic development of the respective regions (Akay, 2021).

As of 2023, a total of 1370 food products have been registered with Geographical Indication in Turkey. Since food and beverage products form the cornerstone of gastronomic tourism, efforts in promotion, marketing activities, and policy implementation at both national and local levels, as well as collaborative efforts between local governments and NGOs, can lead to general and regional development. Food and beverage establishments should pay attention to including registered designation of origin symbols in their menus, as they play a significant role in the region's development.

The promotion of registered designation of origin symbols in different regions should be carried out sustainably. These promotional activities should be supported through the creation of permanent materials such as books and brochures for the protection and promotion of registered or region-specific foods.

References

Akay, E. (2021). Kırşehir İlinin Coğrafi İşaret Almış Ürünlerinin Gastronomi Turizmi Kapsamında Değerlendirilmesi. Sivas İnterdisipliner Turizm Araştırmaları Dergisi, 4(2), 147–159.

Aksoy, M. ve Sezgi, G. (2015). Gastronomi Turizmi ve Güneydoğu Anadolu Bölgesi Gastronomik Unsurlar. Journal of Tourism and Gastronomy Studies, 3(3), 79–89.

Albayrak, Aslı ve Özkul, Emrah (2013). "Y Kuşağı Turistlerin Destinasyon İmaj Algıları Üzerine Bir Araştırma". International Periodical For the Languages, Literature and History of Turkish or Turkic 8(6): 15–31.

Alonso, Abel Duarte ve Liu, Yi (2011). "The potential form arrying local gastronomy and wine; the case of the fortunate islands". International Journal of Hospitality Management, 30, 974–981.

Amira, F. (2009). The role of local food in Maldives tourism: A focus on promotion and economic development. Auckland University, New Zealand Tourism Research Institute (NZTRI).

Barkat, Sidi Mohammed ve Vermignon, V. (2006). Gastronomy Tourism: A comparative study of Two French Regions: Brittany and La Martinique. Paper presented at the Sustainable Tourism with Special Reference to Islands and Small States conference, Malta.

Bayram F., Işın A. (2022). Sürdürülebilir Gastronomi Turizminde Yöresel Yemekler ve Coğrafi İşaretli Ürünlerin Önemi, Gastronomide Akademik Çalışmalar, Detay Yayıncılık, Ankara, 65–83.

Bayram, F., Çiftçi, B. & Kemer, A. K. (2018). Yerel Kalkınmada Gastronomi Turizmi ve Gastronomik Etkinliklerin Önemi, Yerel Kalkınma Sempozyumu (Yekas'2018) Bolu / Türkiye 87–98.

Boyraz, M. (2019). Coğrafi işaretli ürünler: Afyonkarahisar. International Journal Entrep- reneurship and Management Inquiries, 3(4), 26–46.

Çalışkan, Osman (2013). "Destinasyon Rekabetçiliği ve Seyahat Motivasyonu Bakımından Gastronomik Kimlik" Journal of Tourism and Gatronomy Studies, 1(2): 39–51.

Deveci, B., Türkmen, S. &Avcıkurt, C. (2013). Kırsal turizm ile gastronomi turizm ilişkisi: Bigadiç örneği. Uluslararası Sosyal ve Ekonomik Bilimler Dergisi 3(2), 29–34.

Dimara, E., Petrou, A., Skuras, D. (2004). Agricultural policy for quality and producers' evaluations of quality marketing indicators: A Greek case study. Food Policy, 29(5), 485–506

Doğan, B. (2015). Coğrafi İşaret Korumasının Gelişmekte Olan Ülkeler İçin Önemi. E-Journal of New World Sciences Academy NWSA-Social Sciences, 10(2), 58–75.

Du Rand, Gerrie ve Heath Ernie (2006). "Towards a Framework For Food Tourism As An Element Of Destination Marketing", Current Issues in Tourism, 9: 206–234.

Gökdeniz, A., Erdem, B., Dinç, Y. ve Çelik Uğuz, S. (2015). Gastronomi Turizmi: Ayvalık'ta Yerli Turistler Üzerinde Görgül Bir Araştırma, Journal of Tourism and Gastronomy Studies, 3(1), 14–29.

Kalkstein, Carol Ann.-Silkess. (2007). "Food And Food Related Festivals in Rural Destination Branding". Yayımlanmamış Doktora Tezi. Indiana: Purdue University.

Küçükaltan, Gül (2009). "Küreselleşme Sürecinde Gastronomide Yöresel Tatların Turistlerin Destinasyon Tercihlerine ve Ülke Ekonomilerine Etkileri". 3.Ulusal Gastronomi Sempozyumu Bildirisi. Antalya, Alanya: ALTSO: Alanya Ticaret ve Sanayi Odası, 17–18 Nisan.

Long, Lucy. (2004). Culinary Tourism. Lexington. ABD: The University Press Of Kentucky.

Lukose, L. P. (2007). Rationale and prospectsof the protection of geographical indication: An inquiry. Journal of Intellectual Property Rights,12, 212–223.

Mercan, Ş. O., & Üzülmez, M. (2014). Coğrafi İşaretlerin Bölgesel Turizm Gelişimindeki Önemi: Çanakkale İli Örneği. Dokuz Eylül Üniversitesi İktisadi İdari Bilimler Fakültesi Dergisi, 29(2), 67–94.

Orhan, A. (2010). Yerel Değerlerin Turizm Ürününe Dönüştürülmesinde "Coğrafi İşaretlerin" Kullanımı: İzmit Pişmaniyesi Örneği. Anatolia: Turizm Araştırmaları Dergisi, 21(2), 243–254.

Özkaya, E. ve Gündeğer, M. (2014). Yöresel Ürünlerin Gastronomi Turizminde Kullanımı: Ayvacık Örneği. 1. Ulusal Meslek Yüksek Okulları Sosyal ve Teknik Bilimler Kongresi, Zonguldak

Öztürk, S., Işınkaralar, Ö., Yılmaz, D. ve Çılgınoğlu, H. (2021). Bir Kültür Turizmi Destinasyonu Olarak Hasankeyf Örneğinde Ziyaretçi Memnuniyeti. Journal of Bartin Faculty of Forestry. 23(2): 359–369.

Polat, E. (2017). Turizm ve Coğrafi İşaretleme: Balıkesir Örneği. *Meriç Uluslararası Sosyal ve Stratejik Araştırmalar Dergisi*, 1(1), 17–31.

Richards, G. (2002). Gastronomy: an essential ingredient in tourism production and consumption? Hjalager A. M., Richards G. (Ed.), Tourism and Gastronomy (Routledge, London).

Richards, G. (2001). Cultural Attractions and European Tourism (CABI Publishing, London)

Sanchez-Canizares, S. M., & López-Guzmán, T. (2011). Gastronomy as a tourism resource: profile of the culinary tourist. Current Issues in Tourism, 229–245.

Sevimli, Y., Tatlı, M. ve Kızıldemir, Ö. (2021). Göbeklitepe'nin Gastronomi Turizmi Bağlamında Değerlendirilmesi. Güncel Turizm Araştırmaları Dergisi, 5(2), 263–286.

Şengül, S. ve Türkay, O. (2015). "Doğu Karadeniz Mutfak Kültürünün Sürdürülebilirliği Sorunlar ve Çözüm Önerileri". Doğu Karadeniz Bölgesi Sürdürülebilir Turizm Kongresi Bildiri Kitabı, 599–606

Şengül, A. ve Çılgınoğlu, H. (2023). Coğrafi işaretli Yozgat Çanak Peynirinin Gastronomi Turizmi Açısından Değerlendirilmesi. Ağrı İbrahim Çeçen Üniversitesi Sosyal Bilimler Enstitüsü Dergisi, 9(1), 103–118.

Yarış, Ahmet (2014). "Mardin'de Gastronomi Turizmi: Turist Görüşlerine İlişkin BirUygulama". Yayınlanmamış Yüksek Lisans Tezi. Mardin: Mardin Artuklu Üniversitesi.

Yenipınar, U., Köçker, H.& Karacaoğlu, S. (2014). Turizmde yerel yiyeceklerin önemi ve coğrafi işaretleme: Van otlu peyniri. Journal of Tourism and Gastronomy Studies 2(2): 13–23

Yıldız, M. H. (2021). Coğrafi İşaretli Ürünlerin Bölge Halkı Tarafından Bilinirliğinin Gastronomi Turizmi Kapsamında Değerlendirilmesi: Isparta Örneği. (Basılmış Yüksek Lisans Tezi). Kocaeli Üniversitesi.

Yüncü, H. R. (2010). "Sürdürülebilir Turizm Açısından Gastronomi Turizmi ve Perşembe Yaylası" 10. Aybastı-Kabataş Kurultayı Aybastı-Kabataş Kurultayı Yayınları,11: 27–34.

Ecem Akay[*] and Ecem Ince Karaçeper[†]

Chapter 19 An Evaluation on Green Restaurant Consumers' Service and Quality Perceptions Within the Scope of Sustainable Gastronomy

Introduction

In recent years, consumer demand for environmentally responsible products and services has risen significantly, driven by a heightened awareness of the ecological impacts of human activities and a growing preference for options that are healthy, sustainable, and minimally harmful to the environment (Jo, Woo, Taylor, 2024). With the increase in the world population and the growing and changing needs of humanity, some problems have arisen in the service sector as well. The main problems are the destruction of the natural environment, unfair food distribution, unconscious use of existing resources and food waste. In this context, the destruction of food product resources, the formation of unconscious and excessive consumption, wasted food products, the use of environmentally harmful chemicals in food products and food businesses that process these products, the increase in the amount of waste from businesses and the damage caused by these wastes to nature and the damage to water resources have reached serious dimensions (Hilaloğulları, Akdağ, Üzülmez, 2022). Since the late 20th century, sustainability is a concept that aims to improve the socio-cultural structure while addressing issues such as green consumption and gastronomy, ecology and consumers' health (Ince and Öztüren 2021). The discipline of gastronomy aims to present culture, history, and food together by examining the characteristics and uses of ingredients found in food products, and by enhancing the flavor of food that creates similarity and diversity in culinary cultures (Çekal & Doğan, 2021). The concept of sustainable gastronomy is expressed as a concept that emphasizes the concept of environmental sustainability and aims to protect and improve public health and socio-cultural status (Scarpato, 2002). The study aims to

* Lecturer, Doğuş University, Faculty of Art and Design, Department of gastronomy and culinary arts, eakay@dogus.edu.tr

† Assistant Professor, Istanbul Kent University, Faculty of Art and Design, Department of Gastronomy and Culinary Arts, ecem.incekaraceper@kent.edu.tr

explore sustainable gastronomy and green restaurants, evaluate green restaurant criteria, and conduct a bibliometric analysis on consumers' perceptions of service and quality in green restaurants. Bibliometric analysis involves examining publication elements using statistical methods.

Literature Review

The concept of service has various qualities such as intangibility, heterogeneity, non-standardization, simultaneity, perishability and ownership. According to the definitions; intangibility in services means that the service cannot be seen, exhibited, expressed, transported, expressed in any unit of measurement, stored in any way, and standardized (Kurnaz, 2017). When the definitions are considered from a broad perspective, the concept of perceived quality is the judgment that the customer has developed that the product and service purchased is excellent or superior (Gökcek, 2022). When the literature is examined, it is seen that the concept of perceived quality is expressed as an evaluation based on a consumer's individual perceptions of the superiority of a product or the superiority of a service. In the case in question, it can be tangible or intangible while it can be a product or service product. In the dimensions of perceived quality in service-based products, the first dimension is the support tools and the dimension in question is the evaluation of the appropriateness of the tools and methods used for the service. The second is the reliability dimension, which is the determination of the same standard of service at different times. The third is the dimension of competence, which is the assessment of whether the actors in the service the skills have needed in the mentioned services (Gökcek, 2022). It is shaped by the customers' perception of the brand they have received service from and the emergence of customer needs. Expectations consist of individual needs, previous experiences or emotions (Soğancı, Demir, Altun, 2022).

Green Restaurants and Green Restaurant Criteria

Green restaurants are defined as businesses that include dishes prepared with locally grown and locally sourced products with organic certificates; have waste reduction practices and recycling policies; and emphasize the efficient use of energy and water resources and solid waste management policies (İpar, Babaç, Kök, 2020). The Green Restaurant Association guides restaurant businesses and restaurants to provide services in line with sustainable criteria and basically evaluates the relevant businesses based on 7 environmental categories and a total of 100 points. These categories efficiency in water consumption, waste reduction and recyclable applications, sustainable furnishing and building materials, sustainable food, energy consumption, disposable consumables, reducing chemical consumption and pollution (Kurnaz, 2017). When sustainable restaurants in the world are analyzed, it is seen that since 2010, Green Six of the restaurants that best meet the standards set by the Restaurant Association restaurants can be cited as examples. These restaurants are; Chez Panisse, Azurmendi Vizcaya, Geramium Denmark, Manoir aux Quat'Saisons, Brae Melbourne and Mil-Maras restaurants (Şimşek and Akdağ, 2017, akt: Keşkekci and Gençer, 2023).

Methods

Document analysis was used as part of the qualitative research methodology in this study. The purpose is to analyze the publications on service and quality perceptions of consumers of restaurants in the context of sustainable gastronomy between 2017–2024, with specific parameters, considering the chronological order. Within the scope of the research, bibliometric evaluation of the studies published in SCOPUS, Web of Science, National Thesis Center, TR Index, Sobiad, Clarivate and Science Direct databases between 2017–2014 using the keywords "sustainability, green restaurant, service quality, expected quality and perceived quality" was conducted. During the evaluation, the year of publication, keywords, type (master's thesis, doctoral thesis, article, book chapter), sample type, sample size, data collection techniques and relevance of the studies were presented in tables. Bibliometrics is basically expressed as a field of study based on counting, and the concept can be explained as the examination of scientific information sharing tools such as published journals, books, articles, graduate theses, etc. by considering statistical and mathematical techniques (Yılmaz, 2016).

In line with the established objectives of the study, the following questions were addressed: What are the types of studies published in this scope? What is the distribution of the studies published in this scope according to the indexes they are included in? What is the distribution of the studies published in this scope according to their research methods? What is the sampling method and sample size of the studies published in this scope? A total of 53 studies were identified within the scope of the determined criteria. These studies were subjected to content analysis using MaxQda Pro 2020 program.

Findings

During the research process, the 53 studies included were subjected to content analysis and categorized according to predetermined criteria. Accordingly,

Figure 1 shows the criteria used in the analysis process. Considering the journals in which the studies were published, it was observed that there were studies in "Journal of Sustainable Tourism", "International Journal of Gastronomy and Food Science", "Sustainability - MDPI" journals. In addition, it was determined that there were studies published in journals supported in other indexes and in the National Thesis Center database. According to the findings obtained, the authors of the studies, the year of publication of the studies, the name of the studies, the place of publication of the studies, the type of studies (master's thesis, doctoral thesis, article, book chapter) and the relationality in the studies are presented in the tables below.

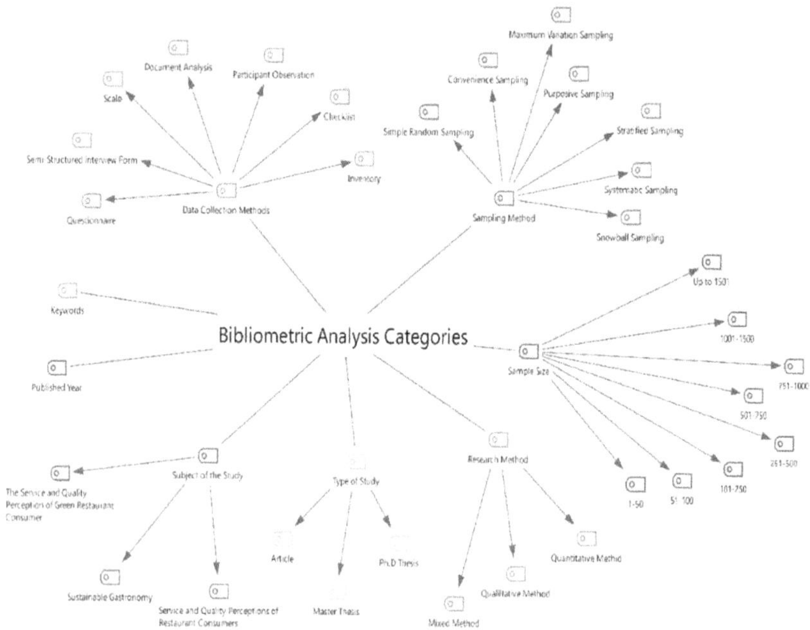

Figure 1. Bibliometric Analysis Categories

As shown in Figure 1, eight different categories have been identified during the research process. These are the topics of the research, types, methods, keywords, publication years, sampling methods, sample sizes, and data collection techniques. Analyses for each context have detailed the strengths and weaknesses of the literature.

Table 1. The Type of Studies by Year

Kod Sistemi	2017	2018	2019	2020	2021	2022	2023	2024	TOPLAM
Type of Study									0
Article									0
Literature Review						1	1		2
Research Article	1	5	5	2	5	8	9	2	37
Ph.D Thesis	1	1				1			3
Master Thesis			1	2	1	6		1	11
Published Year									0
Σ TOPLAM	2	6	6	4	6	16	10	3	53

The table above classifies and presents various types of studies published from 2017 to 2024 in a chronological order. It specifies different types of academic publications such as articles, literature reviews, research articles, PhD theses, and master's theses, and shows the distribution of each type by year. The first column of the table is divided into four main categories under the heading "Type of Study": articles, literature reviews, research articles, doctoral theses, and master's theses. This categorization demonstrates the scope of the study and the diversity of types of academic research.

Table 2. The Subject of Study by Year

Kod Sistemi	2017	2018	2019	2020	2021	2022	2023	2024	TOPLAM
Subject of the Study									0
Service and Quality Perceptions of Restaurant Consumers	1	1		1		3	1		7
Sustainable Gastronomy		4	1	2	3	6	9	2	27
The Service and Quality Perception of Green Restaurant Consumer	1	1	5	1	3	7		1	19
Published Year									0
Σ TOPLAM	2	6	6	4	6	16	10	3	53

The examination of research titles begins with studies on the service and quality perceptions of restaurant consumers. The number of publications in this area is seven, with the year 2021 standing out as a peak year for these studies. During this period, there has been an increased awareness of service quality and customer satisfaction within the restaurant industry. Such interest has led to an increase in the number of studies conducted to understand consumers' changing expectations and how restaurants respond to these expectations. Studies on the service and quality perceptions of green restaurant consumers hold a significant place within this research area, totalling nineteen publications. The year 2022 has been the peak year for this topic, with seven studies making it the period with the most intense research. This situation indicates that consumer expectations for green and sustainable practices in the restaurant sector have increased, and accordingly, academic studies have also seen a rise. The growing interest in the service and quality standards expected from eco-friendly restaurants has prompted academic studies to focus on this area.

Table 3. The Type of The Studies

Kod Sistemi	Literature Review	Research Article	Ph.D Thesis	Master Thesis	TOPLAM
Subject of the Study					0
Service and Quality Perceptions of Restaurant Consumers		5	1	1	7
Sustainable Gastronomy	2	18	1	7	28
The Service and Quality Perception of Green Restaurant Consumer		15	1	3	19
Type of Study					0
Σ TOPLAM	2	38	3	11	54

Table 3 presents the numerical distribution of various types of studies grouped around three different research topics, clearly showing the intensity of academic interest in certain subjects and changes over time. The topics examined are the service and quality perceptions of restaurant consumers, sustainable gastronomy, and the service and quality perceptions of green restaurant consumers. A total of seven studies have been conducted on the service and quality perceptions of restaurant consumers. Most of these studies, five in total, have been realized in the form of research articles. There is also one doctoral thesis and one master's thesis in this area. The absence of a literature review, despite the topic being analysed in depth, indicates that it has not been addressed from a broad perspective and lacks a general evaluation. Overall, this table offers a numerical indicator of the intense interest in certain topics within the academic literature and clearly shows the contribution of each type of research to these topics.

Table 4 displays the frequently used keywords in studies related to sustainable gastronomy and the restaurant industry, showing how these keywords are distributed across three main categories: the service and quality perceptions of green restaurant consumers, sustainable gastronomy, and the service and quality perceptions of restaurant. Environmental sustainability is one of the most common keywords in the table, particularly emphasized in the area of sustainable gastronomy. Green initiatives and tourism are other keywords frequently encountered in the field of sustainable gastronomy. This wide range of keywords and topics illustrates that the restaurant industry and sustainable gastronomy are dynamic and multifaceted fields.

Table 4. Keywords by Research Type

Kod Sistemi	The Service and Quality Perception of Green Restaurant Consumer	Sustainable Gastronomy	Service and Quality Perceptions of Restaurant Consumers	TOPLAM
Keywords				0
Environmental Sustainability	2	1		3
Customer Loyalty		1	4	5
Green Initiatives	5	5		10
Tourism		3		3
Sustainability	4	7		11
Culinary Culture		2	1	3
Organic Practices		2	1	3
Agricultural Sustainability		2	1	4
Sustainability in Cuisine		6		6
Climate Change		2		2
Meat Reduction		2		2
Environmental Impact		3	1	4
Food Innovation	1	7	2	10
Social Responsibility		2		2
Urban Gastronomy		2		2
Sustainable Practices	1	4		5
Local Cuisine		4	1	5
Food Waste		2		2
Food Categories		1		1
Food Choice		1		1
Meat Consumption		2		2
Restaurants	4	4	2	10
Chefs' attitudes		1		1
Augmented Reality			1	1
Intention to Revisit	1		1	2
Food and Beverage Sector		1	2	3
Hospitality Industry		2	2	4
Consumer Satisfaction	1	4	8	13
Dining Experience			2	2
Children's Influence			1	1
Family Decision Making			1	1
Digital Technologies		1	2	3
Customer Preferences		2	3	5
Culinary Innovation		1	2	1
DINESERV			3	3
Multi-criteria Decision Analysis			2	2
Consumer Perceptions		3	4	7
Service Quality	4	8	15	27
Green Restaurants	5	4	2	11
TOPLAM	28	92	54	184

Figure 2. Bibliometric Word Clouds

Figure 2 shows the word cloud created with the MaxQda Pro 2020 data analysis program in line with the findings obtained.

Table 5. Research Methods

Kod Sistemi	The Service and Quality Perception of Green Restaurant Consumer	Sustainable Gastronomy	Service and Quality Perceptions of Restaurant Consumers	TOPLAM
Research Method				0
Mixed Method	1	1		[illegible]
Quantitative Method	4	8	16	28
Qualitative Method	2	13	2	17
Σ TOPLAM	7	[illegible]	[illegible]	47

Table 5 examines the research methods used across three main research topics, illustrating how qualitative, quantitative, and mixed methods are applied. These topics are defined as the service and quality perceptions of green restaurant consumers, sustainable gastronomy, and service and quality perceptions of restaurant consumers. Qualitative methods have been used in seven studies on the service and quality perceptions of green restaurant consumers. These methods aim to deeply understand consumers' experiences, perceptions, and attitudes. Quantitative methods, on the other hand, are widely used, particularly in the field of sustainable gastronomy, with a total of thirteen studies employing this approach. In the field of sustainable gastronomy, quantitative data are used to measure consumer behaviours, market trends, and the impacts of various sustainability practices. Mixed methods combine the advantages of both types of data, offering a more comprehensive perspective.

Table 6. Sampling Types and Sample Sizes by Research Topics

Kod Sistemi	The Service and Quality Perception of Green Restaurant Consumer	Sustainable Gastronomy	Service and Quality Perceptions of Restaurant Consumers	TOPLAM
Sampling Method				0
Maximum Variation Sampling		1		1
Snowball Sampling	1	5		6
Purposive Sampling	2	11	3	16
Convenience Sampling		4	2	[illegible]
Stratified Sampling			1	1
Systematic Sampling				0
Simple Random Sampling	5	9	14	28
Sample Size				0
1-50	1	15	1	[illegible]
51-100			1	1
101-250	1	3	1	[illegible]
251-500	3	5	10	18
501-750	1	2	3	6
751-1000				[illegible]
1001-1500			1	1
Up to 1501	1	1	1	[illegible]
Σ TOPLAM	15	56	[illegible]	[illegible]

Table 6 displays the sampling types and sample sizes used in research related to the restaurant industry across various research topics. A detailed examination is conducted under the headings of green restaurant consumers' service and quality perceptions, sustainable gastronomy, and restaurant consumers' service and quality perceptions. Among sampling types, the most commonly used methods are purposive sampling, convenience sampling, and snowball sampling. Purposive sampling is frequently chosen, particularly in studies concerning restaurant consumers' service and quality perceptions.

Table 7. Data Collection Techniques by Research Topics

Kod Sistemi	The Service and Quality Perception of Green Restaurant Consumer	Sustainable Gastronomy	Service and Quality Perceptions of Restaurant Consumers	TOPLAM
Data Collection Methods				0
Document Analysis	1	5	2	[illegible]
Participant Observation		4	1	5
Semi-Structured Interview Form	3	9		12
Checklist		1		1
Inventory			2	[illegible]
Questionnaire	2	6	8	16
Scale	3	7	11	[illegible]
Σ TOPLAM	9	[illegible]	24	[illegible]

Table 7 illustrates the distribution of data collection methods used in research related to the restaurant industry across three main topics. The frequency of methods used varies depending on the nature of the research topics.

Participant observation is notably used in five studies on sustainable gastronomy. Participant observation is an effective method for understanding consumer behaviours and operational practices within restaurant environments. Semi-structured interview forms are another preferred method, employed in nine studies focused on sustainable gastronomy. This approach provides researchers with the flexibility to delve deeply into specific topics while exploring new issues that may arise during the interviews. Checklists and inventories are also utilized in studies on sustainable gastronomy. Checklists are particularly used to assess sustainability practices in restaurants, offering a quick and objective evaluation based on specific criteria. Inventories are effective tools for systematically recording various resources and materials in restaurants. The questionnaire method is predominantly used in studies addressing the service and quality perceptions of restaurant consumers. This method facilitates the collection of data from a broad group of participants, aiming to achieve generalizable results.

Conclusion and Recommendations

While concerns and studies on sustainability continue globally, many sectors are also working on the issue. Restaurant businesses, which are a labour-intensive sub-branch of the service sector, adopt environmentally friendly approaches and update their policies. For this reason, nature-sensitive practices such as sustainable restaurants or green generation restaurants are on the agenda. Conducting scientific studies, forming scientific approaches on the subject, contributing to the understanding of sustainability and the concept of business management in the sector in the long term are among the developments. In this study, studies published nationally and internationally on sustainable gastronomy management, green restaurants and service and quality perception of green restaurants were compiled and discussed with bibliometric evaluation. Emphasis was placed on the service and quality perceptions of the studies.

Considering the studies examined, the policies implemented in food and beverage establishments within the scope of sustainable gastronomy come to the forefront as environmental sustainability policies such as waste and water management, energy management, pollution prevention. As of 2021, awareness of service quality and customer satisfaction in the restaurant sector has increased in this period when these studies stand out most intensely. This interest has led to an increase in studies to understand the changing expectations of consumers and how restaurants respond to these expectations. Studies on the service and quality perceptions of green restaurant consumers were the most intensive in 2022. This shows that consumers' expectations for green and sustainable practices in the restaurant sector have increased and academic studies have increased accordingly. Increasing interest in the service and quality standards expected from environmentally friendly restaurants has led academic studies to focus on this area. For future studies, it was suggested that the effects of service and quality perceptions on purchasing behavior and customer satisfaction could be studied.

Acknowledgments

The authors are thankful for the employees who participated in this research study.

Funding

No financial support was received for this study.

Authorship Contribution Statement

Ecem Akay, Ecem İnce Karaçeper: Conceptualization, Methodology, Data curation, Writing an original draft. Visualizing, Reviewing, Controlling, Writing – reviewing and editing. Data curation

Declaration of Competing Interest

The authors declare that they have no conflict of interest.

References

Çekal, N. ve Doğan, E. (2021). The Importance of Standard Recipe and Geographical Indications in Sustainable Gastronomy. Journal of Tourism Studies. 4(1):49–60. e-ISSN: 2718-045X.

Gökcek, H., A. (2022). A Qualitative and Quantitative Research To Determine Factors Affecting Perceived Quality: A Restaurant Chain Example. Social Science Development Journal. 7(32):227–240.

Hilaloğulları, M., Akdağ, G., Üzülmez, M. (2022). The Effect of Green Restaurant Practices on Customers' Satisfaction and Behavioral Intention: An Implementation in Istanbul Journal of Tourism and Gastronomy Studies. 10(3):2237–2255.

Ince, E. and Öztüren, A. (2021). Strengths, weaknesses and challenges of municipalities in North Cyprus aspiring to be a sustainable Citta slow tourism destination. Worldwide Hospitality and Tourism Themes, 13(4), 528–534.

İpar, M. S., Babaç, E., & Kök, A. (2020). Evaluatıon Of Customer Revıews For Green Generatıon Restaurants By Content *Analysıs. Journal of Gastronomy, Hospitality, and Travel*, 3(2), 260–269.

Keşkekci, D. ve Gençer, K. (2023). Green Restaurant Applications Within The Scope Of Sustainability. Journal of Silk Road Tourism Research. 3(1):17–25.

Jo, J., Joung, H.-W., & Taylor, J. J. (2023). Impact of Sustainability Attributes of Menu Labels on Restaurant Green Image and Customer Attitudes. *Journal of Marketing Development and Competitiveness*, 17(4). https://doi.org/10.33423/jmdc.v17i4.6646.

Kurnaz, A. (2017). Measurement of Green Restaurants' Service Quality Perception within the Scope of Sustainable Gastronomy: An Example Istanbul. Adnan Menderes University, Institute of Social Sciences, Department of Tourism Management. Doctoral Thesis.

Scarpato, R. (2002). Gastronomy as a tourist product: The perspective of gastronomy studies. Tourism and gastronomy, 11(2), 51–70.

Şimşek, N. & Akdağ, G. (2017). Investigation of Green Genetic Restaurants within Content of Sustainable Gastronomy Tourism. Journal of Academic Social Science Studies. (60):351–368.

Soğancı, F., Demir, A., Altun, Ö. (2022). Technology and Its Importance in Airports. Theoretical and Empirical Current Research in Vocational Sciences. (ed: Daşkın, M., Pala, K.). 142–161. ISBN: 978-605-196-786-8. Çizgi Bookstore. Istanbul.

Yılmaz, G. (2016). Bibliometrics Analysis of Published Papers on Tipping in Restaurants. Journal of Travel and Hospitality Management. 14(2):65–79.

Ayla Solmaz Avcikurt[*] and Yasemin Tekin[†]

Chapter 20 A Healing Center of Antiquity with Faith: Bergama Asklepion

Introduction

Bergama, the starting point of medicine and medical science, is an important tourism center that hosted civilizations, with its natural healing resources and rich cultural heritage. Its history dates back to the Bronze Age (3000 BC). The world's first hospital is the Bergama Asklepion, the healing and treatment center of the ancient age and the first medical school to train physicians such as Hippocrates and Galen who are the pioneering physicians of the medical world. The snake symbol representing medicine and pharmacy also came from the Asklepion. It is the place where medical instruments and many treatment methods used even today first appeared. Bergama Asklepion, which was called the place where death could not approach in the ancient period, is a structure shaped by beliefs and worship, both in its establishment and use. Bergama Asklepion, the temple and treatment center dedicated to Asklepios, the God of Health, who challenged death by healing people, was founded by priests who took part in the treatments alongside the physicians. The Asklepion, where people healed by worshiping in temples and offering gifts to the gods, is very important in terms of medical and tourism sciences. The purpose of this descriptive study is to reveal the origins of the current traditional medicine understanding used for the treatment of many diseases, to create destination awareness, to offer innovative suggestions, and to contribute to the literature of medicine and tourism sciences.

A Healing Center of Antiquity with Faith: Bergama Asklepion

Humankind has always sought healing in medical sciences for many reasons such as fighting disease and epidemics, personal care and beauty, rest, and worship since antiquity. As well as medical interventions, different methods have been used to ensure human health and treatment. Religious beliefs and worshipping

* PhD., Balıkesir University, Medical Faculty, Department of Medical Biology, aylaavcikurt@hotmail.com

† PhD., Balıkesir University, Tourism Faculty, Department of Tourism Management, yasemin.tekin@balikesir.edu.tr

are among the healing and treatment methods. Humankind has sought and tried to maintain a healthy life with various methods such as religious ceremonies, rituals, worshiping, praying, offering sacrifices to the believed god, and believing in the powers of mythological characters.

The structures used for healing and treatment in ancient times were Asklepions. These buildings served as medical schools and hospitals. In addition to medical methods, natural and faith-based treatment methods were used and people were tried to be treated in terms of both physical and mental health in those buildings.

It is claimed that hundreds of Asklepions exist in the wide area that includes Western Anatolia and Greece. Bergama Asklepion is the largest, most famous, and leading health center in the antique world. In this Asklepion, the science of Pharmacology was used for the first time, and people were believed to heal and survive. Other famous Asklepions are the Sanctuary of Asklepios at Epidaurus and Asklepion of the Kos. Due to the cleaning application and treatment and therapy methods with the sound of water, Asklepions were usually built in places with water sources (Çelik, 2018: 1009; Kars, et al., 2020: 113–117; Ökten, 2016: 316). In addition to the Asklepion, Allianoi and Cleopatra Beauty Spa are among the holy waters used for therapeutic purposes since ancient times (Baykal, et al., 2011: 10).

Asklepions are forerunners of modern hospitals. However, the Asklepions are not just a hospital, but a place where superstitious, religious, and scientific methods take place. While people present to a hospital for medical interventions in the modern era, help was sought from the Asklepios, the god of health, and his physicians in ancient times (Erşen, 2007: 98–99). None of the Asklepions are as common and versatile as the Bergama Asklepion. Founded in the Hellenistic era, the Bergama Asklepion demonstrates how far medicine developed in antiquity, by means of suggestion, entertainment, and culture, as well as physical therapies such as healing water, good air, and hot springs. This Asklepion, which has characteristics unique to Anatolia, has no analogues anywhere else, and the treasures it hides have not been fully revealed (Erhat, 2004: 63).

Bergama, the ancient city of Pergamon, in the province of Izmir, survived on a small scale throughout the Middle Ages and was occupied by the Crusaders, Byzantines, Seljuks, and the Ottomans led by Orhan Gazi in 1336, respectively. This city, which is called "Bergama" by the Turks, was revived under the Ottoman rule and became what we see today (Freely, 2008: 25). Bergama is a place where the first parchment paper was made, the first Asian library was established, and the world's tallest and steepest theater was built. In addition, Bergama, which has a great cultural heritage, is a center characterized as the "City of Firsts", where the first big hospital (Asklepion) was built, the first suggestion and natural treatment were applied, the first pharmacology appeared, the first opium model medicine was performed, and the first urban hygiene practices were experienced. Bergama is the starting point of medical science and medicine (Çutuk and Zengin, 2006; Doğan and Sezgin, 2012: 272–277; Süer, 2021: 483).

Bergama Asklepion, where many current medical methods were practiced in ancient times, is the first psychiatry hospital, the first medical school to train famous physicians of the period, and a place of worship dedicated to the god of health (Çelik, 2018: 1009; Kars, et al., 2020: 113).

Bergama Asklepion, which was built in the name of the god of health Asklepios, is a health treatment center established in the middle of the 4th century, outside the city, on the slopes of Geyikli Mountain, in a valley, on the plain where there are water sources believed to provide healing. According to the legend of Bergama Asklepion, Arkhias, a respected person and also the first ruler of Pergamon, injures his foot while hunting in Greece. He is treated in Greece's most famous Epidauros Asklepion. Arkhias, who recovers as a result of the treatment, ensures that the cult of Epidauros Asclepios is established in Bergama to express his gratitude to the god of health (https://kvmgm.ktb.gov.tr/TR-44101/izmir-bergama-muze-mudurlugu.html).

Asklepios, for whom the Bergama Asklepion was built, was described as "mortal" and "great physician" in Homer's Iliad, and later became a god. He is the god of medicine and health in Greek mythology, who continued his art until the end of antiquity (Ağartan, 2009: 49–51; Erhat, 2004: 63; Tanrıver, 2017: 118; Tuna and Özer, 2015: 66). Asklepios is a god who helps and protects people and stands by people against other gods, which distinguishes him from other gods and gives him the image that he is at least as high as Zeus (Erşen, 2007: 101).

According to the legend told by Homer, Koronis, the daughter of the king of the Lapites, marries Apollo, but is killed because of a betrayal. Apollo, on the other hand, cuts the woman's stomach and saves her dying child, and delivers her to the centaur, the priest Chiron. Chiron names this boy Asklepios. Asklepios learns not only to be a surgeon, but also to make medicine, produce a cure from herbs, and resurrect the dead by training with Chiron, who is famous for his healing (Hancı, 2005: 7; Aydın, 2021: 51). After Asklepius starts medicine, he reverses death and life, heals the dying patients, and the number of dead people decreases as he resurrects the dead. This angers Hades, the god of the underworld, and Hades complains about Asklepios to Zeus, and Zeus kills his grandson Asklepios with lightning, thinking that he is disrupting the natural order. Asklepios, who was killed by Zeus, was declared the god of health by Zeus. The temples called Asklepion, which were dedicated to him after the death of Asklepios, the god of health, who challenged death by healing people, were founded by priests. Asklepios married Epione, the goddess of pain relief. Of their daughters, Hygieia had duties about protection from diseases, maintenance of healthy life, cleaning and hygiene, exercise, rest, and diet, Meditrina had a long life, health, and wine-related duties, Aglaea, provided services regarding the healthy body and natural beauty, Iaso and Aceso had duties about healing, and Panacea, as the goddess in charge of the herbal treatment process, provided health services. Telesphorus, Makhaon, and Podaleiros, sons of Asklepios, also took part in the healing process. His daughter Hygieia, who helped Asklepios in his life, his sons Makhaon and Podaleiros, who took part in the Trojan War, and the priests, who were ancient Greek physicians, continued the art of medicine after Asklepios died. In mythology, the god of health Asklepios and his daughter Hygieia were the most popular gods associated with health and medicine, and they were regularly worshiped after the cult of Asklepios (Ağartan, 2009: 50–51; Çelik, 2018: 1008; Erhat, 2004: 62–63; Hancı, 2005: 7; Özcan and Saatçi, 2022: 62; Şahin, 2021: 27–33).

The theater dedicated to Hygieia and many small temples were also located in the Bergama Asklepion (Freely, 2008: 36). Hygieia, the daughter and assistant of Asklepios, is depicted in a long dress feeding a snake from a bowl and is known to have the power to improve health and repel the epidemic, and her name is also included in the Hippocratic Oath (Korkmaz, 2022: 248).

Asklepion, which is reflected in today's ruins with its arrangements made by the Roman Emperor Hadrian in the 2nd century AD, gained importance especially in the Roman Period with its treatment methods and pioneering physicians. The connection of the city of Pergamon with the Asklepion was provided by the 1 km long Sacred Road (Via Tecta). At the large arched door called "Viran Gate" at the beginning of this road, patients were examined by priest physicians (asclepiades), and those who were about to die and pregnant women were not admitted. According to the rumor, there was an inscription "Death is forbidden in Asklepion for the blessing of all gods" on this entrance door (https://kvmgm.ktb.gov.tr/TR-44101/izmir-bergama-muze-mudurlugu.html). With this inscription at the entrance, patients who came to Asklepion for treatment were made to feel better psychologically from the very first moment. The patients who came to Bergama Asklepion were examined and rested right after they entered the main gate, and they were prepared for treatment by drinking the holy water that came out of the rock between the temple of Asklepios and the theater, washed in this water, and dressed in a white garment. The patients received both physical and mental treatments at Asklepion. The most important of the applied treatment methods is the treatment by suggestion. Physicians, who addressed the patients in the sleeping areas of the hospital, continued their suggestions during sleep. Thus, psychotherapy sessions were conducted by interpreting the dreams of the patients who slept with the effect of what they heard. If the patient is lucky, god will make him heal in his dream. What is expected from the patient is to receive some signs from God in his dreams. Priests and physicians will interpret the dream of the patient and suggest remedies for him. The first condition of treatment in Asklepion is faith. Religious treatment methods with holy waters, medicinal herbs, and ointments were applied in the temples. There is an underground tunnel connecting the Sanctuary in the center of the hospital and the treatment area. The sunlight leaking into this tunnel, the sounds of running water, religious songs called hymnos, and the suggestions of physicians were also used for treatment purposes. After their treatment in the tunnel, the patients who reached the round treatment area bathed in hot springs, healing water baths, and mud baths. People who took a mud bath with the order of God ran around the temples three times and cleaned the mud on the holy fountain. After the water treatment, the patients were supported to sunbathe on the specially built terrace. In Asklepion, where support was received from faith, priests as well as physicians were at the forefront in the treatment of patients. In addition to faith, sports and music were also used to find healing in Asklepion where there are structures such as theater and libraries to keep the morale of the patients high and entertain them. In Asklepion, patients were treated with walking, jogging, and gymnastics, and sports were used as an effective means of rehabilitation and recreation. Individuals listening to music for healing, spending time in the library, and socializing in these areas also healed by praying in temples. If their condition worsened, patients were taken out of the Asklepion because death was not accepted (Çelik, 2018: 1008–1010; Doğan and Sezgin, 2012: 274; Erşen, 2007: 100–101; Freely, 2008: 36; Kars, et al., 2020: 114–116; Çutuk and Zengin, 2006).

The word "patient" was never used for those who came to seek healing, but was expressed as "guest" during their stay in Asklepion. Money was taken according to the diseases and financial conditions of the guests, and only the villagers of Bergama were treated free of charge. Those who recovered as a result of the treatment visited the Temple of Asklepios, gave financial aid, and left a small model of his recovered organs made of bronze, marble, or mud while leaving Asklepion. It was ensured that those who healed encountered new guests on the holy road while leaving Asklepion, and thus the newcomers gained high morale and hope for recovery. The treatment methods applied in Asklepion and information about dreams were learned by reading the votive stones and inscriptions unearthed during excavations (Menekay, 2009: 2–4; Tuna and Özer, 2015: 70–72).

In Bergama Asklepion, there are various areas for the treatment of patients, as well as for their worship and socialization. Some of these structures are a hospital, physical therapy center, surgical center, psychiatry center, healing water pools and mud centers, sleeping rooms, toilets, tunnels, sports fields, temples and places of worship, and a theater, library, and bazaar dedicated to Asklepios, the god of health, and Hygieia, the goddess of health. The relaxing and healing effect of entertainment was also utilized in Bergama Asklepion through plays, festivals, and bazaars organized. There were shops where patients and visitors could buy their needs, gifts, or votive offerings to the god on the road that starts from the entrance door, about one kilometer, and is lined with columns. In the covered galleries in the courtyard, physicians and priests practiced suggestion by being in constant communication with the patients (Çelik, 2018: 1009; Doğan and Sezgin, 2012: 274; Erşen, 2007: 101; Freely, 2008: 36; Kars, et al., 2020: 116).

Asklepios Temple is known as the most impressive structure in Bergama Asklepion. It is mentioned that there are statues of healing gods associated with this sanctuary in the niches in the temple and the statue of Asklepios in the middle (Freely, 2008: 35).

The most important physicians of Anatolian medicine were born, trained, and educated in Bergama. The pioneer of pharmacy, physician Galen (Galenos), known as the "Prince of Physicians", was born in Bergama. He received training in Bergama and different centers and improved his anatomy and professional knowledge in asklepions. He returned to Bergama and worked as a physician for gladiators, especially in the field of orthopedics and surgical diseases. Many studies of Physician Galen emerged in the library of Bergama, which is thought to have about 200,000 works. More than 20 of Galen's works in the fields of surgery, psychiatry, anatomy, physiology, philosophy, ethics, pharmacology, and botany have survived. Hippocrates, one of the leading physicians of medical science, was trained in Asklepions, evaluated diseases with a rational point of view, not supernatural reasons, and contributed to medical science by determining the professional ethics rules that physicians should have (Doğan and Sezgin, 2012: 272–277; Freely, 2008: 24, 34; Kars, et al., 2020: 116–117; Kuşkonmaz, 2016: 149). In the books of Galen, who was the physician of the Bergama Gladiator Barracks, information about life in the ancient period and the population of Bergama was also included. Accordingly, the population of Bergama was expressed as 160,000, of which 40,000 were locals and citizens, and 120,000 were foreigners and slaves (Tuna and Özer, 2015: 68).

The discovery of bronze bone and silver tools found in Bergama excavations indicates the existence of surgical interventions such as fistula opening and curettage. The bones found in the excavations revealed that the first brain drilling operation (trepenation) performed in Anatolia in ancient times was in Bergama and that the person lived for a while after the operation (https://kvmgm.ktb.gov.tr/TR-44101/izmir- bergama-muze-mudurlugu.html; Doğan and Sezgin, 2012: 272–274; Kuşkonmaz, 2016: 148).

The Hippocratic Oath, one of the oldest texts of the Greek language, is known as the first text in medical ethics, and it is seen in the first sentences that the mythological creators are sworn in. In the original Hippocratic Oath, physician gods and goddesses are kept as witnesses to the oath taken. These divine powers are responsible for health, illness, life, and death (Korkmaz, 2022: 247; Tanrıver, 2017: 121);

> *"I swear by Apollo the physician, and Asklepios, and Hygieia and Panacea and all the gods and goddesses as my witnesses…"*
>
> *(Hippocrates)*

Representing the healing power, Asklepios, the god of health, is depicted with a male statue with a snake wrapped around his staff. It is believed that Asklepios holds a snake in one hand and a bowl in the other which contains a healing mixture of wheat, honey, and oil. The snake and the bowl were combined and used as the symbol of pharmacists (Ağartan, 2009: 51; Doğan and Sezgin, 2012: 274).

While visiting the Asklepion ruins, the snake symbol can be seen in many works. The Serpent Column, located at the entrance of the Asklepion, is thought to have been built by Galen. The snake symbol representing medicine and pharmacy also emerged from Asklepion. According to the legend, when a patient who come to the Bergama Asklepion during the time of Galen, known as the pioneer of pharmacy, cannot be treated, he is taken to the entrance door of the temple, thinking that he will not be cured, and his relatives are told to take him. The patient, who is left in front of the door, sees two snakes drinking milk from the same container and leaving their poison in the milk while fighting over the milk. The patient desperately drinks milk from this container to commit suicide and falls asleep right there. When the patient recovers and stands up, Galen tells the recovered patient, "I thought snake venom was curative, but I did not dare to try it in the treatment of patients. You justified my thinking. From now on, the symbol of Asklepion will be the double serpent." He had a column erected as an offering to the god of health, Asklepios, on which there were two snake figures drinking milk from the same container and leaving poison in the container. The staff in Asklepios' hand represents the tree of life and the snake represents youth (Hancı, 2005: 6–7; Kuşkonmaz, 2016: 149; Tanrısever, 2019: 436). The snake symbolizes youth and immortality with its ability to change its skin (Patacı, 2016: 167), and thus, in some cultures, witch doctors ate snakes to obtain this skill of the snake (Korkmaz, 2022: 249). According to Asklepios, the physician should be silent like a snake, keep secrets, and work patiently and calmly. The staff, on the other hand, represents that medicine cannot be learned in a short time, but the physician needs learning and experience until he gets old and leans on the staff. The staff, staff-snake, or two coiled snakes have been used as protective and healing symbols in various countries over the years. Since it lives underground, the belief that the snake is the creature that knows best the healing power in medicinal plants has also been influential and the snake has been accepted as a symbol of medicine since the cult of Asklepios (Hancı, 2005: 6–8).

The fact that Bergama Asklepion belongs to the history of medicine was the work called "Sacred Stories", in which Aelius Aristides wrote his memoirs during the period he spent 17 years in the Bergama Asklepion and was treated for 4–5 years. In these memoirs, an important work on temple medicine, Aristides included information about his illnesses and his presentation to gods and physicians for the treatment of his diseases, and especially mentioned the healing baths. While some of the inscriptions that provide evidence for the relationship between Bergama Asklepion and medicine are located in the field, some were moved to the Bergama Museum. In these inscriptions, the complaints that caused the diseases, the treatment methods, and the thanks of the patients who were satisfied with their treatments and healed are included (Akça, 2013; Soyöz, 2020: 104).

Many treatment methods such as treatments applied for the suggestion, consolation and relieving boredom, putting patients to sleep with suggestions in sleeping rooms, interpretation of dreams seen, drinking healing waters from the source, bathing and purifying in healing, hot and cold waters, mud bath and cleansing, sea bathing on the seaside, treatment with the sound of water, treatment with music, treatment with drugs, treatment with medicinal herbs, nutrition, and special diet methods, emptying of the intestines, blood draw (hijama), massages with vegetable oil or ointments, benefiting from sunlight and sunbathing therapies, walking barefoot, socializing in the library, theater and market, making the patients have fun and releasing their emotions and relaxing with theatrical performances, psychodrama treatments in the theater and surgical interventions were applied in the Bergama Asklepion, which is believed to heal people and keep them away from death. The inscription " Death is forbidden in Asklepion for the blessing of all gods " on the entrance door of the Asklepion is perhaps the first and most effective in terms of creating a positive start to the treatment and healing process by directing the thoughts and feelings of people who see, know and seek healing.

Conclusion and Recommendations

Bergama, a medical, social, cultural, sportive, and recreational center with many structures such as its library, pool, theatre, gymnasium, stadium, bazaar, festival, and temple area, offering healing to the body and mental health with its Asklepion, is an important tourism destination in Türkiye.

Considering the existence of the Bergama Asklepion and the variety of treatment methods, it can be said that this structure pioneers and directs various branches of science today, especially medical science. When evaluated in terms of tourism, Asklepions can be considered as structures where various types of tourism such as faith tourism, health tourism, cultural tourism, tourism sociology, thermal tourism, festival tourism, and aromatherapy tourism are experienced both in the past and today.

In order to preserve the value of Bergama Asklepion in medicine and tourism sciences and to increase tourism mobility in this destination, various investments can be made in this area by giving importance to the promotion, accessibility, and sustainability of the destination.

Since the destination is valuable for the fields of medicine and tourism, the number of national and international congresses to be held can be increased and made traditional. Particularly, the participation of the great physicians and tourism professionals in the field at the congresses to be held in the destination and their visits to the region can raise awareness in the medical and tourism sciences and sectors, increasing the quantity and quality of scientific studies to be carried out in this field, the value of the destination, and the number of visitors.

Appropriate physical conditions can be prepared and financial support can be provided for students and academicians studying in various disciplines such as archeology, sociology, fine arts, conservatories, history, literature, and sports sciences, especially medicine and tourism, to visit the destination and carry out their education in this atmosphere.

This destination can be promoted with many projects, especially the "Travel lover Project", initiated under the auspices of the Ministry of Youth and Sports, allowing young people between the ages of 18–30 to stay in state dormitories free of charge and thus planned to keep Türkiye's tourism alive, and enabling young people to visit the destination and see the riches there.

In order to revive third-age tourism in this destination, entrance to the area can be free of charge or discounts can be applied. It may be beneficial to make the necessary investments to encourage disabled people to visit the destination. Ergonomic walking routes can be arranged for disabled individuals. Texts and signs can be prepared in Braille alphabet for visually impaired individuals or headphone service that transmits information in different languages can be offered. Sign language-speaking professionals can be employed for individuals with hearing loss.

Themed festivals can be organized to revive the ancient Bergama and Asklepion life and culture. Souvenirs that reflect the present and ancient life of the destination can be produced and diversified. These products can be sold in the destination area, at cultural and scientific events, and at road rest stations.

Due to the interest and intensive use of technology today, a QR code can be created for each of the buildings in the destination. When this QR code is read by the visitor with his smart device, texts and videos sharing information with different language options can be accessed. Within a structure to be built in the Asklepion area, with tools such as augmented reality applications and virtual reality glasses, the life and treatment methods of that period can be experienced by the visitors and thus they can gain a different and unforgettable experience.

Bergama and Bergama Asklepion can be featured on the virtual museum platform (https://sanalmuze.gov.tr/) created by the Ministry of Culture and Tourism during the pandemic period. Similarly, special internet pages and social media accounts can be created for Bergama and the Bergama Asklepion, thereby increasing the interest in the destination.

While preparing this review study, we noticed that the number of research on Bergama Asklepion in medical science and tourism science is low. For this reason, scientific studies on various subjects such as researching the visitor profile, expectations, and satisfaction of visitors to the region can contribute to the literature.

References

Ağartan, C. A. (2009). A living myth in artworks: Asklepios, god of health, and his daughter Hygieia. *Zeynep Kamil Medical Bulletin*, 40(1), 49–52. Retrieved from https://dergipark.org.tr/tr/pub/zktipb/issue/22071/236965.

Akça, T. (2013). *Evaluation of Pergamon Asklepion from a medical perspective.* Çukurova University Health Sciences Institute, PhD Thesis, Adana, 2013.

Aydın, F. (2021). The effect of roman mythology on medical terminology and the origins of the names given to diseases. *Eskişehir Medical Journal*, 2(1), 48–52.

Baykal, F., Emekli, G. & Ataberk, E. (2011). The role of geographical factors in integrating Dikili and Bergama tourism. *Aegean Journal of Geography*, 20(1), 1–16. Retrieved from https://dergipark.org.tr/tr/pub/ecd/issue/4871/66891.

Çelik, S. (2018). Musical therapy in ancient Greece: the example of Bergama Asklepion. *Proceedings of The 1st International Congress of Humanities and Social Sciences Research*. Antalya. Karabuk University Press, pp. 1007–1010.

Çutuk, S. & Zengin, E. (2006). Investigation of physical education and sports culture and management strategies in Pergamon (Bergama) civilization. *Journal of Sports Management and Information Technologies*, 1(1). Retrieved from https://dergipark.org.tr/tr/pub/sybtd/issue/11436/136519.

Doğan, S. & Sezgin, F. (2012). How aware is Türkiye of the importance of Bergama in the history of medicine? *Journal of Social and Human Sciences*, 4(2), 271–279. Retrieved from https://dergipark.org.tr/tr/pub/sobiadsbd/issue/11354/135715.

Erhat, A. (2004). Dictionary of mythology. 13th Edition. Remzi Bookstore, Istanbul.

Erşen, D. (2007). Guide to traveling and understanding ancient cities - Aegean. Ekin Publishing, Istanbul.

Freely, J. (2008). Turkish civilizations guide 3 - Aegean Coasts. 4th Edition. Translated by Tuncay Birkan, Gürol Koca, Aslı Biçen. Yapı Kredi Publications, Istanbul.

Hancı, I. H. (2005). Snake story. *Journal of Continuous Medical Education*, 14(8), 6–9. Retrieved from https://www.ttb.org.tr/STED/sted0805/hikaye.pdf.

Kars, N., Bolat, N., Sirer, E. & Akaydın, A. (2020). The first treatment center in antiquity: Asklepion. In: Aydin, A. R., Keskin, I., Yelçe, N. Z. (Eds.), History of Disability Writings. Istanbul University Press, Istanbul, pp. 109–118. doi: 10.26650/B/AA09.2020.007.07.

Korkmaz, V. (2022). Footprints of greek mythology in medical terminology. *Eskişehir Medical Journal*, 3(2), 246–253. doi: 10.48176/esmj.2022.81.

Kuşkonmaz, Ş. M. (2016). A 2,000-year-old hospital and a physician in Bergama: Asklepion and Galen. *Konuralp Medical Journal*, 8(2), 147–150.

Menekay, S. (2009). Asklepion Bergama. *Dirim Journal of Medicine*, 1–4.

Ökten, A. İ. (2016). Mythology and neurosurgery. *World Neurosurgey*, 90, 315–321. doi: 0.1016/j.wneu.2016.02.106.

Özcan, S. & Saatçi, G. (2022). Hygieia as identity; "goddess? nurse?". *Kirsehir Ahi Evran University Journal of Health Sciences*, 6(1), 61–72.

Patacı, B. (2016). Asclepius as a mythological figure between deification and demonization. *Milele and Nihal*, 13(2), 154–182. doi: 10.17131/milel.284851.

Soyöz, U. (2020). The healing touch: sacred healing journey to the roman temple of Asclepion in Kos (island), *METU JFA*, 37(2), 103–126. doi: 10.4305/metu.jfa.2020.2.5.

Süer, S. (2021). Use of cultural heritage for tourism purposes: the example of Bergama district. *Journal of Humanities and Tourism Research*, 11(3), 479–502. doi: 10.14230/johut1088.

Şahin, Z. (2021). *Cult of Asklepius and Asklepions in Anatolia*. Gaziantep University Social Sciences Institute, M.A. Thesis, Gaziantep, 2021.

Tanrısever, C. (2019). Galen forgotten in Asklepion. *IV. International Culture Symposiums: Proceedings of International Health Society and Culture Symposium*. KIBATEK, Ankara, pp. 431–441.

Tanrıver, D. A. (2017). Cult of Asclepius in Smyrna. *Journal of Atatürk University Fine Arts Institute*, 38, 117–130. Retrieved from https://dergipark.org.tr/tr/pub/ataunigsed/issue/29520/301303.

Tuna, A. & Özer, M. C. (2015). Music as a healing method in Bergama Asklepion. *Journal of EU State Conservatory of Turkish Music*, 7, 63–73. Retrieved from https://dergipark.org.tr/tr/pub/konservatuvardergisi/issue/46521/584046.

https://kvmgm.ktb.gov.tr/

https://sanalmuze.gov.tr/

Hüseyin Yaşa* and Merve Onur†

Chapter 21 An Evaluation on the Relationship Between Social Media and Gastronomy Tourism

Introduction

With the developments in new information and communication technologies, especially the widespread use of mobile internet and smartphones, the use of social media platforms is increasing. Social media, one of the most effective mass communication tools today, offers important opportunities for gastronomy tourism. Gastronomy tourism generally includes domestic and international tourism trips to places visited to discover local food and beverages, to observe the process from production to consumption, to interact with local people, and to realize experiences such as festivals and restaurants. Therefore, social media, which is effective in almost all areas of daily life, has become a tool that plays an active role in determining their own preferences based on the comments of other individuals who have experienced restaurants, allowing individuals to share their dining experiences by enabling the discovery of local dishes or restaurants belonging to new or different cultures. On the other hand, while social media creates positive opportunities for gastronomy tourism, it can also create some negative opportunities. Individuals who strive for ostentatious consumption for other users on social media can transfer superficial visual information by moving away from the detailed experiences of local cultural products. Therefore, this study is important both in terms of understanding the effects between the two fields and making certain inferences on the potentials in these fields, and in terms of guiding researchers who want to conduct research between the two fields. The aim of this study is to examine the relationship between social media and gastronomy tourism. In this context, the literature on social media and gastronomy tourism was reviewed and the relationship between the two fields was evaluated through literature review method.

* Dr., Independent Researcher, hsynyasa@gmail.com
† Dr., Research Asistant, merveonur@osmaniye.edu.tr

Social Media Concept and Historical Development Process

From a terminological point of view, the concept of "social" is defined as the instinctive need of individuals to communicate with each other, a situation that has been going on since the beginning of mankind. The concept of "media" can be defined as a tool that enables individuals to interact (Safko & Brake, 2009). These two concepts have come together to create the concept of "social media".

Social media can be defined as interactive virtual environments where individual(s) come together in web-based environments without time and space restrictions and share their ideas, feelings and thoughts with other individual(s), including multimedia (text, visual, video, audio, etc.) features. Therefore, social media consists of the communication and multimedia (text, visual, video, audio, etc.) sharing of users with each other (Yaşa, 2023).

The concept of social media has undoubtedly come into our lives with the emergence of the Internet. The foundations of the Internet were first laid in 1969 with the name ARPANET as a project based on a simple interconnected network of the US Department of Defense (Gerla & Kleinrock, 1982). Although the birth of social media environments has become widespread with the emergence of web 2.0 infrastructure technologies, there are different ideas in the literature on this subject. Social media was first introduced in 1969 by the online service provider CompuServe. In 1971, the first electronic mail was sent between computers (Dao, 2015). In 1978, two technology enthusiasts, Ward Christensen and Randy Suess, found a way to share information and stay in touch with other friends through a software called BBS (Bulletin Board System), which allowed computer users to connect to the Internet using their modems. This system is known as the first "social network" example (Sanlav, 2014). On the other hand, according to most opinions, the foundations of social media were laid in 1979 by Tom Truscott and Jim Ellis at Duke University with a worldwide discussion system called "User Network", also known as "Usenet", which was open to everyone and allowed many individuals to send mass messages at the same time and publish messages in the formthey wanted. Later, the online diary platform called "Open Diary", founded by Susan and Bruce Abelson in 1989, is another social media environment that brings together users around the world in a single environment on a daily basis(Kaplan & Haenlein, 2010). In the following years, the first website "SixDegrees.com" was established, which combined the features of the existing social media environments up to that time and the current social media concept. This platform emerged as a website where users could create profiles, add friends to these profiles and review them (Boyd & Ellison, 2007).

Starting with the foundations of the internet in the 1960s, Facebook, founded by Mark Zuckerberg and his close friends at Harvard University in 2004, YouTube in 2005, Twitter in 2006, Instagram, founded by Mike Krieger and Kevin Systorm in 2010 under the name Burbn and later known by its current name, and TikTok, which emerged in 2016, have become one of the most preferred social media environments today with their unique structure, usage and target audience.

Characteristics of Social Media

Social media, an output of Web 2.0 technology, offers unique experiences to its users by incorporating many features. One of the most basic features of social media environments is that users have the opportunity to interactively communicate and share with other users regardless of time and space. Another important feature of social media is that it enables instant and fast communication. Users can interact quickly and instantly, such as commenting on and liking other users' posts. In addition, users can quickly send a private message to another user simultaneously/independently, as well as make a voice or video call.

Increasingly diverse social media environments have the potential to increase the level of interaction and communication between individuals or individuals or societies. Therefore, it is important for institutions, organizations or brands to take active action by representing the role of an easy and effective communication tool for the target audience that they have difficulty in reaching and to deliver their own message messages to the target audience quickly (Yaşa, 2022).

Social media, which offers different experiences to its users thanks to its features, has the following characteristics (Mayfield, 2008): - Participants: social media is open to the participation of users. Social media encourages participation and is shaped and developed with the feedback of the participants. - Openness: every user can comment, give feedback and like all posts on social media environments. All comments and posts are accessible to all users. - Conversation: unlike traditional media, social media allows two-way communication and eliminates time-space boundaries, which is a great advantage. - Community feature: social media can enable individuals to quickly and effectively form communities or groups around any topic. In this way, individuals can shape society by staying in dialogue with themselves and others outside the community regardless of time and space. - Interconnectedness: users who share content on social media accounts can link to other users for ease of use, and this provides an opportunity for communication to be intertwined.

Importance and Effects of Social Media

In today's world, social media environments stand out as the most powerful means of mass communication. With the increasing use of social media environments day by day, the importance of these environments increases and attracts the attention of many individuals.

Before social media environments came into the lives of individuals, content produced by experts and professionals was delivered to consumers through traditional media. The content transfer to businesses, institutions and organizations regarding the expected product or service choices of consumers could not be provided with rapid feedback. This situation can cause a negative picture for both consumers and businesses and institutions. Therefore, changes due to factors such as user-based content production, increased participation and freedom, decreasing importance of physical spaces, time and speed have increased the importance of social media (Singh & Diamo, 2019).

The effects of social media can be categorized into different categories in terms of social interaction. In this context, some effects such as complementary, debilitating and empowering can be mentioned. The fact that it provides an easy and inexpensive way to communicate between individuals, regardless of space and time constraints, falls within the scope of empowering effects. It has a weakening effect in terms of weakening the relationships that individuals will establish with their family and friends. In addition, it can be mentioned as a complementary effect in the sense that it can create an appointment stage before meeting face-to-face through social media environments (Heres & Thomas, 2007). On the other hand, the negative effect of social media, which can be mentioned as the negative effect of social media, allows its users to express hostility and hatred against the "other" directly and unlimitedly, thus enabling the production or circulation of hate speech (Yaşa & Öksüz, 2020).

Social Media Usage in the World and Turkey

According to January 2024 data by the research group "We are social" (We Are Social, 2024); the number of active social media users is 5.04 billion. This constitutes 62.3 % of the world's total population. The number of social media users in the world increased by 5.6 % compared to 2023. Since the beginning of 2023, 266 million new users have started using social media. This translates to 8.4 new social media users per second in 2023. The report shows that social media users spend 2 hours and 23 minutes a day on social media environments. The reasons for using social media were found to be the highest at 49.5 % for communicating with family and friends and the lowest at 19.7 % for sharing information about their own lives. The most used social media environments worldwide are Facebook, YouTube and WhatsApp. On the other hand, the world's favorite social media environment is Instagram.

According to the social media usage data of the research group "We are social" in Turkey (We Are Social, 2024); the total number of active social media users is 57.50 million. The average daily time spent on social media is 2 hours and 44 minutes. Social media users in Turkey use social media to communicate with family and friends at the highest rate (53 %) and at the lowest rate (19.5 %) to find like-minded and like-minded communities and to follow celebrities and influencers. Instagram is the most used social media environment in Turkey, followed by WhatsApp and Facebook. Instagram is one of the most popular social media platforms. Other social media platforms following Instagram are WhatsApp in second place and X (Twitter) in third place.

Gastronomy Concept and Historical Process

Gastronomy is a combination of the Greek words gaster (stomach) and nomas (law) (Özdemir & Dülger Altıner, 2019). Gastronomy entered the literature at the end of the 9th century with Joseph Bercholux's Gastronomie ou L'Homme des champs a Table (Gastronomy or Man from Field to Table) in 1801. The French author, who first introduced the term gastronomy, quoted Berchous' work in 1804, and the literature was strengthened by the first official study by the French Jean Anthelme Brillat Savarin with the study of "taste physiology" (Kivela & Crotts, 2005).

With the discovery and control of fire, human beings have made radical changes in their lifestyle. One of the most important of these changes is the realization that the flavors and nutritional values of food change when cooked. With the realization of this change, they started the adventure of gastronomy in history. In this context, it is known that the first in history to have a separate section in their homes for cooking, a kitchen-like structure was unearthed in the Neolithic life excavation sites of Jericho (7,700 BC) in the Jordan Valley and Çatalhöyük (7,000 BC) in Turkey (Akdeniz, 2019; Karataş, Aksu & Deniz, 2023).

Gastronomy is a science (Santich, 2004; Zengin, Uyar, & Erkol, 2015). Gastronomy is a terminology that has recently experienced a fashion trend (Yüksel, 2019) and there are many definitions of the term gastronomy in the literature. The term gastronomy is a new word that has passed from French to Turkish after the 1990s. In the dictionary of the Turkish Language Institution, gastronomy is described with two definitions: "the curiosity of eating food well", and secondly "a well-organized, pleasant and delicious kitchen, food order and system that is suitable for health" (Samancı, 2020). At the same time, Yüksel (2019) defines gastronomy as "the sociological, biological and historical culture of taste".

Gastronomy is a part of geography and history because it defines and classifies foodstuffs; of physics and chemistry because it examines and analyzes the composition and properties of foodstuffs; of the art of cooking because it adapts dishes and makes them pleasing to the sense of taste; and of economic policy because it seeks methods of selling the best product profitably at the most affordable price (Açıkkol, 2006 as cited in Samancı, 2020; Yurt & Sağır, 2023).

Gastronomy Tourism

Tourism is one of the world's most value-generating industries. Tourism destinations compete with each other to increase their tourist numbers and revenues. In this competitive environment, countries stand out not only with their history, sea and beaches but also with their culinary cultures (Bekar & Belpınar, 2015). Gastronomy tourism emerges as a result of the search for alternative tourism (Emekli, 2006).

The concepts of gastronomy and tourism, which is a new term in the relevant literature, have been evaluated together since 1983 (Belisle 1983; Hobsbawm & Ranger, 1983) and have been used to express that other societies learn about their culture through their food (Sarıışık & Özbay, 2015). UNWTO defines gastronomy tourism as "the totality of recreational experiential activities related to food and beverages for recreational purposes to gastronomic destinations" (UNWTO, 2017).

Everyone who is closely interested in gastronomy is interested in local and traditional dishes, high class dishes and a special dish of that country. Those who travel within the scope of tourism formed by those who are interested in all of these three categories of food at the appropriate place and time are considered as gastro tourists (Hatipoğlu, 2010). Gastronomy tourism; visiting food producers, gastronomy museums, food festivals, restaurants and special areas to taste a special type of food or to see the production of a food, and eating from the hands of a famous chef are considered within this scope (Hall et al., 2003). Gastronomy tourism, which is a rapidly growing tourism type, can be divided into many niche sub-fields such as wine tourism, beer tourism, chocolate tourism and cheese tourism.

Gastronomy Tourism in the World and Turkey

Gastrotourism is growing as more people choose to visit and learn about local cuisine and beverages. Through gastronomy tourism, there is a communication between the tourists and the local people and thus an emotional bond is established (Çavuş et al., 2018; Sormaz et al., 2023).

Among the findings of a study, 59 percent of respondents stated that food and beverage is now much more important than it was five years ago. It is also stated that 83 % of individuals believe that food influences their travel experience and leaves a lasting impression. Accordingly, when we look at the food and beverage-based expenditures of tourists, similar results are encountered (Everett & Aitchison, 2008; Horng & Tsai, 2010).

Tourists spent a total of $1.1 trillion in 2019, of which $278.9 billion was spent on food and beverage activities. The amount spent on food and beverage activities is expected to reach 1,796.5 billion dollars by 2027, with a Growth Rate of 16.8 % from 2020 to 2027. Within gastronomy tourism, it is stated that food festivals make the biggest contribution to the sector (Kiteski, 2019). In the world, many of the food-oriented events that attract the most tourists are held in the USA (Arıkan, 2016).

Gastronomy tourism, which constitutes an important part of tourism income in many destinations around the world, is an area waiting to be developed in Turkish destinations. At the same time, Turkey has extremely rich values in terms of gastronomy tourism potential.

According to TurkStat data, the per capita expenditure of tourists coming to Turkey in 2021 was $834 and the average food and beverage expenditure of tourists within this expenditure was $183. Food and beverage activities have the highest share of 22 % in tourism revenues generated in Turkey. When the accommodation sector is compared with the gastronomy sector in terms of employment, the added value of the gastronomy sector to the country's economy is 70 %, while the added value of the accommodation sector is only 30 %. These data reveal the importance of gastronomy tourism (Küçükkömürler, Şirvan & Sezgin, 2019; European Comission, 2014). Therefore, gastronomy forms the basis of tourism and economy in today's conditions. With the impact of gastronomy tourism, both the country's economy and society are changing and developing. In addition, gastronomy tourism also forms the basis of a culture transfer environment. Therefore, the necessity of activating gastronomy tourism, which is stated to be within the sustainable tourism targets, should be emphasized (Sormaz, Yılmaz & Pala, 2023).

Social Media and Gastronomy Tourism Relationship

Today, compliance with technological innovations is very important for organizations and businesses in a competitive environment. (Yurt & Sarı, 2022). It can be said that social media has a great place in the field of gastronomy as in every field today. Especially social media environments such as (Instagram, Facebook, Twitter, Pinterest, etc.) have a high potential to influence the development of destinations by highlighting gastronomic elements (Vukolic, Gajić, & Penic, 2022). When making travel plans, people look at cuisine and food cultures when choosing destinations. This situation actually shows the interaction between social media platforms and gastronomy tourism (Keskin, Örgün, & Akbulut, 2017; Horng & Tsai, 2012; Lange-Faria & Elliot, 2012).

Yuan et al. (2022) concluded that Chinese tourists plan their vacation plans using Instagram, Facebook & other networks before traveling. Social media platforms that are effective on travel plans are also effective on gastronomy travel. In the study conducted by Çelik (2023), it was determined that the use of social media is effective on gastronomy tourism. Another finding obtained in the same study was that most of the participants used Instagram to obtain gastronomic information.

Considering the speed of spread of social media, the values of cities in the field of gastronomy culture can also be promoted effectively (Faraç & Cinnioğlu, 2021). Therefore, businesses can create advertisements and content for gastronomy tourism at low costs and develop effective marketing strategies (Nadda et al., 2015).

On the other hand, tourists can share information about their experiences on social media, and many food and beverage businesses share their customers' experiences and promotional posts on their social media accounts (Cox et al., 2009). As a result, customers can be emotionally affected by experience stories, comments, photos and videos shared on social media (Koban, 2020). This affects users' decisions positively or negatively when choosing a restaurant (Banerje, Bhattacharyya & Bose, 2017).

Again, with the food and table sharing on social media platforms, food from different cultures has been recognized by other people (Çaycı, 2019). In this context, web pages for food and beverage establishments or gastronomy cities serving on social media serve as a bridge between the destination, the business and the target customer (Tezcan, 2017). When the literature is evaluated, the use of social media in the decision-making of people who will participate in gastronomy activities can be defined in three stages (Dwityas & Briandana, 2017). First, the pre-travel stage is the stage before individuals go on a trip. In this stage, individuals use social media to generate the desire or demand to travel and to collect and evaluate information about tourism image, product image and tourism activities. The second is the travel stage. This stage includes the period when travelers experience gastronomy tourism experiences and consume tourism-related services such as food. This stage also involves the search for additional information on tourism products. The third is the post-trip phase. This is the stage in which travelers complete their activities and return home. After the trip, travel experiences are evaluated and travelers determine their level of satisfaction with certain products, activities and information sources based on the decisions made during the trip. These evaluations are important as they will affect individuals' tourism behaviors in the future. Therefore, it is seen that travelers actively benefit from social media before, during and after gastronomy tourism. It will guide their future travel plans and the experiences they have will provide guidance for future travels.

Conclusion

Social media, which has a significant impact on individuals and societies, is of great importance for businesses in different sectors. In this context, one of the sectors highly preferred by social media users is gastronomy tourism. Considering the existing studies in the literature on gastronomy and media, the effect of media on participation in gastronomy courses (Cankül & Abiş, 2023), the effect of media on consumers' interest in learning to cook (Worsley et al, 2014), the effects of gastronomy and travel-based television programs on behavioral intention in tasting (Yılmaz & Göksu, 2017), the relationship between television cooking programs and gastronomy tourism (Taşpınar & Temeloğlu, 2018), the relationship between gastronomy and social media (Yuan, Ping & Yen 2023).

When we look at the studies based on social media and gastronomy tourism, it is seen that there are not many studies on topics such as Generation Z, tourist guides, promotion of web pages, customer communication, media relations, corporate identity, design, culture, local cuisine and its effect on travel intention for gastronomy (Gökdeniz, et al., 2015; Selda, Albayrak & Güdük, 2016; Uluçay, 2017; Cizrelioğulları, Altun & Barut 2020; Çelik, 2023).

This study aims to conceptualize the impact of social media on participation in gastronomy tourism. As a result of the literature review, it was observed that there is a limited number of conceptual frameworks addressing the relationship between social media and gastronomy tourism. Looking at the studies that have been conducted in general, it is seen that the use of traditional media and social media based on gastronomy is a technological tool that causes significant effects by both the consumer and the user, and that it has both advantages and disadvantages. Consumers' sharing of local foods or food culture specific to a destination on social media platforms is an important tool for consumers to discover new dishes and stories in their gastronomy travel to the destination. To summarize its advantages;

- In gaining a gastronomy image of the destination,
- Marketing of gastronomic values and culinary branding,
- In the formation of gastronomy travel intentions,
- Increasing interaction between gastronomic values and the target audience,
- Future travel plans and the experiences gained to guide future trips,
- Tourists have an effect on the tasting intention of an unfamiliar local food,
- Increased participation in gastronomy-based courses,
- Provides information on how a local food is cooked,
- Guidance on where to eat and what to eat,
- It is effective in increasing participation in gastronomy-based tourism activities (gastronomy festivals, local food markets, restaurants, gastronomy museums, etc.).

References

Akdeniz, D. (2019). Alternatif bir gastronomi tarihi incelemesi: resim sanatında yemek hazırlıkları ve mutfak sahneleri. *Selçuk Üniversitesi Sosyal Bilimler Meslek Yüksekokulu Dergisi*, 22(2), 490–507.

Arıkan, A. D. (2016). Gastronomi turizmi. Yüksek, G. (Eds.). Alternatif Turizm İçinde (Ünite 8). Eskişehir: Anadolu Üniversitesi Yayınları.

Banerjee, S., Bhattacharyya, S., & Bose, I. (2017). Whose online reviews to trust? Understanding reviewer trustworthiness and its impact on business. *Decision Support Systems*, *96*, 17–26.

Bekar, A., & Belpınar, A. (2015). Gastronomi turizmine ilişkin turist görüşlerinin kültürler arası karşılaştırılması. *Yaşar Üniversitesi E-Dergisi, 10*(38), 6519–6530.

Belisle, F. (1983). Tourism and food production in the caribbean. *Annals Of Tourism Research*, 10: 497–513

Boyd, D. M., ve Ellison, N. B. (2007). Social network sites: Definition, history, and scholarship. *Journal of computer-mediated Communication*, 13(1), 210–230.

Cankül, D., & Abiş, A. D. (2023). Gastronomi kurslarına katılımda medyanın etkisi. *Tourism And Recreation*, *5*(2), 58–65.

Çavuş, A., Işık, M. F., & Yalçın, C. (2018). Gastronomi turizmi çekiciliği açısından Zigana yöresi. *Atatürk Üniversitesi Sosyal Bilimler Enstitüsü Dergisi, 22*(Özel Sayı 3), 3071–3085.

Cizrelioğulları, M. N., Altun, Ö., & Barut, P. (2020). Sosyal medya uygulamalarını kullanan Z kuşağının gastronomi turizmi açısından incelenmesi. Gastronomi Araştrmaları içinde (ss. 41–71). Ankara: İksad Yayınevi.

Cox, C., Burgess, S., Sellitto C. & Buultjens, J. (2009). The role of user-generated content in tourists' travel planning behavior. *Journal Of Hospitality Marketing & Management, 18*, 743–764.

Çaycı, E., A. (2019). Sosyal medya'da dijital yemek kültürü. *International Journal Of Cultural And Social Stuides*, 5(1), 120–136.

Çelik, M. (2023). Sosyal medya kullanımının gastronomi turizmine etki düzeyi üzerine bir inceleme. *Journal Of Recreation And Tourism Research, 10*(4), 1–13.

Dao, D. V. (2015). Social media classification scheme in online teaching and learning activities: a consideration for educators. *International Journal of Education and Social Science*, 2 (4), 85–94.

Dwityas, N. A. & Briandana, R. (2017). Social media in travel decision making process. *International Journal Of Humanities And Social Science, 7*(7), 193–201.

Emekli, G. (2006). Avrupa Birliği Bölgesel Politikasında Turizmin Yeri ve Türkiye Turizmi. Içinde Iv. Ulusal Coğrafya Sempozyumu. Ankara.

European Comission (2014). Home Page/Labour Productivity, Personnel Costs, https://commission.europa.eu/index_en (Accessed 28.04.2024).

Everett, S., & Aitchison, C. (2008). The role of food tourism in sustaining regional identity: a case study of cornwall, South West England. *Journal Of Sustainable Tourism, 16* (2), 150–167.

Faraç, M., & Cinnioğlu, H. (2021). Türkiye'deki Unesco gastronomi şehirlerinin sosyal medya hesapları üzerine bir araştırma. *Journal Of Tourism Intelligence And Smartness*, *4*(3), 323–340.

Gerla, M., & Kleinrock, L. (1982). *Flow control protocols. In Computer Network Architectures and Protocols*. Boston, MA: Springer US.

Gökdeniz, A., Erdem, B., Yakup, D., & Uğuz, S. Ç. (2015). Gastronomi turizmi: Ayvalık'ta yerli turistler üzerinde görgül bir araştırma. *Journal of Tourism & Gastronomy Studies*, *3*(1), 14–29.

Hall. M. C., Sharples, L., Mitchell R., Macionis, N. & Cambourne B. (2003). Food Tourism Around The World: Development. *Management And Markets, Butterworth-Heinemann*. Elsevier.

Hatipoğlu, A. (2010). *İnançların gastronomi üzerine etkileri bodrumdaki beş yıldızlı otellerin mutfak yöneticilerinin görüşlerinin belirlenmesine yönelik bir araştırma.* Yayımlanmamış Yüksek Lisans Tezi. Sakarya, Sakarya Üniversitesi, Sosyal Bilimler Enstitüsü.

Heres, J. & Thomas, F. (2007). "Civic participation and ICTs", Information and Communication Technologies in Society E-living in a digital Europe, By B. Anderson (Ed.). MalcolmBrynin. JonathanGershungandYoelRaban, Routledge, s. 175–188.

Hobsbawm, E. & Ranger T. (1983). The Invention Of Tradition. Cambridge: Cambridge University Yayınları.

Horng, J. S., & Tsai, C. T. S. (2010). Government websites for promoting East Asian culinary tourism: A cross-national analysis. *Tourism management, 31*(1), 74–85.

Horng, J. S., & Tsai, C. T. (2012). Constructing indicators of culinary tourism strategy: an application of resource-based theory. *Journal Of Travel & Tourism Marketing,* 29(8), 796–816.

Kaplan, A. M. & Haenlein, M. (2010). Users of the world, unite! the challenges and opportunities of social media. *Business Horizons,* 53 (1), 59–68.

Karataş, M., Aksu, F. B., & Deniz, G. (2023). Unesco yaratıcı şehri Gaziantep'in gastronomi turizmi cazibe unsurları. *Turizm Ve İşletme Bilimleri Dergisi, 3*(1), 105–125.

Keskin, E., Örgün, E., & Akbulut, B. A. (2017). Gastronomi kavramının kelime ilişkilendirme testi aracılığıyla analizi. *Journal Of Tourism And Gastronomy Studies, 5*(3), 255–267.

Kiteski, D. (2019). Food Tourism; Definition, Statistics, And Everything You Need To Know (2019). Https://Passportsymphony.Com/Food-Tourism-Definition-Statistics-And-Everything-You-Need-To-Know/. (Accessed: 26.05.2024).

Kivela, J., & Crotts, J. C. (2005). Gastronomy tourism: A meaningful travel market segment. *Journal of Culinary Science & Technology,* 4(2-3), 39–55.

Kivela J., & Crotts C. J. (2008). Gastronomy tourism. *Journal Of Culinary Science & Technology, Vol.4* (2/3).

Koban, E. (2020). Gastronomik Turizm - Pazarlama ve Rekabet Yaklaşımıyla. Ekin Kitapevi.

Küçükkömürler, S., Şırvan, N. B., & Sezgin, A. C. (2019). Dünyada ve Türkiye'de gastronomi turizmi. *Uluslararası Turizm Ekonomi ve İşletme Bilimleri Dergisi,* 2(2), 78–85.

Lange-Faria, W., & Elliot, S. (2012). Understanding the role of social media in destination marketing. *Tourismos,* 7(1),193–211.

Mayfield, A. (2008). What is social media. https://indianstrategicknowledgeonline.com/web/mayfield_strat_for_soc_media.pdf (Accessed: 05.05.2024).

Nadda, K. V., Dadwal, S. S., Mulindwa, D. & Vieira, R. (2015). Role of Social Media in Tourism, Handbook of Research on Global Hospitality and Tourism Management içinde (142–162. ss.). United States: IGI Global Book Series.

Özdemir, G. & Dülger Altıner, D. (2019). Gastronomi kavramları ve gastronomi turizmi üzerine bir inceleme. *Erzincan Üniversitesi Sosyal Bilimler Enstitüsü Dergisi*, Cilt 12, Sayı 1, Ss. 1–14

Safko, L., & Brake, D. K. (2009). The social media bible: tactics, tools, and strategies for business success. John Wiley & Sons.

Samancı, Ö. (2020). Gastronomy: *An interdisciplinary meeting. Anatolia: Journal of Tourism Research, 31(1)*, 92–95.

Sanlav, Ü. (2014). *Sosyal medya savaşları*. İstanbul: Hayat Yayınları.

Santich, B. (2004). The study of gastronomy and its relevance to hospitality education and training. *Hospitality Management, 23*, 15–24

Sarıışık, M., & Özbay, G. (2015). Gastronomi turizmi üzerine bir literatür incelemesi. *Anatolia: Turizm Araştırmaları Dergisi*, *26*(2).

Selda, U., Albayrak, A., & Güdük, T. (2016). Edirne iline özgü gastronomi kültürünün pazarlanmasinda sosyal medyanin rolü ınstagram'da Edirne iline yönelik etiketlemeler üzerine bir içcrik analizi. *International Journal of Social and Economic Sciences*, 6(2), 71–80.

Singh, S., & Diamo, S. (2019). *Sosyal Medya Pazarlaması* (Çev: Başdaş, V.). Ankara: Nobel Akademik Yayıncılık.

Sormaz, Ü., Yılmaz, M., & Pala, K. (2023). An evaluation of the gastronomy tourism potential of amasya from a sustainable tourism perspective through swot analysis. *Journal Of Tourism & Management Research*, *8*(2), 1169–1179. Https://Doi.Org/10.5281/Zenodo.8278914

Sormaz, Ü., Yılmaz, M., Özata, E., & Büyükyıldırım, C. (2023). Burdur ilinin gastronomi turizmi potansiyelinin swot analizi ile değerlendirilmesi (Evaluation of the gastronomic tourism potential of Burdur province by swot analysis). *Journal Of Tourism & Gastronomy Studies, 11*(2), 1262–1279.

Taşpınar, O. & Temeloğlu, E. (2018). Influence of tv cooking shows on the behavioral ıntention of participating in gastronomic tourism. *Journal of Tourism and Gastronomy Studies* 6(2), 24–40.

Tezcan, E. T. (2017). Bir halkla ilişkiler uygulama alanı olarak sosyal medya kullanımı: change.org zeytin hayattır imza kampanyası. *E-Journal Of New Media*, 1(1), 47–52.

Uluçay, D. M. (2017). Gastronomi turizmi işletmelerinin web siteleri üzerine karşılaştırmalı bir değerlendirme. *Erciyes İletişim Dergisi*, 5(2), 196–208.

UNWTO (2017). Published By The World Tourism Organization, Madrid, Spain May. Second Global Report On Gastronomy Tourism. Unwto,2017 file:///C:/Users/User/Desktop/Gastronomy_Report_Web%20unwto.Pdf. (Accessed: 26.05.2024).

URL (2023). Food Tourism; Definition, Statistics, And Everything You Need To Know (2019). Https://Passportsymphony.Com/Food-Tourism-Definition-Statistics-And-Everything-You-Need-To-Know/. (Accessed: 26.05.2024).

Vukolic, D., Gajić, T., & Penic, M. (2022). The effect of social networks on the development of gastronomy–the way forward to the development of gastronomy tourism in Serbia. *Journal Of Tourism Futures.*

We are Social (2024). Digital 2024, your ultimate guide to the evolving digital world. https://wearesocial.com/uk/blog/2024/01/digital-2024/. (Accessed: 10.05.2024).

Worsley, A., Wang, W., Ismail, S. & Ridley, S. (2014). Consumers interest in learning about cooking: The influence of age, gender and education, *International Journal of Consumer Studies, 38*(3). DOI:10.1111/ijcs.12089.

Yaşa, H. (2022). Çevre(cilik) hareketi olarak sosyal medyada sıfır atık hareketi. *Selçuk Üniversitesi Sosyal Bilimler Enstitüsü Dergisi* (49), 212–230.

Yaşa, H. (2023). *Kanaat önderliği bağlamında influencerlar ve içerikleri üzerine fenomenolojik bir araştırma.* Anadolu Üniversitesi Lisansüstü Eğitim Enstitüsü, Yayınlanmış Doktora Tezi, Eskişehir.

Yaşa, H., & Öksüz, O. (2020). Nefret söyleminin inşasında sosyal medyanın rolü: ekşi sözlük örneği. *Erciyes İletişim Dergisi*, 7(2), 1383–1408.

Yılmaz, H., & Göksu, P. (2017). The effect of gastronomy travel based tv programs on behavioral tasting ıntention. Interdisciplinary Conference of Management Researchers Empowering Sustainable Tourism.

Yuan, L., Ping C. & Yen, T. (2023). Building wom from hedonic experience of social media in the field of gastronomy. *Asian Research Journal of Arts & Social Sciences.* 19(1), 13–25

Yuan, Y., Chan, C. S., Eichelberger, S., Ma, H. & Pikkemaat, B. (2022). The effect of social media on travel planning process by chinese tourists: the way forward to tourism futures, *Journal Of Tourism Futures*, 1–20, Doi: 10.1108/Jtf-04-2021-0094.

Yurt, İ., & Sağır, Y. E. (2023). The Effect Of Staff Behaviours in Food and Beverage Establıshments On Customer's Eating and Drinking Experience. *Gastroia: Journal of Gastronomy and Travel Research, 7*(2), 448–460.

Yurt, İ., & Sarı, Ö. (2022). Teknolojik gelişmeler ve insan kaynağına etkisi: yiyecek içecek [Technologıcal developments and ıts ımpact on human resources: a revıew of the food and beverage ındustry]. *Journal of Gastronomy, Hospitality and Travel*, 5(4).

Yüksel, A. (2019). Gastronomi Ne Değildir?. *Seyahat ve Otel İşletmeciliği Dergisi*, *16*(1), 186–190.

Zengin, B., Uyar, H., & Erkol, G. (2015). Gastronomi Turizmi Üzerine Kavramsal Bir İnceleme. *Ulusal Turizm Kongresi*, *1*, 16.

Atif Akkil[*] and Vedat Kayiş[†]

Chapter 22 Relationship Between Current Trends and Food Culture

Introduction

Globalisation is a process in which technological, economic and sociological changes are effective worldwide. This process also leads to significant changes in the food and beverage sector. In particular, the diversification of gastronomic elements can attract the attention of individuals (Bozok & Yalın, 2018). This situation encourages people to preserve their own cultural identity and heritage, and at the same time to have new experiences by interacting with other cultures. In addition, with the impact of factors such as globalisation and migration, the popularity of food from different cultures has increased worldwide (Garcia, 2020). Therefore, under the influence of globalisation, although a kind of homogenisation is observed among food cultures, efforts are also made to preserve and promote cultural differences (Beşirli, 2012: 11) local flavours and traditional dishes. For this reason, the unique food culture and flavours of each region still exist.

Throughout history, cuisine and culinary culture have been the subject of research in various aspects (Hande & Akoğlan Kozak, 2015). The cuisine of each society has basic characteristics; these are generally associated with religion, beliefs, geographical factors and local resources (Şavkay, 2000: 11). At the same time, food culture harbours meanings shaped by geographical conditions and cultural factors and evolves with social, economic and political changes over time (Kanık, 2016: 10). While this change can sometimes be seen in the meanings of food and consumption habits, it can sometimes manifest itself in changes in table manners.

Changes and transformations in food culture may differ according to geographical regions and are often influenced by changes in the economic and cultural structures of societies and shape culinary culture (Fendal, 2014: 158). For example, economic growth and increased prosperity in a region can affect the

* Lecturer., Düzce University, Akçakoca Tourism and Hotel Management School, Department of Gastronomy and Culinary Arts, atifakkil@duzce.edu.tr

† Lecturer., Düzce University, Akçakoca Tourism and Hotel Management School, Department of Gastronomy and Culinary Arts, vadatkayis@duzce.edu.tr

eating habits and preferences of local people. Individuals with more income may visit restaurants and food and beverage outlets more frequently to enjoy a variety of dishes. In addition, migration movements and cultural interactions can also shape the food culture of a region. The coming together of different cultures can lead to the creation of new recipes and the differentiation of existing dishes.

Eating and drinking activities are considered not only for nutritional purposes but also as a part of social behaviour and culture (Beşirli, 2010). In particular, elements such as the selection, preparation, presentation and consumption of meals reflect the values, traditions and lifestyle of a society (Çaycı, 2019). These habits are passed down from generation to generation, creating a cultural heritage. At the same time, the attractiveness of a destination is not limited to its historical and natural beauties; the food culture of that destination can also be intriguing for tourists (Ardıç Yetiş, 2015).

Today, the proliferation of social media platforms has a significant impact on food culture (Smith, 2021). The easy sharing of restaurant experiences and food posts has enabled new food trends to spread rapidly. In addition, with the increase in healthy living trends, interest in organic and natural products has also increased (Jones, 2019). This has led to a change in eating and drinking habits and the prominence of healthy eating preferences.

Food Culture and Social Change

In the age of empire, one of the strong indicators of societies was the ability to change the world order and the course of history. From today's perspective, it is observed that the emperors and/or rulers who make conquests struggle to include the world in their sphere of sovereignty (Sağır & Memiş, 2014). Social structure is a set of institutions that ensure the continuation and functioning of a community. These rules are the rules that individuals find ready in society and are passed down from generation to generation in a traditional way (Sever, 2019). Food is an important factor in defining human communities.

Food culture bears the traces of the society in which it is found. Therefore, food culture can be used to define people, because food culture is shaped by many factors such as social interaction, geography, religion, settlement pattern, population growth and structure (Çaycı, 2019). At the same time, with the social, economic and political changes in human history, the codes of the way of eating, i.e. food culture, also change (Kanık, 2016: 10). However, these explanations about food may be considered too ambitious, because there are different parameters related to food. Therefore, it is not enough to define people by tracing food cultures. When eating a good meal or having a gastronomic experience, individuals tend to take photos of the meal and share them on social media platforms (Çaycı, 2019).

In Turkish societies, meat and dairy products form the basis of dietary habits and have a historically important place. However, there are significant differences between rural and urban life. People living in rural areas generally prefer meat and dairy products from natural and organic sources, while people in cities tend to consume more processed and ready- to-eat foods (Abdurrezak, 2014: 5).

Eating and drinking habits are closely related to the economic structure of a society. Individuals with higher income levels generally have access to healthier and more varied foods, while those with lower income levels may turn to lower-cost and less healthy options. This is one of the important factors determining the dietary habits and lifestyles of individuals within the society. In addition, the communicative characteristic of eating and drinking habits plays an important role in social interactions and social relations. Especially in Turkish culture, hospitality has a great importance and the food served during the hospitality of the guests shows the hospitality and financial status of the host. Therefore, eating and drinking habits are a factor that determines not only dietary preferences, but also relationships and status in society.

The Effect of Technological Developments on Food Culture

In recent years, there has been a rapid technological transformation across the world. This transformation is seen as a fundamental factor affecting gastronomy at every stage from food production to service and final consumption (Ertaş & Kırmızıkuşak, 2021). Starting towards the end of the 18th century and triggered by the use of steam engines, technological developments accelerated with the changes in software and communication fields and started to be called Industry 4.0 (Mutlu Öztürk, 2020). The elements that make up Industry 4.0 include the proliferation of robots, the Internet of Things, artificial intelligence, sensors, cognitive technologies, nanotechnology, internet services, quantum informatics, technological clothing and jewellery, augmented reality, smart signage, smart robots, big data, 3D printers and smart networks (Mutlu Öztürk, 2020) The impact of technological developments on the food and beverage industry has increased significantly in recent years. In addition, we see that crisis situations necessitate the digitalisation of the sector. The World Health Organisation's (WHO) declaration of Covid-19 as a global pandemic has made it imperative for many sectors to develop various strategies to combat this crisis (Sohrabi et al., 2020). During the Covid- 19 pandemic, the widespread use of cloud-based kitchen networks is an example of the effects of technology on food delivery ().

In this context, with the rapid advancement of technology, many innovations are expected to be realised in the field of gastronomy. In today's world, the globalisation process, which has multidimensional economic, technological, cultural, environmental and educational effects, is realised through the interaction of factors such as developments in communication technologies, economic integrations, tourism movements and cultural interactions (Çetinkaya & Gül, 2023).

It is predicted that many more innovations will be realised in kitchens with technology. (Yumurtacı and Keçebaş, 2009). Technology plays an important role not only in kitchens but also in the digitalisation and dissemination of food culture. Social media platforms enable the spread of food cultures by creating a bridge between different cultures, especially by sharing food photos, videos, recipes and restaurants. In this way, through technology, it is possible for local food cultures belonging to a certain region to be discovered by different societies and for these cultures to be recognised on a global scale (Philipov, 2017).

The possibilities offered by technology contribute to the development and universalisation of food culture in a digital environment by making it easier for people to share, learn and interact with food-related experiences. In addition, emerging technologies such as artificial intelligence and big data analytics enable restaurants and food businesses to better understand customer preferences and offer personalised experiences. For example, AI-powered systems can make recommendations based on customers' past orders or offer dishes that are suitable for a specific diet or nutritional requirements. While this increases customer satisfaction, it can also improve the efficiency of businesses and provide a competitive advantage.

The Role of Social Media on Food Culture

Human beings are social beings and desire both to acquire knowledge and to share their knowledge with others. This sharing is not only limited to information; it also includes elements such as one's location, activities, food and beverages consumed, and accommodation. People tend to share any information that makes them feel valuable and that they want others to be aware of. At this point, social media platforms function as tools that fulfil this need for sharing for many people (Eryılmaz & Şengül, 2016). Scientific studies have shown that social media has an impact on local cultures as well as consumption habits. With the advancement of communication technology, it has become possible for a post made anywhere in the world to spread to other parts of the world within seconds. This situation plays an important role in placing mass media at the centre of globalisation (Taylan & Arklan, 2008).

Social media platforms are used as a tool for individuals to exhibit status, prestige and ostentatious consumption behaviours (Çetin & Yaylı, 2019). In particular, they are in an effort to show themselves to other individuals by sharing luxury brand preferences, travelling and different dining experiences (Sabuncuoğlu, 2015). These platforms also play an important role in marketing the products of food and beverage businesses and shaping consumer behaviour by influencing people's eating habits. Social media is the general name of virtual environments where users create and share their own content and share it with large masses (Eryılmaz & Zengin, 2014). According to Weber , "social media are online platforms where people with common interests come together to share their thoughts, comments and ideas" (Weber, 2009). The spread of social media platforms affects businesses and customers in many areas, including the food and beverage industry (İbiş & Engin, 2016). It is known that social media enables reaching the target audience at low costs in terms of reaching consumers, which is the main purpose of marketing strategies. Through these platforms, consumers share their eating and drinking experiences, influence their environment and want to experience different foods (Seçer & Boğa, 2017). With the development of technology, the screen resolution quality of smartphones is increasing, which allows a closer and more detailed look at the images and videos shared on social media. The attractiveness of visuals increases the degree to which people are affected. Therefore, the attractiveness of the content shared through social media is important (Sağdıç & Şapcılar, 2023).The content that individuals share on their social media accounts shapes their social identities. This visibility on social platforms is generally aimed at displaying consumption habits. Therefore, individuals tend to share and relive their experiences (Mak, Lumbers, Eves, & Chang, 2012).

The use of social media also affects consumers' choice of restaurant or destination. Today, it is an important indicator not only in terms of determining eating habits, but also reflecting the social status and socio- economic status of the individual. The choice of restaurant or destination not only shows social status, but also gives information about the socio- economic status of the individual. When eating a good meal or having a gastronomic experience, individuals tend to take photos of the meal and share them on social media platforms (Çaycı, 2019). Especially eating in stylish and luxurious restaurants is a behaviour that symbolises social status for many people. Therefore, individuals are in search of social status and prestige in their restaurant or destination choices. However, when a good meal is eaten or a gastronomic experience is experienced, individuals are more likely to take a photo of the meal and share it on social media platforms.

As a result, it points to the digitalisation of food culture and the increasing role of social media in sharing food experiences. People have the opportunity to express their food preferences and cultural tastes through the food photos and experiences they share. Social media platforms increase cultural diversity by enabling users to recognise and explore food from different cultures. In this way, social media transcends the boundaries of food cultures and becomes a global sharing and interaction area. In addition, food posts on social media have become an important tool for the promotion and marketing of restaurants and food businesses. In particular, restaurants are trying to attract customers by promoting themselves on popular social media platforms and attracting attention with visually impressive food presentations. food posts on social media, along with the choice of restaurant or destination, have become an important factor in expressing individuals' social status and cultural identity, recognising different cultures and influencing the food industry. This accelerates the process of digitalisation and globalisation of food culture. This is becoming an important factor for consumers to experience and learn about different food cultures.

Health Trends and Nutrition Habits

The industrialisation process, which started in the 1800s, has led to radical changes in people's eating habits (Özdemir, 2020). With the development of communication and technology that started in this process, overpopulation in our country and all over the world, unplanned industrialisation, excessive use of artificial fertilisers and pesticides have brought environmental pollution to the agenda at a worldwide level. In recent years, since it is aimed to obtain more products to feed the increasing population, the use of growth hormones and antibiotics to plants and animals, and the addition of food additives to foods have made healthy nutrition difficult.(Akın, Özkoçak, & Gültekin, 2015). In parallel with this, the increasingly widespread cultivation of Genetically Modified Organisms (GMO) gives the impression that it is almost impossible to obtain healthy and natural food from these plants and animals. Therefore, the basic conditions of healthy, quality and long-lasting life (Adızel, Özdemir, Durmuş, & Akın, 2010). In order to provide healthy life trends in current trends should be adopted.

Thanks to these trends, consumers are turning towards healthy options in their food experiences. Food and beverage activities have become a process of entertainment, leisure time utilisation and unforgettable experiences for individuals. In this context, gastronomy trends play an important role. The gastronomy trends examined in recent years show that eating and drinking habits are becoming more organic, consumers are turning towards agriculture and natural colours and adopting a healthy eating approach (Özdemir, 2020). The gastronomy trends that have become popular in recent years for healthy nutrition are as follows (Sarıışık & Kardeş, 2019):

- Black foods
- Purple foods
- Bowl food
- Edible flowers
- Healthy eating
- Raw food
- Organic restaurants
- Plates and dishes with stories
- Natural foods

The Raw Food trend, which forms the basis of healthy nutrition, is a method also known as germination or sprouting and has been used since ancient times. Throughout history, people have obtained new products by preserving plant seeds in moist environments or used these seeds in their meals (Kabacik & Bayram, 2023). Today, this method is frequently used to form the basis of healthy nutrition in order to maintain and improve its nutritional and functional properties. Sprouts obtained by germination method are widely used especially in Far Eastern cuisine (Kabacik & Bayram, 2023).

In developed countries, policies towards consumers are generally focussed on consumer education and information, while in developing countries, consumer protection efforts are at the forefront. In developed countries such as the USA, Canada and Switzerland, consumers have reached the stage of being interested in and having a say in political, technological, social and economic problems in society. In developing countries, on the other hand, consumer movements are not yet organised or have just started to be organised. Therefore, the issue of healthy nutrition in the food culture of consumers has started to take place in gastronomic trends and the tendency to return to natural and organic products has increased. With the development of technology, these trends have taken their place in catering practices and people have started to offer takeaway services. In the takeaway process, some conditions must be controlled in order for consumers to consume healthy products (Babaoğul, Şener & Buğday, 2007). These conditions are as follows:

- Delivery of the personalised meal hot and ready to eat.
- The food is not mixed together during the packaging process.

- Carrying out production in accordance with health and hygiene rules.
- Labelling of spices and ingredients used.
- The dishes are harmonious in terms of appearance, taste and consistency.
- Employment of personnel specialised in takeaway services.
- To take into account the dietary habits of individuals with special nutrition
- Paying attention to food harmony in the preparation of menus.
- Preparation of satisfyingly large portions.
- Serving an audience that cares about healthy nutrition
- Packages are suitable for recycling

Health trends are among the issues that should be at the centre of all trends. In order to transfer today's scarce resources to future generations by using them actively and efficiently and to create healthy eating habits, consumers need to be informed about this issue. Otherwise, activities such as the use of pesticides by the increasing population and GMO changes will lead to changes in food culture as well as threatening healthy nutrition.

Conclusion

Food is not only a biological need that ensures the continuation of life, but also a means of expressing identity as a symbol of membership of a community or religion. Food is not only simply a food but also an important element in the analysis of social relations (Gürhan, 2017). Food is not only an action that meets the physiological needs of humans; it is also a complex phenomenon with cultural, social and psychological dimensions. Eating habits are passed down from generation to generation as part of cultural heritage and have an important role in determining the identities of societies (Appadurai, 1988). Food culture is evaluated in a wide range from the selection of ingredients to cooking methods, from presentation to consumption rituals and has a deep relationship with factors such as history, geography and economic status of the society.

Nowadays, food culture is increasingly changing under globalisation and commercial interaction. The spread of fast food chains may lead to the homogenisation of local food cultures and the forgetting of traditional dishes. However, this situation can be reversed. Supporting local markets, small farmers and traditional products is an important step to preserve local food cultures. It is also important to promote sustainable agricultural practices and criticise industrial food production.

Food culture also has a major impact on health and nutrition. Eating habits play an important role in determining the risk of many chronic diseases such as obesity, diabetes and heart diseases (Popkin & Adair, 2012). Therefore, promoting healthy eating and popularising balanced eating habits are critical for improving the overall health of societies. Food also functions as a means of social interactions and personal relationships. Environments such as restaurants, cafes and dinner parties enable people to come together to eat together, communicate and strengthen their relationships. Therefore, it is important to focus on the social dimension of food culture in order to reinforce societies' sense of unity and solidarity.

Today, food culture is influenced by current trends such as technological developments and changing consumer preferences. In particular, with increasing environmental and health concerns, people's eating habits and dietary preferences are changing. Plant-based diets such as veganism, vegetarianism and flexitarianism are becoming a preferred lifestyle for more and more people. This trend has led to the enrichment of traditional food cultures with vegetarian and vegan options and the popularisation of plant-based foods. In addition, with the development of technology, digital platforms such as online food ordering and cookbooks are also influencing eating habits. People can now peruse the menus of many different restaurants online and easily order the food they want. In addition, cookbooks and food blogs allow people to improve their cooking skills at home and try dishes from different cultures.

All these current trends show that food culture is constantly evolving and adapting to changing societal needs and consumer preferences. Food culture not only fulfils the need for nutrition, but is also recognised as part of social and individual identity. Therefore, it is important to preserve cultural diversity and promote healthy eating, taking into account the impact of current trends on food culture

References

Abdurrezzak, A., O. (2014). İşlevsel Teori Bağlamında Yemek Kültürünün İletişimsel Yönü. Electronic Turkish Studies, S. 9 (11), s. 1–16.

Adızel, Ö; Özdemir, K.; Durmuş, A.; Akın, G. (2010). Genetiği Değiştirilmiş Organizmaların (GDO) Doğa ve İnsana Etkileri, Yüzüncü Yıl Üniversitesi Fen Bilimleri Enstitüsü Dergisi, Cilt: 15, Sayı: 2, Sayfa: 148–153.

Akın, G., Özkoçak, V., & Gültekin, T. (2015). Geçmişten Günümüze Geleneksel Anadolu MutfakKültürününGelişimi.Antropoloji (30), 33–52. https://doi.org/10.1501/antro_0000000319

Appadurai, A. (1988). How to make a national cuisine: Cookbooks in contemporary India. Comparative Studies in Society and History, 30(1), 3–24.

Ardıç Yetiş, Ş. (2015). Kapadokya yemek kültürü ve Mustafapaşa Beldesi (Sinasos) örneği.

Babaoğul, M., Şener, A., & Buğday, E. B. (2007). Tüketici yazıları I. Ankara: Hacettepe Üniversitesi Hastaneleri Basımevi.

Beşirli, H. (2010). Yemek, kültür ve kimlik. Millî folklor, 22(87), 159–169.

Beşirli, H. (2012). Yemek sosyolojisi. Yiyeceklere ve Mutfağa Sosyolojik Bir Bakış, Phoenix, 1, 184.

Bozok, D., & Yalın, G. (2018). Gastronomide Yeni Trend: Siyah Yiyecekler. Güncel Turizm Araştırmaları Dergisi, 2(Ek1), 251–261.

Çaycı, A. E. (2019). Sosyal Medya'da Dijital Yemek Kültürü. Uluslararası Kültürel Ve Sosyal Araştırmalar Dergisi, 5(1), 120–136.

Çetin, M. & Yaylı, A. (2019). Restoran Tercihinde Gösterişçi Tüketim Eğilimi, Turizm Akademik Dergisi, 6(2), 227–238.

Çetinkaya, T., & Gül, K. (2023). Küreselleşme Bağlamında Ekonomik ve Teknolojik Boyutların Türk Mutfağı Algısına Etkisi. Türk Turizm Araştırmaları Dergisi, 7(4), 577–596.

Ertaş, A., & Kırmızıkuşak, D. (2021). Geleceğin Mutfak Kültürü. Academia. edu, 14–35.

Eryılmaz, B., Zengin, B. (2014). Butik Otel İşletmelerinin Sosyal Medya Kullanımına Yönelik Bir İnceleme: Facebook Örneği, Kastamonu Üniversitesi İİBF Dergisi, Sayı: 4, s. 42–59.

Eryılmaz, B., & Şengül, S. (2016). Sosyal medyada paylaşılan yöresel yemek fotoğraflarının turistlerin seyahat tercihleri üzerindeki etkisi. Uluslararası Türk Dünyası Turizm Araştırmaları Dergisi, 1(1), 32–42.

Fendal, D. (2014). "Türkiye'deki Kahve ve Mutfak Kültürünün Dönüşümü Üzerinden Küreselleşme Sürecinde Küresel ve Yerel Kültürün Etkileşim ve Eklemlenişi", Galatasaray Üniversitesi İletişim Dergisi, ss. 147–180.

Garcia, M. (2020). Globalization and the Popularization of World Cuisines. Food and Culture Review, 15(4), 312–327.

Gürhan, N. (2017). Toplumsal Değişme Ve Yemek Kültürü Üzerine Sosyolojik Bir Çözümleme: Mardin Örneği. Journal of International Social Research, 10(54).

Hande, T., & Akoğlan Kozak, M. (2015). Türk Mutfağının Gelişiminde Gurmelerin Rollerine Yönelik Algılamalar. Anatolia: Turizm Araştırmaları Dergisi, 26(2), 207–220.

İbiş, S. & Engin, Y. (2016). Öğrencilerin Yiyecek ve İçecek İşletmesi Seçiminde Sosyal Medyanın Rolünün Belirlenmesi. Mehmet Akif Ersoy Üniversitesi Sosyal Bilimler Enstitüsü Dergisi, 8(17), 322–336

Jones, R. (2019). The Rise of Interest in Organic and Natural Foods. International Journal of Gastronomy and Food Science, 8(3), 210–225.

Kabacik, M. & Bayram, F. (2023). "Raw Food Practices in Gastronomy". Ö. Yayla, A. Işın, İ. Yazıcıoğlu ve F. Bayram (Ed.). Gastronomy Attractions and Practices in Tourism (s. 165–175) içinde. Berlin: Peter Lang.

Kanık, İ. (2016). Gastro Gösteri-Popüler Kültür Ürünlerinde Yemeğin Kültürel Gösterisi, İstanbul: Ayrıntı Yayınları.

Mak, A., Lumbers, M., Eves, A. & Chang, R. (2012). Factors Influencing Tourist Food Consumption. İnternational Journal of Hospitality Management, 31: 928–936.

Michael Mussa, 2000. Factors drivingglobaleconomicintegration, Proceedings - Economic Policy Symposium - Jackson Hole, Federal Reserve Bank of Kansas City, pages 9-55.)

Mutlu Öztürk, H. (2020). Teknolojik Gelişmeler ve Gastronomi Alanına Yansımaları: Gastronomi 4.0. Güncel Turizm Araştırmaları Dergisi, 4(2), 222–239. https://doi.org/10.32572/guntad.703872

Özdemir, B. (2020). Gastronomi Akımlarında Sağlıklı Mor Yiyecekler. Gsı Journals Serie B: Advancements in Business and Economics, 3(1), 16–30. https://doi.org/10.5281/zenodo.4030210

Philipov, M. (2017). Becoming Food, Eating Media. Geoforum, 84, 241–242.

Popkin, B. M., & Adair, L. S. (2012). Global nutrition transition and the pandemic of obesity in developing countries. Nutrition Reviews, 70(1), 3–21.

Sabuncuoğlu, A. (2015). "Sosyal Medyanın Bir Gösteriş Tüketimi Mecrası Olarak Kullanımı" İletişim Çalışmaları, 2015.369–380. Sakarya: Burak Ofset.

Sağdıç, E. & Şapcılar, M., C. (2023). Sosyal Medyada Yapılan Yiyecek İçecek Paylaşımları ile Yiyecek İçecek Tüketim Motivasyonları Arasındaki İlişkinin İncelenmesi: UNESCO Gastronomi Şehirleri Üzerine Bir Araştırma. Journal of Tourism & Gastronomy Studies, 11(1), 523–545. https://doi.org/10.21325/jotags.2023.1205

Sağır, A., & Memiş, P. (2014). Kuşatıcı Bir Toplumsal Değişme: Küreselleşme Ve Bu Süreçte Farklılık Ve Özgürlük Arayışları. Akademik İncelemeler Dergisi, 1(2), 19–33.

Sarıışık, M. & Kardeş N. (2019). Gastronomi Akımları ile Renklerin İlişkisi Üzerine Bir İnceleme. International Conference On Eurasian Economies (pp. 430–438) Famagusta.

Seçer, A., & Boğa, M. (2017). Sosyal Medyanın Tüketicilerin Gıda Ürünleri Satın Alma Davranışına Etkisi. KSÜ Doğa Bilimleri Dergisi, 20(4), 312–319. https://doi.org/10.18016/ksudobil.280469

Sever, M. (2019). 1980 Sonrasında Türkiye'de Toplumsal Değişme. Uluslararası Halkbilimi Araştırmaları Dergisi, 2(3), 2–10.

Smith, J. (2021). The Impact of Social Media on Food Culture. Journal of Food Studies, 10(2), 45–58.

Sohrabi, C., Alsafi, Z., O'Neill, N., Khan, M., Kerwan, A., Al-Jabir, A., & Agha, R. (2020). World Health Organization declares global emergency: A review of the 2019 novel coronavirus (COVID-19). International Journal of Surgery, 76, 71–76.

Şavkay, Tuğrul (2000). Osmanlı Mutfağı, Şekerbank Yayınları, İstanbul.

Taylan, H. H., Arklan, Ü. (2008). "Medya ve Kültür: Kültürün Medya Aracılığıyla Küreselleşmesi", Afyon Kocatepe Üniversitesi Sosyal Bilimler Dergisi. 10(1), 85–97.

Weber, L. (2009). Marketing To The Social Web: How Digital Customer Communities Build Your Business,

Yumurtacı, M. ve Keçebaş, A. (2009). Akıllı ev teknolojileri ve otomasyon sistemleri. 5. Uluslararası İleri Teknolojiler Sempozyumu (IATS'09), 13–15 Mayıs, 2009, Karabük, Türkiye

Serkan Gün[*]

Chapter 23 Globalization, Mass Tourism, and Sustainability

Introduction

The significance of tourism as a global industry is widely acknowledged. As an industry, tourism represents a significant source of income for any country. By generating revenue, it can invigorate the national economy. Furthermore, the tourism industry provides employment opportunities and contributes to the country's foreign exchange earnings.

In consideration of these three factors, the concept of "mass tourism" is of paramount importance. Mass tourism is defined as the act of visiting an entertainment destination by many people at the same place and at the same time. While mass tourism is a significant contributor to the socio-culture, economy and environment of every country, it has direct impacts on these areas. Thus, alongside the positive effects of mass tourism, it can also have some negative, destructive impacts (Banerjee and Khana, 2021).

Tourism management, especially for mature and mass tourism destinations, is increasingly focusing on sustainability (GSTC, 2013). Sustainability involves prioritizing the responsible use of natural resources, reducing the environmental impact of activities, minimizing waste production, and alleviating pressure on water, soil, and biodiversity. Furthermore, the document emphasizes the utilization of renewable energy sources, the conservation of cultural heritage, and the protection of natural resources. These measures are designed to ensure the long-term cultural integrity of destinations. According to Bramwell (2004), mass coastal tourism faces various threats including deteriorating infrastructure, environmental degradation, and diminishing profitability due to increased competition among similar vacation destinations. Additionally, tourist expectations have shifted as many reject the idea of package holidays in favor of personalized travel forms. Consequently, local policymakers are prompted to diversify tourism products and develop policy responses to cope with the decreasing economic returns of standard "sun and sea" tourism. The underlying idea of this policy is to

* Assistant Professor, Siirt University, School of Tourism Management and Hotel Management, Department of Gastronomy and Culinary Arts, serkan.gun@siirt.edu.tr

develop specialized products that are thought to better accommodate changing tastes and the specific interests of tourists (Bernini, Urbinati, and Vici, 2015).

It is considered particularly important to explore new trends in tourism, especially the phenomena of alternative tourism and sustainability, compared to mass tourism. It should not be forgotten that the phenomenon of tourism is associated with human activities that have an impact on natural ecosystems and resources (Belias et al., 2017).

Globalization and Tourism

Globalization is seen as the highest level of internationalization, primarily facilitated by international trade and the commodification of information through media communication. The main actors most affected by globalization include mega corporations, typically multinational companies, and consumers themselves. From an economic perspective, the term "globalization" refers to the increasing interdependence of production and markets between different countries, facilitated by trade in goods and services, international strategic alliances, mergers, cross-border capital flows, cooperation and technology exchanges. Smeral (1998) also defines globalization as an increase in the international division of labor through the international fragmentation of production and a political tendency towards a more liberal economic order (Peric, 2005).

As a global phenomenon, globalization has impacted nearly every aspect of social life, particularly in the post-war period. As a specific dimension of globalization, the global economy has triggered various changes, including liberalization, international trade, competition, the free movement of capital and labor, and investments. A unique feature of this phenomenon is the removal of barriers to the flow of capital, goods, finance, and investments (Vukadinović, Knežević, and Mizdraković, 2013). Tourism, as one of the most significant economic activities during the globalization process, is particularly associated with mass tourism, which is seen as a visible manifestation of globalization. Thus, the development of tourism can be observed

in every economic, social, cultural, and even political dimension of these processes (Holowiecka, Grzelak-Kostulska, & Kwatkowski, 2011). In research and theory, tourism stands out as a strong example of globalization's impact. Key features of globalization's influence on tourism development include advancements and technology transfers, effective transportation of tourists, open borders, and other factors. With the increasing significance of the internet, fast communication, and lower costs of air travel, globalization has opened new opportunities for the development of tourism through advances in electronic technology, communication, and transport. The internet has significantly reduced costs by diminishing or eliminating intermediaries and has become one of the most sought-after amenities for comfortable accommodation. The globalization process has shaped the formation and operation of a global tourism market, where competitors from any origin expect to compete on an equal basis (Čerović, Vukadinović, and Knežević, 2015).

Despite varying approaches to the phenomenon of globalization, most definitions recognize that both the economy and society are significant factors underlying the globalization of the world economy. Kosolapov (2001) defines globalization as an intensification of crises, a weakening of the role of the national state, a trend towards the concentration of power, the global expansion of consumer goods, the resolution of local problems within a global community context, and the flexible and multicentric nature of international relations. As outlined by Dreher (2006), the process of globalization can be defined as a phenomenon that erodes national borders, integrating national cultures, economies, governance and technologies while simultaneously producing complex interdependencies (Gulcemal, 2020).

Identifying the drivers of globalization in the contemporary context is challenging due to the lack of a systematic and coherent explanation (McGrew 2011). In the extant literature, the forces behind globalization are predominantly divided into four interconnected categories: economy, technology, culture, and politics (Dwyer et al. 2009; Fletcher and Westlake 2006; McGrew 2011; Mussa 2000). Given that tourism is an integral aspect of the globalization process, it is unsurprising that these forces are also evident in the tourism sector. As Cohen (2012) outlines, tourism has facilitated the development of a modern global transportation system that makes remote destinations accessible with minimal effort, while also contributing to the growth of establishments such as hotels, airports, and resorts. Governments have increasingly streamlined official procedures to facilitate the processing of an increasing number of tourists. Concurrently, the globalization of the tourism industry has resulted in the fragmentation of the tourism production system and the internationalization of marketing regulations, knowledge transfer, ownership structures, services and outsourcing (Song, Li, and Cao, 2017);

- **Technology:** The role of technology in the explanation of globalization is of paramount importance. The advent of modern transportation and communication technologies has created a "shrunken" world, compressing space and time (McGrew 2011). Since the 1980s, developments in communication technologies have enabled the dissemination of information worldwide at much lower costs (Azarya 2004; McCann 2008). Transportation and information technologies are of great importance for the tourism sector. The industry sells faith-based products, and service providers are geographically dispersed.

- **Economy:** Globalization is a phenomenon that emerges from the perspective of capitalism and market dynamics, which are widely discussed in economic geography (McGrew 2011). The logic of market dynamics posits that globalization is a direct consequence of market competition. In theory, free trade allows countries to maximize their welfare based on their comparative advantage, while global competition and market forces ensure that goods and services are produced efficiently at minimum cost.
- **Policy:** The term "politics" is most associated with the ideological foundations of globalization. As McGrew (2011) notes, nearly all explanations of contemporary globalization refer to the ascendance and prevalence of neoliberal ideology in the OECD world, accompanied by the implementation of policies such as deregulation, liberalization, and privatization (Mahon 2010). The prevailing political trend among OECD countries since the 1970s has resulted in the relaxation of restrictions on capital mobility and the liberalization of national economies.
- **Culture:** The concept of culture also plays a pivotal role in the process of globalization. The exponential growth in population since the Second World War has created a demand for a vast array of economic products. This has led to the emergence of trade opportunities, particularly in light of the unequal distribution of population between countries (Fayed and Fletcher, 2002). Cultural factors, including the influence of media sources, have contributed to a certain degree of homogenization.

Globalization has certain driving forces that have clear impacts on the tourism, travel, and hospitality industries. The liberalization of air transport and trade in services, widespread adoption of information and communication technologies, and extensive use of the internet in the sale and marketing of travels and tourism packages contribute to the continual growth of the tourism industry (ILO 2001). This situation is evident in the calls for better coordination between national governments and international and regional industry organizations regarding air transport, airport infrastructure, tourism investments, and tourism development policies. Such improvements enhance the contribution of the travel and tourism industry to national economies and regional development. Countries, both developed and developing, which are vitally dependent on tourism as a primary source of wealth and job opportunities and seek foreign investment to better utilize and manage their natural, historical, or cultural resources for tourism purposes, are currently at the forefront of advocating for the liberalization of air transport (IVC 2006). UNESCO (1999) also notes that electronic technology has facilitated this growth by providing access to fare and hotel information and online booking services, another form of advancement globally utilized (Mustafa, 2010).

Globalization has brought significant benefits to the tourism industry. It contributes to overall economic growth by facilitating trade, the movement of people and capital, and the creation of new jobs in both emerging and developed economies. Furthermore, it encourages the development of tourism-related infrastructure in underdeveloped regions, thereby improving their position in the international market. Tourism benefits from globalization and develops in accordance with global principles of cultural, ecological, socio-economic and sustainable development. However, globalization also presents new challenges in marketing a tourism destination. Segmentation and cultural differentiation are important for the future competitiveness of a destination. Marketing professionals must be aware of local habits, cultural needs and local market characteristics of specific markets to utilize them when developing marketing plans for destinations in the international market. Knowing how to distinguish between global, international, and sub-national cultures is essential. Therefore, destinations must build their cultural identities to better position themselves in the market (Pindzo and Radulovic, 2017).

Advantages of Globalization in Tourism:

- Increase in capital, international trade and human flows,
- Providing growth and creating new employment in both developing and developed economies,
- Providing investment from foreigners in touristic places and hotels,
- Emergence of new activities and business forms (marketing etc.),
- Formation of multinational companies (global players) and strategic alliances (hotels, airlines and tour operators),
- Global management within the scope of tourism,
- Focusing on sustainable development and quality as a priority goal,
- Providing new experiences and benefits, low cost and advanced technology for tourists,
- Creating a global market (glocalization)

Disadvantages of Globalization in Tourism

- Loss of local characteristics and creation of a global village,
- Cultural changes that can lead to a homogeneous culture associated with globalization,

- ➢ Environmental and ecological degradation,
- ➢ Negative effects of climate change on tourism destinations,
- ➢ Negative effects of global warming on tourism,
- ➢ Causing loss of identity, closure and operational difficulties for local SMEs and national companies,
- ➢ Increasing competition,
- ➢ Disappearance of local products and standards, standardization of tourism products becomes difficult,
- ➢ It is difficult to standardize and internationalize the tourist infrastructure,
- ➢ Cultural and linguistic barriers,
- ➢ Endangering the business of (local) tourist intermediaries due to the emergence of new technologies (Internet, CRS, GDS).

Mass Tourism

The term 'mass tourism' is used to describe the movement of large numbers of organized tourists to popular holiday destinations for recreational purposes. It is characterized by mass consumption and the use of standardized packaged products. Conceptually, this type of tourism includes standardized packaged experiences and entertainment products intended for mass tourist consumption (Poon 1993). The origins of mass tourism can be traced back to various changes that occurred in industrial societies in the second half of the 20th century. Indeed, the history of mass tourism can be traced back to Thomas Cook's inaugural group tour to the Great Exhibition in London in 1851 (Page 2012). Bramwell (2004) posits that the significance and growth of leisure travel can be attributed to a confluence of factors, including increased personal mobility, spending power, internationalization and the advancement of public transportation in modern societies (Naumov and Green, 2015).

In the early years of the twentieth century, tourism continued to expand as a result of increased wealth, interest, outgoing attitudes, and advancements in people's transportation. These transportation advancements have enabled mass travel. Briefly, mass tourism refers to pre-planned tours for groups of people traveling together under the organization of tourism professionals for similar purposes (recreation, sightseeing, etc.). Over time, mass tourism has been mistakenly used or confused with holiday tourism. However, mass tourism is the complete opposite of 'individual tourism,' although both encompass holiday, timeshare, and other tourism markets (Sezgin and Yolal, 2012).

Mass tourism, involving tens of thousands of tourists traveling to the same destination in the same period of the year, is one of the most famous forms of tourism worldwide (Cooper et al., 2005). Mass tourism is particularly popular in 'All-Inclusive' packages due to their low cost and standardized services. The services offered in all-inclusive packages include air and land transportation, food and beverages, accommodation, morning and evening tea and treats, unlimited local drinks and, in some cases, excursions (Samarathunga and Gnanapala, 2016). As a result of the ever-increasing demand for tourism products and services, tour operators, hotels, destination management companies, cruise companies, airlines and other service providers can offer their products and services at lower prices, creating large economies of scale (Samarathunga and Gamage, 2020).

The emergence of alternative tourism can be attributed to the negative effects of mass tourism on destination regions and the intense criticism of mass tourism. Alternative tourism encompasses a range of forms, including small-scale tourism, soft tourism, nature tourism, green tourism and integrated tourism. Alternative tourism has been proposed as a means of reconciling social, natural and community values with the pursuit of economic benefits, as it is believed to have fewer negative impacts on destinations, populations and the environment than mass tourism, while still generating positive economic impacts (Smith and Eadington, 1992). Moscardo (2001) emphasized the role of alternative tourism in promoting the preservation of natural, social and historical assets in tourist destinations. Consequently, it has been identified as a pivotal factor in the evolution of tourism. The advent of alternative tourism has led to the emergence of sustainable tourism as a primary objective in the development of tourism (Christou, 2012).

The concept of sustainable tourism emphasizes the necessity of all tourism development to contribute to the socio-economic development of the destination region and increase natural riches. As a result, sustainable tourism has increased awareness of ecological and environmental awareness in every tourism activity (Dodds and Butler, 2010). In contrast, mass tourism aims for quick economic returns, thus paying little attention to social and environmental impacts and sustainable development. Mass tourism activities largely depend on climate and seasonal conditions. Consequently, the term "sea-sun-sand tourism" was coined to describe this phenomenon. Mass tourism activities are seasonal and reach their highest capacity and peak in the summer months. Consequently, there is a large influx of tourists to popular mass tourism destinations during the summer months, which results in heavy congestion and high population movements. Mass tourism is based on fixed programs (predetermined packages) driven by travel agencies and tour operators. Dodds (2005) posits that alternative tourism is based on families, small groups and even individuals, and that alternative tourism activities can be carried out throughout the year to create average capacity (Christou, 2012).

Sustainability and Tourism

The inaugural definition of sustainable development was set forth in the 1987 United Nations report of the World Commission on Environment and Development, titled "Our Common Future," also known as the Brundtland Report. This report defined sustainable development as development that meets the needs of the present without compromising the ability of future generations to meet their own needs (Brundtland, 1987). The National Sustainable Development Strategy also defines sustainable development as a long-term, goal-oriented, synergistic and comprehensive process that affects living conditions and all aspects at all levels, and that meets spiritual, material and biological needs. The social needs and interests of people must be met (iii) by eliminating or significantly reducing interventions that endanger, destroy or damage their way of life and conditions, (iv) without becoming a burden on the country, (v) by helping to protect resources and (vi) by protecting cultural heritage and values (Ozili, 2022).

In the past decade, the tourism sector has undergone significant developments. Alongside the positive effects of creating growth in destinations, there can also be negative impacts that significantly affect the environment, culture, politics, society, and economy. Sustainable development requires changes that aim to increase the satisfaction of users regarding specific environmental decisions to achieve set goals. The growing need for sustainability is also a result of increasing knowledge and concerns about the impact on tourism and environmental issues in general (Minciu and Stanciu, 2010). Developing tourism without a sustainability awareness can lead to severe disruptions not only for society but also for the environment and the entire tourism industry (Dutescu, Popa, and Ponorîca, 2014).

The concept of 'sustainable tourism' first entered the language of tourism development policy approximately twenty years ago. This development, which generally reflects the emergence and widespread adoption of sustainable development, is seen as an appropriate response to the challenges arising from the scope, scale and consequences of tourism development. Since the mid-1960s, there have been increasing calls for restrictions on tourism, particularly in response to the rapid growth of international mass tourism and the relentless global expansion of the so-called "pleasure circuit" (Turner and Ash, 1975). It has long been argued that sustainable tourism principles represent more than a micro solution to a macro problem (Wheeller, 1991). There is a paucity of evidence to suggest that the principles of sustainable development or sustainability are being adopted at the level of the tourism and travel industry, individual tourism businesses, or indeed at the grassroots level. However, sustainable tourism remains an enjoyable global policy goal. The concept of sustainable development has been expanded to include all types of economic activities and development, including tourism. Coccossis (1996) examines sustainable tourism from four different perspectives: The concept of sustainable tourism encompasses four distinct areas: economic sustainability of tourism, ecologically sustainable tourism, sustainable tourism development, and sustainable development. The latter is particularly relevant in that it focuses on the long-term viability of the industry, despite the clear needs for environmental quality, and promotes sustainable tourism development (Yazdi, 2012).

The concept of sustainability is becoming increasingly prominent in the context of tourism development and planning. The demand for sustainability in tourism is based on a number of interconnected processes. Since the 1960s and 1970s, global tourism has grown rapidly, exerting significant impacts on the local and global tourism system. These effects, especially the perceived negative consequences of tourism growth in destination areas, have made it necessary to limit and direct tourism growth in many places (Britton, 1991; Butler, 1991; Wolfe, 1982). Additionally, the social context of the late 1960s and 1970s supported a focus on global inequalities and environmental concerns. Furthermore, this increased the political conceptualization of sustainable development, together with the policy needs and objectives to limit the impact of tourism, as presented in the Brundtland Commission's 1987 report, "Our Common Future". In the early 1990s, the idea of sustainability was incorporated into tourism development with an increasing number of indicators and conceptualizations (Cater, 1993; WTO, 1993). The transformation of production and consumption towards post-Fordist production in Western societies also supported the emergence of alternative and 'green' consumption trends in the tourism sector. This resulted in the emergence of various new forms and terms, including ecotourism and responsible tourism, which are often seen as more environmentally friendly products. As more environmentally friendly products emerged (Mowforth and Munt, 1998; Wheeller, 1993), so too did the concept of responsible tourism (Hughes, 2004; Poon, 1993). Indeed, the tourism industry has become responsible for various aims and effects on both global and local scales. Generally, responsible tourism refers to tourism development principles and practices aimed at making places better for people to live in and visit. It aims to promote ethical consumption and production among all stakeholders, minimizing the negative social, economic, and environmental impacts of tourism on destination communities and environments, and maximizing its positive social, economic, and environmental effects (Saarinen, 2014).

Principles of Sustainable Tourism

Many destinations now follow strategies aimed at implementing a sensitive approach to tourism. Below is a list of important principles for national governments, destinations, and organizations guided by the ethics of sustainable and responsible tourism (Begum, 2021):

- **Participation:** Members of a community must maintain control over the development of tourism by participating in the process of identifying resources to be developed and sustained, establishing a community tourism vision, and establishing strategies and goals for the management and development of tourism.
- **Stakeholder Participation:** The development of tourism initiatives should be informed by a broad-based social input. Potential participants may include voluntary service groups, local non-governmental organizations (NGOs), women, the poor, economic development departments, municipal governments, tourism associations, town partnerships, visitors bureaus, regional representatives of tourism development, and other parties that may be affected or involved.
- **Local Ownership:** It is evident that tourism development should provide quality employment for community residents. Indeed, providing employment should be regarded as an integral part of any local tourism development strategy. Historical evidence indicates that policies that focus on providing education and training to local people and allowing local entrepreneurs and businesses to access finance are of vital importance.
- **Establishing Local Business Connections:** It is recommended that links be established between local businesses in the tourism sector with the objective of ensuring that tourism expenditures remain within the destination. Furthermore, it is proposed that tourists be enabled to purchase imported services and products.
- **Sustainability of the Resource Base**: It is the responsibility of communities to ensure that the resource base is not diminished to a greater extent than that which was inherited. In addition, it is important to consider the impact of actions that result in the irreversible dispersal of resources, with a view to avoiding such consequences.
- **Community Objectives:** It is essential that the needs of society and tourists are aligned. This can be achieved through the support of the wider community,

which ensures a balance is struck between the development of the social, economic and cultural industries and the protection of the environment.

- **Collaboration:** It is of paramount importance that local attractions, tourism operators and other businesses collaborate with one another, as the quality and performance of one can directly affect the other.
- **Monitoring and Evaluation:** It is imperative that communities implement monitoring and conservation strategies if they are to protect the resources that underpin the tourism industry and the environment on which it depends.
- **Accountability:** The utilization and administration of public assets, such as communal lands and water sources, must be accompanied by a clear accountability framework to prevent the misappropriation of these resources.
- **Education:** The advancement of sustainable tourism necessitates the implementation of educational and training programs that will enhance the comprehension of business and public stakeholders. These programs should encompass courses in hotel management, tourism, the operation and establishment of small businesses, and other related subjects.
- **Positioning:** The objective of sustainable tourism development is to encourage activities and uses that reduce poverty and enhance and leverage the sense of place, character of the landscape, opportunities for the local community and the community's sense of identity.

It is imperative that tourism considers the global and local environmental impacts. It should seek to promote the protection of habitats, natural areas and wildlife, and to minimize damage to them. One of the keys to sustainable and successful tourism is to achieve a distinct sense of differentiation from other competing destinations. Lasting and successful attractions are truly about local history, culture, industry, natural resources and lifestyle. Reflecting social values means dynamically representing the past, present and future aspirations of the local community. It is of the utmost importance to comprehend the expectations and requirements of specific market segments and broader market trends. The incorporation of distinctive features, or "packaging," enhances the appeal and probability of a destination being visited. The augmentation of existing attributes serves to diversify the local economy and provides a more comprehensive tourism experience. The narrative of a destination is of paramount importance in providing a more rewarding experience, which ultimately contributes to the preservation of the destination. The design of a destination should respect and preserve resources, reflect the values of the community, and be effective in storytelling. While tourism is a means of social and economic development, it must also consider the benefits desired by both tourists and host communities. To be considered a good tourism business, it is essential that it engages with the community, collaborates with other stakeholders and businesses, and helps to build local capacity (Sharpley, 2009).

Ensuring the sustainability of the tourism sector and addressing seasonal problems requires a balanced distribution of tourism activities both regionally and seasonally. In this context, diversifying tourism activities, that is, creating alternative tourism types, is of great value (TÜSİAD, 2014). The demand for alternative tourism has evolved over time, and as countries began to market their touristic potential as alternative tourism activities, both scientists and public authorities began to adopt different approaches regarding the types and scope of alternative tourism. For example, alternative tourism types in Turkey include hunting tourism, river tourism, mountaineering, air sports, golf tourism, Silk Road tourism, faith tourism, winter sports tourism, bird watching, congress tourism, health-thermal tourism, cave tourism, yacht tourism, underwater tourism, diving tourism and plateau tourism (Yıldız and Gürhan, 2020).

References

Azarya, V. (2004). Globalization and international tourism in developing countries: Marginality as a commercial commodity. *Current Sociology, 52*(6), 949–967.

Banerjee, S., & Khana, D. (2021). Mass tourism, its challenges and sustainability: A study on Santiniketan. *IJARIIE-ISSN(O), 7*(1), 219–223.

Begum, F. M. (2021). Sustaınable Tourısm. *International Journal of Tourism and Hospitality Management.*

Belias, D., Velissariou, E., Kyriakou, D., Vasiliadis, L., Roditis, A., Varsanis, K., & Koustelios, A. (2017). The differences on consumer behavior between mass tourism and sustainable tourism in Greece. *5. International Conference on Contemporary Marketing Issues (ICCMI)*, 176–197.

Bernini, C., Urbinati, E., & Vici, L. (2015). Vısıtor Expectatıons And Perceptıons Of Sustaınabılıty In A Mass Tourısm Destınatıon. *ToSEE – Tourism in Southern and Eastern Europe* (3), 1–17.

Bramwell, B. (2004). Mass Tourism, Diversification and Sustainability in Southern Europe's Coastal Regions. *Coastal Mass Tourism: Diversification and Sustainable Development in Southern Europe* (s. 1–31). içinde Clevedon: Channel View.

Britton, S. (1991). Tourism, capital and place: Towards a critical geography of tourism. *Environ. Plan. Soc. Place* (9), 451–478.

Brundtland, G. H. (1987). What is sustainable development. *Our common future, 8*(9), 1–27.

Butler, R. (1991). Tourism, environment, and sustainable development. *Environ. Conserv*(18), 201–209.

Cater, E. (1993). Ecotourism in the third world: Problems for sustainable development. *Tourism Manag*(14), 85–90.

Čerović, S., Vukadinović, P., & Knežević, M. (2015). The Influence Of Globalızatıon On Tourısm And Impact Of Tourısm On Other Actıvıtıes Wıth An Emphasıs On Greenfıeld Investments In Tourısm. *Sitcon*, 47–52.

Christou, L. (2012). Is it possible to combine mass tourism with alternative forms of tourism: the case of Spain, Greece, Slovenia and Croatia. *Journal of Business Administration Online*, 1–8.

Coccossis, H. (1996). Tourism and Sustainability: Perspectives and Implications. *Sustainable Tourism? European Experiences.* içinde U.K.: Cab International.

Cohen, E. (2012). Globalization, global crises and tourism. *Tourism recreation research, 37*(2), 103–111.

Cooper, C., Fletcher, J., Gilbert, D., Fyall, A., & Wanhill, S. (2005). *Tourism: Principles and practice*. London: Pearson Education.

Dodds, R. (2005). New Tourism Realities for Sustainable Tourism Policy? Implications and Recommendations for Stakeholders. *TTRA Canada Conference.* Kelowna: Journal of Business Administration.

Dodds, R., & Butler, R. W. (2010). Barriers to Implementing Sustainable Touism *Mass Tourism Destinations Tourismos Journal.*

Dreher, A. (2006). Does globalization affect growth? Evidence from a new index of globalization. *Applied Economics, 38*(10), 1091–1110.

Dutescu, A., Popa, A. F., & Ponorîca, A. G. (2014). Sustainability of the Tourism Industry, Based on Financial Key Performance Indicators. *Amfiteatru Economic Journal, The Bucharest University of Economic Studies, 16*(8), 1048–1062.

Dwyer, L., Edwards, D., Mistilis, N., Roman, C., & Scott, N. (2009). Destination and enterprise management for a tourism future. *Tourism management, 30*(1), 63–74.

Fayed, H., & Fletcher, J. (2002). Report: Globalisation of economic activity: issues for tourism. *Tourism Economics, 8*(2), 207–230.

Fletcher, J., & Westlake, J. (2006). Globalisation . *International handbook on the economics of tourism* (s. 464–480.). içinde Cheltenham: Edward Elgar Publishing.

GSTC. (2013). *Sustainable tourism destination criteria.* http://www.gstcouncil.org. adresinden alındı

Gulcemal, T. (2020). The Globalızatıon Indıcators And Tourısm Development: A Dynamıc Panel-Data Analysıs For Medıterranean Countrıes. *Journal of Business, Economics and Finance (JBEF), 9*(3), 253–261.

Holowiecka, B., Grzelak-Kostulska, E., & Kwatkowski, G. (2011). Impact of Globalization on Tourist Preferences and Activiti. *Scale of globalization* (s. 55–62). içinde Ostrava: University of Ostr.

Hughes, G. (2004). Tourism, Sustainability and Social Theory. *A Companion to Tourism* (s. 498–509). içinde Oxford, UK: Blackwell.

ILO. (2001). *Human Resources Development, Employment and Globalization in the Hotel, Catering and Tourism Sector, Report for discussion at the Tripartite Meeting on the Human Resources Development, Employment and Globalization in the Hotel* . Geneva: International Labor Organization Office.

IVC. (2006). *The Economic Impact of Air Service Liberalization, Inter Vistas- Consulting.*

Kosolapov, N. .. (2001). The international political organization of the globalized world: models for a medium-term prospect. *Social sciences and modernity*(6), 141–143.

Mahon, R. (2010). After neo-liberalism? The OECD, the World Bank and the child. *Global Social Policy, 10*(2), 172–192.

McCann, P. (2008). Globalization and economic geography: the world is curved, not flat. *Cambridge Journal of Regions, Economy and Society, 1*(3), 351–370.

McGrew, A. (2011). *The logics of economic globalisation" In Global political economy.* New York: Oxford University Press.

Minciu, R., & Stanciu, P. (2010). Bukovina's tourism perspectives – a strategic approach. *Journal of Tourism, 10*(10), 81–90.

Moscardo, G. (2001). Cultural and Heritage Tourism: The Great Debates in Tourism. *Tourism Twenty-First Century.* içinde London: Continuum.

Mowforth, M., & Munt, I. (1998). *Tourism and Sustainability: A New Tourism in the Third World.* Routledge: London, UK.

Mustafa, M. H. (2010). Tourism and Globalization in the Arab World. *International Journal of Business and Social Science, 1*(1), 37–48.

Naumov, N., & Green, D. (2015). Mass tourism. *Encyclopedia of Tourism.* içinde New York: Springer.

Ozili, K. P. (2022). *Sustainability and sustainable development research around the world.* Munich Personal RePEc Archive-MPRA.

Page, S. (2012). *Tourism Management: An Introduction.* Oxford: Butterworth-Heinemann.

Peric, V. (2005). Tourısm And Globalızatıon. *e 6th International Conference of the Faculty of Management Koper Congress Centre Bernardin.* Slovenia: Managing the Process of Globalisation in New and Upcoming EU Members.

Pindzo, R., & Radulovic, D. (2017). The Impact Of The Globalızatıon On Tourısm Industry. *Organısatıonal Behavıor And Types Of Leadershıp Styles And Strategıes In Terms Of Globalızatıon.* içinde United Kingdom: Compass Publishing.

Poon, A. (1993). *Tourism, Technology and Competitive Strategies.* Wallingford, UK: CAB International.

Saarinen, J. (2014). Critical Sustainability: Setting the Limits to Growth and Responsibility in Tourism. *Sustainability, 6*(1), 1–17.

Samarathunga, W. H., & Gamage, D. (2020). *Alternative Tourism as an Alternate to Mass Tourism during the Post-COVID-19 Recovery Phase: the Case of Sri Lanka.* Sage Submissions.

Samarathunga, W., & Gnanapala, W. K. (2016). Impact of All-inclusive Tour Packages with Special Reference to Bentota tourist Resort. *Sustainable Tourism in the Global South: Communities Environments and Management.* içinde Cambridge: Cambridge Scholars Publishing.

Sezgin, E., & Yolal, M. (2012). Golden Age of Mass Tourism: Its History and Development. *Visions for Global Tourism Industry - Creating and Sustaining Competitive Strategies.* içinde intechopen.

Sharpley, R. (2009). Sustainable Tourism Development: A Critique. *Tourism Development and the Environment: Beyond Sustainability?* içinde London, Sterling: Professor Richard Anthony John Sharpley.

Smeral, E. (1998). The impact of globalization on small and medium enterprises: new challenges for tourism policies in European countries. *Tourism management, 19*(4), 371–380.

Smith, V. L., & Eadington, W. R. (1992). *Tourism Alternatives: Potentials and Problems In The Development Of Tourism.* New York: John Wiley.

Song, H., Li, G., & Cao, Z. (2017). Tourism and Economic Globalization: An Emerging Research Agenda. *Journal of Travel Research.*

Turner, L., & Ash, J. (1975). *The Golden Hordes: International Tourism and the Pleasure Periphery.* London: Constable.

TÜSİAD. (2014). *Evaluation Towards an Alternative Tourism Development in Turkey.* Istanbul: TÜSİAD Publications.

UNESCO. (1999). *Tourism and Culture: Rethinking the Mix, the UNESCO Courier.*

Vukadinović, P., Knežević, G., & Mizdraković, V. (2013). Globalizacija istrane direktne investicijeu Srbiji. *Globalizacija i savremeno poslovanje* (s. 57–64). içinde Bijeljina: Univerzitet Sinergija.

Wheeller, B. (1991). Tourism"s troubled times: responsible tourism is not the answer. *Tourism Management, 12*(2), 91–96.

Wheeller, B. (1993). Sustaining the ego. *J. Sustain. Tourism*(1), 121–129.

Wolfe, R. (1982). Recreational travel: The new migration revisited. *Ont. Geogr*(19), 103–122.

WTO. (1993). *Sustainable Tourism Development: Guide for Local Planners.* Madrid, Spain: World Tourism Organization .

Yazdi, S. k. (2012). Sustainable Tourism. *American International Journal of Social Science, 1*(1), 50–56.

Yıldız, Ü., & Gürhan, N. (2020). The Effects of Alternative Tourism Activites on Sustainable Tourism Potential and Employment. *Revista Argentina de Clínica Psicológica, 29*(5), 1567–1581.

Dönüş Çiçek* and Şerife Üsküdar†

Chapter 24 Reflections of Turkish Bath Traditions on Bath Museums

Introduction

One of the important elements of cultural promotion, museums, where the material and spiritual heritage of humanity takes place, exhibited and protected, are nowadays diversified and opened with different themes reflecting the society's daily life. One of them is the bath museums. The baths, which have been built with the realization of the healing power of hot waters, have been important in keeping the unique traditions of each civilization alive from ancient times to the current times, and have found their place in different disciplines such as painting, sculpture, medicine, music, cinema and tourism.

It has been recognized that Turkish baths have an essential place in Turkish culture from the past to the present. When the historical trace of the baths is examined, the tradition of the baths in Anatolia can be said to be a continuation of the Roman bath culture, which has changed and developed into a unique building by changing and transforming over time (Yegül, 2006; Özdal, 2023).

Apart from washing, which is the main function of the baths, it became a part of social and social life during the Seljuk and Ottoman periods (Ürük, 2016). Besides the understanding of cleanliness from the past, the acceptance of Islam and the understanding of cleanliness foreseen by the religion have been effective in the formation of the Turkish bath culture. This culture, which is associated with Turkish society and has become a tradition by developing new features over time, has continued its existence throughout history (Güler & Yasa, 2021). Baths have been an essential part of social life and a symbol of a culture passed down from generation to generation. As Yegul (2017; 2006) states, they have become representatives of a social organization.

When the literature is investigated, it is remarkable that there are many research on the cultural functions of Turkish baths (Ünver, 1973; Bozok, 2005;

* Assoc. Prof. Dr., Anadolu University, Eskisehir Vocational School, dcicek@anadolu.edu.tr

† PhD, Sivas Provincial Health Directorate, Directorate of Public Health Services, suskudar@hotmail.com.tr

Uluumay, 2009; Özgen, 2016). Baths have been featured in travelogues (Sezen, 2021), miniatures, painting (Büyükkol & Arda, 2016), poetry (Kaplan, 2011) and sometimes in a film scene. In these research, different subjects such as health, bath preparation process, bath traditions, bath etiquette, materials used in baths, bath sets have been discussed. The preservation of the bath culture, which has a rich cultural content, is crucial for the protection of cultural heritage. Therefore, it can be said that bath museums have undertaken this mission. In this part of the book, Turkish bath culture and Turkish bath museums, which are one of the reflections of this culture on tourism, are presented.

Turkish Bath Traditions

Turkish baths, which also take place as a social living space in Turkish culture, have been places where traditions with rich cultural contents are implemented, where many important events from birth to weddings are experienced and blessed (Özgen, 2016). The most popular of these traditions is the bridal bath. The bridal bath, also called henna bath, is usually organized a few days before the wedding or at the henna night. This tradition in which the guests are previously invited, the expenses of the bath are covered by the groom. Depending on the wealth of the groom, the groom can either close the bath completely or pay for a certain number of kurnas (Büyüktanır, 2009). The bride's bundle is important in the bridal bath. In this bundle, which is carefully packed by the groom's side, there are bath items (loincloth, towel set, comb, mirror, bath bowl, kildan, pouch, soap, etc.) and the bride uses these items. The regions may have different practices in the preparation of the bridal bundle (Emiroğlu, 2010; Demir, 2013). Another bath tradition that women organize is the Loğusa Hamam. In some regions, this tradition, which is also called forty-removal bath, heralds the end of the mother's puerperium period and the baby can be visited. Because the mother who gives birth does not go out with the baby for forty days, and guests are not accepted except for close relatives (Büyüktanır, 2009). The rituals performed in the puerperal bath, which is being performed by the mother's close family members, differ according to the regions, as in the bridal bath (Atlı, 2020). Another tradition in women's baths is to look for a bride-to-be in the bath. This tradition, which is called girl liking bath, is crucial to have an idea about the health and beauty of the young girl, and to get some information about the family of the liked girl from the bathhouse staff (Tuzcu, 2019). Another of the bath traditions is the circumcision bath. According to this tradition, the day before the circumcision, the child is taken to the bathhouse by his father and relatives, washed and cleaned, and dressed in his clothes. In some regions, henna is burnt on the right hand of the child after bathing. Prayers, beautiful words and folk songs are used to relieve the child's restlessness. In places where there is a custom of kirvelik, all costs of the bath are paid by the kirve (Uysal, 2016). The groom's bath is another bath tradition. It is a tradition for entertainment purposes in which the groom-to-be is taken to the hammam by his relatives and friends before the wedding, where he is massaged, eaten and drunk (Türkyılmaz, 2001). In addition to these traditions, bath traditions such as soldier's bath, Eid (Arefe) bath, votive bath, engagement bath, seven baths also have an significant role in Turkish culture (Türkyılmaz, 2001).

Turkish Bath Museums in Turkey

Some of the baths that have been maintained and preserved in the historical process have served their original functional purpose, while many of them have been restored and given new functionalities (Akalın, 2023). For instance, Şengül Hamam in Kütahya has been converted into a geological museum, Istanbul Tahtakale Hamam into a touristic bazaar, and Istanbul Ördekli Hamam into a cultural center. In addition, some baths have been restored and turned into bath museums to keep the bath culture alive and to pass on the traditions to new generations. Museums, which are mediators in the promotion of cultural values, are centers that represent the cultural identity and cultural heritage of a region and bring tourism mobility to the region. Museums are considered to be the safeguards of cultural heritage (Stylianou- Lambert, Boukas &Yeralia, 2014) and museums collect, preserve and exhibit things that form the basis of cultural sustainability and tangible connections between past, present and future (Davies & Wilkinson, 2008.).

The International Cauncil of Museums (ICOM), which Turkey is a member of, defined the concept of museum as follows: "*A museum is a permanent non-profit institution at the service of society that research, collects, preserves, interprets and exhibits tangible and intangible heritage. Public, accessible and inclusive museums promote diversity and sustainability. It provides a wide range of experiences for education, entertainment, reflection and sharing of knowledge, operates and communicates ethically, professionally and with the participation of communities*" (ICOM, 2022). The museums are places that provides information about the history and cultural values of a society, is effective in national and international promotion and provides intergenerational transfer (Çekal, Doğan & Aktürk, 2022). Starting with archaeology and ethnography museums, museum formation has diversified over time and started to be established with different themes. One of these is the Turkish bath museums, which focus on Turkish baths (Özdal, 2023), one of the places where cultural production and transmission takes place, and important traditions are preserved. In Turkey, a totally of seven Turkish bath museums have been determined (Table 1).

Table 1. Turkish Bath Museums

Name of Museum	Province	Foundation Year	Building Type
Beypazari Turkish Bath Museum	Beypazarı (Ankara)	2012	Historical Bath
Gaziantep Bath Museum	Şahinbey (Gaziantep)	2015	Historical Bath
Bayezid II Museum of Turkish Bath Culture	Fatih (İstanbul)	2015	Historical Bath
Tahtalı Bath Museum	Battalgazi (Malatya)	2018	Historical Bath
Eyüpsultan Tomb Bath Museum	Eyüpsultan (İstanbul)	2022	Historical Bath
Eskisehir Bath Museum	Odunpazarı (Eskişehir)	2022	New Construction
Zeyrek Tiled Bath Museum	Fatih (İstanbul)	2023	New Construction

Source: Created by the authors.

Beypazari Turkish Bath Museum: The third in the world and the first in Turkey, this museum, which presents the bath culture, has been established in 2012 with the support of Ankara Development Agency and Living Museum with the conversion of Rüstem Pasha Bath, which had been built by Rüstem Pasha, the vizier of Magnificent Suleiman in Beypazarı, into a museum (Demir, 2013). The bath consists of two parts. The first section, the coldness section, exhibits silver embroidered horseshoes and hamam bowls, silk loincloths, gülabdans, sürmedans, hamam bundles and censers. In this part of the museum there are horseshoes used by the mother of the bath, which visitors are allowed to try on. In the second section, which is the warmth section; mother-of-pearl and silver inlaid horseshoes as well as handmade olive oil soaps are also exhibited. In the halvets in the temperature section of the bath, there are exhibitions themed as bridal bath and groom's bath. In the museum, different exhibition techniques are used to enable visitors to experience the henna bath tradition in its entirety with its instrument, word, clothing and attire. Razors, styles, barber basins are exhibited in the shaving section of the museum (Demir, 2013).

Gaziantep Bath Museum: The bath is in the city center of Gaziantep and is part of the Lala Mustafa Pasha Complex. Lala Mustafa Pasha Bath, which represents the Ottoman era bath architecture, was constructed in 1564–1565. The bath has been built half buried in the ground to keep the heat inside. The bathhouse, which consists of coldness, warmth, warmth and ashen sections, has been restored by the Department of Conservation Implementation and Inspection Office (KUDEB) under the leadership of Gaziantep Metropolitan Municipality, was opened as a museum in 2015 and obtained private museum status in 2021.

There are three wax sculptures at the entrance of the museum. One of the sculptures is depicted distributing mırra, the traditional coffee of the region. Beside it is a depiction of a person in a loincloth resting in the cold section of the bath. The other sculpture is of a peshtamal maker waiting to hand a peshtamal to the bathers. The wall paintings in this section include images of people from the baths. In the warmth section; there are soaps and bowls in double-sided wall-mounted showcases, and in the middle of the section there is a fountain with cold water flowing. In the museum, the traditions of *groom's bath, bride's bath* and *puerperal bath* are depicted with sculptures. In one of the private bathing rooms in this section there is an oral history area and in the other there is a three-dimensional model of the Topkapı Palace baths (Ateş & Üzüm, 2021). The museum organizes workshops in different categories. One of the traditions of Gaziantep baths, şamşırak (fragrant water) is explained in this workshop, the prepared şamşırak is put in small jars and presented to the visitors attending the workshop. The museum also organizes puerperal emi (paste) workshops and scented soap workshops for adult and children (Yetkin, 2024).

Bayezid II Turkish Bath Culture Museum: It is located in the complex founded by Bayezid II in 1505. The bath, which was added to the complex later (Baltacı, 2005), was built by Gülbahar Hatun, the wife of Bayezid II, in 1507–1508 as a double bath for public use. The bath, which served until the 1930s, was decided to be reorganized as a museum with the projects prepared in 2011 (Beyazıt II Bath Restoration and Cultural Heritage Museum Organization Projects). The bath was restored in accordance with the original and registered as a museum in 2015 (Tozduman Terzi, 2017). In the museum, which is entered from the cold section of the women's bath, a door was opened later, allowing access to the cold section of the men's bath. This place was organized as a temporary exhibition area, and the warmth and temperature areas in both the women's and men's sections were evaluated as permanent exhibition areas. In the permanent exhibition area, Ottoman period bath items are exhibited. The warmth section of the women's bath is the area where archaeological stone artifacts from the Byzantine and Ottoman periods are exhibited. These artifacts are registered in the inventory of the Istanbul Archaeological Museums Directorate and were found in and around the baths. Another feature of the museum is that it is organized in a way that allows meetings and seminars with speakers from different fields to be held. Furthermore, the museum organizes various workshops with students during the academic year. (URL-1).

Tahtalı Bath Museum: The bath in Battalgazi district of Malatya is dated to the 19th century depending on the architectural features it bears. The bath, which consists of undressing, warmth, temperature, material and water tank sections, is of great importance in terms of reaching the present day intact (Akalın, 2023). The bath, which was restored by Battalgazi Municipality in 2014, was opened as Tahtalı Bath Museum in 2018. In the museum, there are scenes such as *groom's bath, bride's bath, girl's bath, soldier's bath, circumcision bath, puerperal bath, rich bath*. In addition to these scenes, which are animated with wax sculptures and conveyed to the audience, there is an oral history room in the bathhouse, and the history of the bathhouse is presented to visitors in electronic environment. Also on exhibition are items used in Turkish baths. In the bath, where the conception of audience- oriente interaction is dominant rather than artifact-oriented exhibition in the traditional sense (Keleş, 2003), traditions and oral culture elements are conveyed to visitors through costumed museum experts, amateur theater performances, conferences and audio exhibitions.

Eyüpsultan Tomb Bath Museum: Located in Fatih, Istanbul, the bath is known as Eyüp Sultan Mosque Kebir Bath and Fatih Sultan Mehmet Bath. The Tomb Baths of Eyüp Sultan Complex is one of the first baths built in Istanbul by Sultan Mehmet the Conqueror (Yörük, 2016). Haskan (1993) suggests that this bath may have been destroyed in the earthquake of 1509 and was later rebuilt by Mimar Sinan in the form of a double bath. The bath was restored to its original state and opened as a museum in 2022 (Altun, 2023). On the lower floor of the museum, which consists of two floors, items belonging to the bath culture are displayed. Concurrently, information about the baths given by experts and the restoration stages of the Eyüpsultan Türbe Bath are shown to visitors through videos. The foundations of the bath, which were uncovered during the restoration phase, were covered with a glass floor. There are also visuals of the old state of the bath on the walls. On the upper floor, which was formerly used as a dressing room, the expression and objects of water culture are included (URL-2).

Eskişehir Bath Museum: Located in the historical Odunpazarı region of Eskişehir province, Eskişehir Bath Museum was designed and built in a modern way based on traditional Turkish bath architecture.

It was opened by Eskişehir Metropolitan Municipality in 2022. The museum is Turkey's first and only Turkish bath museum that focuses on the bathing culture from Antiquity to the Republican period. In addition to the characters of the natır, meydancı, tellak, külhanbeyi, bath mother, sherbet seller and Eskişehir's famous halvacı, barbers are also included in the museum. The museum features bath scenes depicted in wax sculptures. In the museum where information about bath culture is given with digital panels, soap, soap boxes, soap seal prints, clogs, cologne bottles, barber materials, gülabdanlıklar, combs, bath tubs, towels, loincloths, mirrors, censer holders are exhibited (URL-3).

Zeyrek Çinili Bath Museum: The bath, also called Hayreddin Pasha Bath, Kaptan Pasha Bath, Tezgâhâhçılar Bath, was built by Mimar Sinan between 1540–1546 as a double bath. As a result of the rise in the road level over time, the bath is descended by stairs (Kuruçay, 2011). The bath was purchased and restored by Ali Bey between 1833–1850 and put into service (Kabaoğlu, 2013: 1178). Nevertheless, the bath, which was closed in the following years, was purchased by The Marmara Group in 2010, and its restoration was completed in 2023 as a result of the Çinili Hamam Conservation-Repair and Rehabilitation Project. Designed as an additional building in harmony with the architecture of the bath, the museum exhibits artifacts from the Roman, Byzantine and Ottoman periods unearthed during excavations carried out under the supervision of the Istanbul Archaeological Museums. One of the rooms in the museum exhibits approximately 3000 pieces of blue-white Iznik tiles discovered during restoration and excavations and preserved inside the bath. In the 16th century, there is a representative panel explaining the process from the production stage of these tiles specially produced for the Çinili Hamam to their exhibition in various museums around the world and showing the representative arrangement of the tiles at the time the bath was built. Another room is used for the exhibition of historical bath objects used in the baths (horseshoes, towels, bowls, etc. decorated with precious metals and mother-of-pearl). The bath also include schematics and models explaining the water and heating systems designed by Sinan the Architect (URL-4).

Conclusion

Baths are important cultural spaces used by people to meet their needs for cleaning, healing, socializing and entertainment (Kasapoğlu Akyol, 2020). With the innovations brought by modern life, the importance of baths and bath culture is rapidly forgotten and takes its place in history within social development (Güler & Yasa, 2021). Bath Museums are important in terms of introducing the place and importance of water and baths in daily life, how the change in bath culture has taken place in the historical process, and the bath traditions from the past to the present to new generations. In the meantime, these museums aim to document, preserve and transmit the Turkish bath culture to the future.

In Turkey, where the first examples of Turkish baths in the world were found, the first Turkish bath museum opened was the Beypazarı Turkish Bath Museum. Gaziantep Turkish Bath Museum, Bayezid II Turkish Bath Culture Museum, Tahtalı Turkish Bath Museum, Eyüpsultan Türbe Turkish Bath Museum, Eskişehir Turkish Bath Museum and Tiled Turkish Bath Museum were opened respectively. Eskişehir Bath Museum is the first museum designed in accordance with the traditional Turkish Bath architecture. The Tiled Bath Museum was designed as an additional building in accordance with the architecture of the bath. The bath buildings where other bath museums were established have been restored while preserving their original architecture. These museums are important in terms of preventing the damage of bath buildings by remaining idle, as well as the proper utilization of cultural heritage.

Different practices are carried out in bath museums to make cultural transfer more understandable. In this context, in the bath museums in Beypazarı, Gaziantep, Malatya and Eskişehir, it is important to explain the bride's bath, groom's bath traditions, which are part of Turkish culture and traditions, using sculpture and technology, and to exhibit the items used in the baths to help new generations understand the culture. Furthermore, the presentation of traditions in museums through interactive techniques rather than classical presentation techniques takes museum visitors out of the status of being an audience. Thus, it can be said that culture transfer is more effective.

References

Akalın, F. B. (2023). Malatya tarihi tahtalı hamamın plastik eser yansımaları. *Socıal Mantalite and Researcher Thınkers Journal (Smart Journal)*, 9(78), 5352–5360.

Altun, S. (2023). Eyüpsultan'da kültür turizmi bağlamında miras rotası önerisi, *Uluslararası Beşerî ve Sosyal Bilimler İnceleme Dergisi*, 2(2), 87–97.

Ateş, H. & Üzüm, İ. H. (2021). Toplumsal mirasın aktarılmasında müzelerin rolü, Gaziantep büyükşehir belediyesi'ne bağlı müzeler üzerine bir değerlendirme, *International University Museums Association Platform Journal of Cultural Heritage UNIMUSEUM*, 4(2), 82–95.

Atlı, S. (2020). Hayatın geçiş dönemlerinde hamam ve işlevleri, Social *Science Development Journal*, 5(22), 240–252

Baltacı, C. (2005). *XV–XVI. yüzyıllarda osmanlı medreseleri*, İstanbul: M.Ü. İlahiyat Vakfı Yayınları.

Bozok, D. (2005). Türk hamamı ve geleneklerinin turizmde uygulanışı: Bursa merkez ilçede bir araştırma, *Balıkesir Sosyal Bilimler Dergisi*, 8(13), 62–86.

Büyükkol, S. & Arda, Z. (2016). Türk kültüründe hamam geleneği ve resim sanatına yansımaları, *idil*, 5(27), 2047–2062.

Büyüktanır, F. (2009). Geçmişten günümüze Sivas'ta hamam kültürü, Sivas: Sivas Cumhuriyet Üniversitesi Sosyal Bilimler Enstitüsü Yüksek Lisans Tezi.

Çekal, N., Doğan, E. & Aktürk, H. (2022). Türkiye'de gastronomi müzeleri ve özellikleri, *Sosyal, Beşerî ve İdari Bilimler Dergisi*, 5(11), 1655–1673.

Davies, M. & Wilkinson, H., 2008. Sustainability and museums: Your chance to make a difference. London: Museums Association. https://archive-media.museumsassociation.org/sustainability_web_final.pdf

Demir, Z. S. (2013). Türk hamamı müzesi ve kına hamamı sergisi, *Journal of Ankara Studies*, 1(2), 48–55.

Emiroğlu, İ. (2010). Konya'da hamam kültürü ve kullanım eşyaları. Konya: Selçuk Üniversitesi Sosyal Bilimler Enstitüsü Yüksek Lisans Tezi.

Güler, F. & Yasa, G. (2021). İzmir ili hamam geleneği ve hamam takımları giyim ve süsleme özellikleri, *19 Mayıs Üniversitesi Güzel Sanatlar Fakültesi Seramik Cam Bölümü Uluslararası Sempozyum ve Sergi Anadolu'nun Geleneksel El Sanatları "Sanatçı ve Zanaatkarlar Buluşması"* Bildiri Kitabı, ISBN 978-625-7464-15-4, 236–249.

Haskan, M. N. (1993). *Eyüpsultan tarihi*. İstanbul: Eyüpsultan Vakfı Yayınları.

ICOM (2022). *Museum definition* international cauncil of museums, https://icom.museum/en/resources/standards-guidelines/museum-definition/ (Erişim tarihi: 27.05.2024)

Kabaoğlu, C. (2013). Çinili hamam projesi, *kargir yapılarda koruma ve onarım semineri V, Bildiriler Kitabı*, İstanbul: İBB-KUDEB Yayınları.

Kaplan, Y. (2011). Türk hamam kültürünün divan şiirine yansımaları, *Atatürk Üniversitesi Türkiyat Araştırmaları Enstitüsü Dergisi* (44), 132–155.

Kasapoğlu Akyol, P. (2020). Suyla arınma'dan modern zamanların "gelin hamamı"na: Tarihi Karacabey hamamı'nda gelin hamamı. *Folklor/Edebiyat,* 26(103), 691–706.

Keleş, V. (2003). Modern müzecilik ve Türk müzeciliği, *Atatürk Üniversitesi, Sosyal Bilimler Enstitüsü Dergisi*, 2(2), 1–17.

Kuruçay, A. (2011). *İstanbul'un 100 hamamı*, İstanbul: İstanbul Büyük Şehir Belediyesi Kültür A.Ş. Yayınları.

Özdal, A. (2023). Gaziantep kent merkezinde hamam kültürü. *Motif Akademi Halkbilimi Dergisi*, 16(43), 1000–1023.

Özgen, Ö. (2016). Kültürel miras kapsamında "Türk hamamı" üzerine bir inceleme. *İletişim Kuram ve Araştırma Dergisi*, 42,11–138.

Sezen, G. (2021). Batılı seyyahların izleniminden hamamlar. *Türkiyat Mecmuası*, 31(1), 301–322.

Stylianou-Lambert, T., Boukas, N., & Christodoulou-Yerali, M. (2014). Museums and cultural sustainability: stakeholders, forces, and cultural policies, *International Journal of Cultural Policy*, 20(5), 566–587. https://doi.org/10.1080/10286632.2013.874420

Tozduman Terzi, A. (2017). *Gravürlerle Fatih*, İstanbul: Fatih Belediyesi.

Tuzcu, G. (2019). Geçmişten günümüze Merzifon hamam kültürü. Ankara: Gazi Üniversitesi Sosyal Bilimler Enstitüsü, Yüksek Lisans Tezi.

Türkyılmaz, D. (2001). Türk kültüründe hamam geleneği ve Eskişehir hamamları. Ankara: Hacettepe Üniversitesi Sosyal Bilimler Enstitüsü, Yüksek Lisans Tezi.

Uluumay, Esat (2009). *Eski hamam, eski tas*. İstanbul: Yapı Kredi Yayınları.

URL-1, https://muzeyum.istanbul.edu.tr/tr/content/muzeler/ii.-bayezid-turk-hamam-kulturu-muzesi (Erişim tarihi: 28.05.2024).

URL-2, https://eysam.org/eyupsultan_turbe_hamami_muzesi/ (Erişim Tarihi: 28.05.2024).

URL-3, http://eskisehirhamammuzesi.com/?sayfa=kurumsal (Erişim tarihi: 29.05.2014).

URL-4, https://zeyrekcinilihamam.com/tr/mimar-sinanin-basyapitlarindan-zeyrek-cinili-hamam-kapsamli-restorasyonun-ardindan-ozgun-islevine-kavusuyor (Erişim tarihi: 29.05.2024).

Uysal, Y. (2016). Adıyaman halk kültüründe sünnet merasimi ve kirve kültürünün topluma yansımaları, *Turkish Studies* (Elektronik), 11(4),1015–1024, 11(4), 1015–1024.

Ünver, S. (1973). Türk hamamı. *Belleten,* 37 (145), 87–94.

Ürük, Z. F. (2016). Medeniyetler içinde hamamın gelişimi ve kültürel olarak mekân analizleri, *Akademik Sosyal Araştırmalar Dergisi*, 28, 185–209.

Yegül, F. K. (2006). *Antik çağ'da hamamlar ve yıkanma*, İstanbul: Homer Kitabevi

Yegül, F. K. (2017). Anadolu'da hamam kültürünün evrimi, *Z Dergisi*, 2 (Güz), 163–169.

Yetkin, T. (2024). Çağdaş müzecilik üzerine bir değerlendirme Gaziantep hamam müzesi örneği, Gaziantep: Gaziantep Üniversitesi Sosyal Bilimler Enstitüsü, Yüksek Lisans Tezi.

Yörük, S. (2016). XVIII. yüzyıl Eyüp Kazası hamamları ve hamam çalışanlarının demografik yapısı, *BELLETEN*, 80(289), 823–856.

Mehmet Beyazgül*

Chapter 25 Using Machine Learning Algorithms in Restaurant Cost Prediction

Introduction

In the dynamic landscape of the culinary world, restaurants constantly endeavor to balance quality, affordability, and sustainability. Many difficulties arise in predicting and controlling operational costs for the sustainable growth and profitability of restaurants. In this respect, using machine learning algorithms is a new, promising approach to estimating restaurant costs in a way that leads to better-informed decisions and optimizing resources in this regard.

Despite the paramount value of the cost forecast in sustainable restaurant management, the conventional approach often lacks both accuracy and flexibility. Labor-intensive manual forecasting and elementary models are not only error-prone but also inadequate in capturing the complex dynamics of the restaurant environment. Consequently, managers of restaurants grapple with uncertainty, risk of financial instability, and inefficiency in their operations (Ceballos-Palomares, Nava-Jiménez, Caballero-Morales, & Cano-Olivos, 2021; Kostakis, Boskou, & Palisidis, 2011; Yang, Pan, & Song, 2013). Therefore, this research suggests that algorithms developed in machine learning to predict restaurant costs, making a data-driven approach more far-reaching than traditional methodologies. Machine learning models can uncover complex patterns and derive insights from data in large sets of datasets comprising variables such as menu offerings, customer demographics, seasonality, and market trends (Mortazavi et al., 2016; Rajkomar et al., 2018). In predictive analytics, restaurants can predict their cost fluctuations, optimize the inventory management, and devise strategic price strategies to enhance the financial resilience (Sakib, 2023; Şahinbaş, 2022).

Using machine learning algorithms to predict restaurant costs uses artificial intelligence techniques to analyze large data related to restaurant operations, pricing, and customer behavior. Several studies have delved into the application of machine learning to predict various aspects such as review helpfulness, price, financial distress, bankruptcy, revenue, visitor patterns, and sales

* Assist. Prof. PhD., Harran University, Tourism Faculty, Department of Gastronomy and Culinary Arts, mehmetbeyazgul@harran.edu.tr

forecasting in the restaurant industry. Lee, Kwon, and Back (2021) worked on the development of prediction models for review helpfulness by the application of machine learning (ML) algorithms to analyze available big restaurant data, which proved the potential for ML in customer preference and decision making. Similarly, Şahinbaş (2022) noted the use of different machine learning modeling algorithms for restaurant price prediction, highlighting the effectiveness of techniques like XGBoost, Random Forest, and Artificial Neural Networks in this domain. Furthermore, studies by S. Y. Kim and Upneja (2014) and H. Kim and Gu (2006) show that machine learning algorithms may forecast restaurant financial difficulties and insolvency. These models help to avoid crises or identify factors or symptoms early enough that can lead to financial instability, ensuring timely measures. Additionally, Bortan and Baisakov (2021) have explored the application of machine learning to anticipate income and the number of future guests in the restaurant industry and emphasized the importance of forecasting in operational planning and decision-making. Furthermore, in the literature on menu engineering and cost analysis in restaurants, machine learning has been used to activity-based costing (ABC), as demonstrated by Raab, Mayer, Kim, and Shoemaker (2009). In the study, it is argued that ML makes it possible to understand its cost structure better and, hence, make well-informed decisions about prices based on actual metrics of the restaurant's profitability. In addition to cost prediction, machine learning algorithms were widely adopted for sales prediction and operational efficiency optimization in restaurants. Gómez-Talal (2024), introduced an ML model using multiple correspondence analysis for sales forecasting, demonstrating the potential of data-driven approaches in improving revenue projections and inventory management. Mhlanga (2018), highlighted the factors impacting restaurant efficiency, emphasizing the role of data envelopment analysis in identifying economic risks and enhancing operational performance.

More specifically, the application of machine learning algorithms in restaurant cost prediction is a data-driven approach that helps restaurants in the optimization of their pricing strategies, forecasting financial outcomes, and, therefore, increasing the operational efficiency within their system. Leveraging advanced analytics and AI techniques, restaurants can derive valuable insights into customer behavior, cost structures, and revenue streams to improve profits and thus gain more significant competitive advantages. The aim of this study is to make a comprehensive examination of the use of machine learning methods in restaurant cost predictions. Although machine learning techniques are very important, there is little information on how to apply them for estimating restaurant costs.

This study aimed to cover this gap by providing details regarding the application of ML algorithms for cost prediction in restaurants.

Utilizing Machine Learning Algorithms for Cost Prediction

Accurate estimates in the field of restaurant cost prediction can be obtained by utilizing diverse machine learning methods. The kind of data, the difficulty of the task, and the intended alternatives between interpretability and prediction performance are often taken into consideration when choosing a particular algorithm. In this context, some frequently used algorithms are as follows:

Decision Trees: Decision trees are structures that resemble trees, with each leaf node representing a class label and each inside node representing a choice based on a feature. Decision trees are interpretable and appropriate for tasks involving regression and classification because they divide data according to attribute values using a top-down induction technique (Zhou & Wang, 2012). Decision trees are appropriate for changing datasets because they gradually learn from data and update the tree structure to accommodate new knowledge (Wang, Chu, Xie, Wang, & Sun, 2018). Since decision trees are a flow chart, they provide an advantage to decision makers in terms of easy interpretation and understandability (Chien & Chen, 2008).

Random Forests: Another ensemble learning method called Random Forest builds several decision trees during training and produces the average prediction for regression tasks or the class mode for classification problems (Breiman, 2001). By building a forest of trees with randomized feature subsets, Random Forest lessens overfitting and boosts accuracy.

Gradient Boosting: Gradient boosting is another ensemble learning technique that combines numerous weak predictive models to generate a robust predictive model. It scans a subset of the complete dataset to select the data point that produces the most disagreement, therefore boosting the model's predicted performance (Lu & Bongard, 2009).

K-Nearest Neighbors: K-nearest neighbors is a non-parametric approach that predicts the value of a new data point using the values of its k nearest neighbors from the training data. Data points are classified based on the majority class of their k-nearest neighbors, making it simple and effective for pattern recognition tasks (Zhang, Chen, Liu, & Xi, 2020). In other words, the algorithm classifies data using analogy based on its close link to past data.

Naive Bayes: Naive Bayes is a probabilistic algorithm that predicts the probability of an event occurring based on the presence of certain features (Serag et al., 2024). It is based on the assumptions of independence, normality between variables and that all attributes should be equally important (Bayes, 1968; Serag et al., 2024).

Support Vector Machines: Support Vector Machines (SVM) are models that use data to perform classification and regression tasks. SVM is an ideal search for a hyperplane that divides classes in high dimensions with the biggest margin between them (Poojitha & Kanagasabai, 2022). Support vector machines are beneficial when there are a large number of potential predictors. They are commonly employed for classification problems and can deal with both linear and nonlinear interactions between variables.

Artificial Neural Networks: Artificial neural networks (ANN) are a sort of deep learning algorithm that can discover intricate patterns and relationships in data. They consist of interconnected layers of artificial neurons that process and transmit information. ANNs are computational models based on the human brain's neural structure. ANNs are made up of interconnected nodes structured into layers. Each node processes information and transmits it to the next tier. ANNs learn by modifying weights to reduce output mistakes, making them extremely effective at identifying complicated patterns (Hsu, 2012).

In summary, these machine learning algorithms can predict costs by analyzing historical data, identifying patterns, and making predictions based on the relationships between input features and cost outcomes. Each algorithm has its unique characteristics and applications, making them valuable tools for cost prediction in various domains. The choice of algorithm(s) depends on various factors, including the size and structure of the dataset, the presence of temporal dependencies, the need for interpretability, and the computational resources available.

Results

This paper explores the application of machine learning algorithms in predicting restaurant costs, emphasizing the significance of accurate cost estimation for sustainable restaurant management. Traditional cost prediction methods often lack precision and adaptability, leading to challenges in financial stability and operational efficiency for restaurants. By leveraging machine learning algorithms, restaurants can enhance their cost forecasting capabilities, enabling informed decision-making and resource optimization. Thus, restaurant companies can gain significant contributions in terms of financial health and operational efficiency. Various machine learning models, such as decision trees, random forests, gradient boosting, and artificial neural networks, offer a data-driven approach to analyze complex patterns and derive insights from diverse datasets encompassing factors like menu offerings, customer demographics, and market trends.

Restaurants can optimize inventory management, budgeting, and pricing strategies by leveraging machine learning algorithms to predict fluctuations in costs related to ingredients, labor, utilities, and other operational expenses. Healthy stock management may be made possible by forecasts that consider the demand volume, seasonality, and market trends into consideration. Also, for predictive maintenance of equipment and facilities to minimize downtime, reduce maintenance costs, and ensure seamless operations machine learning can be used. Thus, by integrating machine learning algorithms into cost prediction processes, restaurants can enhance financial resilience, operational efficiency, and profitability in a competitive industry landscape.

While prior studies have investigated many areas of restaurant management and predictive analytics, this study stands out for its specific focus on cost prediction and the use of cutting-edge machine learning techniques. While some research has used basic statistical models like regression analysis, the use of complex algorithms like neural networks, decision trees, and ensemble approaches is a novel approach. This study aims to help better understand the effects of machine learning algorithms that have unparalleled levels of accuracy and granularity in cost forecasting for restaurants.

While the chapter discusses the importance and potential applications of machine learning in the restaurant industry, it does not delve into the challenges or obstacles that may arise during the implementation process. Furthermore, the sample outcomes of machine learning applications in the restaurant business are not thoroughly explored. More study is needed to solve these constraints and gain a better knowledge of the practical uses and efficacy of machine learning algorithms in predicting restaurant costs.

Future studies in the realm of machine learning applications for the restaurant industry could focus on conducting comparative analyses of different machine learning algorithms for cost prediction to determine the most effective approach. Longitudinal studies could be undertaken to assess the long-term impact of machine learning implementation on cost-forecasting accuracy and operational efficiency in restaurants.

References

Bayes, T. (1968). Naive bayes classifier. *Article Sources and Contributors*, 1–9.

Bortan, A., & Baisakov, B. (2021). Predicting Future Visitors in the Restaurant Business Using Machine Learning. *Herald of Kazakh-British Technical University, 18*(3), 36–41. doi:10.55452/1998-6688-2021-18-3-36-41

Breiman, L. (2001). Random Forests. *Machine Learning, 45*(1), 5–32. doi:10.1023/A:1010933404324

Ceballos-Palomares, L.-N., Nava-Jiménez, A.-B., Caballero-Morales, S.-O., & Cano-Olivos, P. (2021). Forecast Methods and Periodic Review Inventory Model for Supply Planning to Reduce Food Waste. *Asian Journal of Economics Business and Accounting*, 24–40. doi:10.9734/ajeba/2021/v21i530378

Chien, C. -F., & Chen, L. -F. (2008). Data mining to improve personnel selection and enhance human capital: A case study in high-technology industry. *Expert Systems with Applications, 34*(1), 280–290. doi:https://doi.org/10.1016/j.eswa.2006.09.003

Gómez-Talal, I. (2024). Avoiding Food Waste From Restaurant Tickets: A Big Data Management Tool. *Journal of Hospitality and Tourism Technology, 15*(2), 232–253. doi:10.1108/jhtt-01-2023-0012

Hsu, K. W. (2012, 12–14 July 2012). *Hybrid ensembles of decision trees and artificial neural networks.* Paper presented at the 2012 IEEE International Conference on Computational Intelligence and Cybernetics (CyberneticsCom).

Kim, H., & Gu, Z. (2006). Predicting Restaurant Bankruptcy: A Logit Model in Comparison With a Discriminant Model. *Journal of Hospitality & Tourism Research, 30*(4), 474–493. doi:10.1177/1096348006290114

Kim, S. Y., & Upneja, A. (2014). Predicting Restaurant Financial Distress Using Decision Tree and AdaBoosted Decision Tree Models. *Economic Modelling, 36*, 354–362. doi:10.1016/j.econmod.2013.10.005

Kostakis, H., Boskou, G., & Palisidis, G. (2011). Modelling Activity-based Costing in Restaurants. *Journal of Modelling in Management, 6*(3), 243–257. doi:10.1108/17465661111183676

Lee, M., Kwon, W., & Back, K. J. (2021). Artificial Intelligence for Hospitality Big Data Analytics: Developing a Prediction Model of Restaurant Review Helpfulness for Customer Decision-Making. *International Journal of Contemporary Hospitality Management, 33*(6), 2117–2136. doi:10.1108/ijchm-06-2020-0587

Lu, Z., & Bongard, J. (2009). Exploiting Multiple Classifier Types With Active Learning. doi:10.1145/1569901.1570228

Mhlanga, O. (2018). Factors Impacting Restaurant Efficiency: A Data Envelopment Analysis. *Tourism Review, 73*(1), 82–93. doi:10.1108/tr-07-2017-0109

Mortazavi, B. J., Downing, N. S., Bucholz, E. M., Dharmarajan, K., Manhapra, A., Li, S. X., . . . Krumholz, H. M. (2016). Analysis of Machine Learning Techniques for Heart Failure Readmissions. *Circulation Cardiovascular Quality and Outcomes, 9*(6), 629–640. doi:10.1161/circoutcomes.116.003039

Poojitha, M. R. S., & Kanagasabai, M. (2022). Decision Tree Over Support Vector Machine for Better Accuracy in Identifying the Problem Based on the Iris Flower. doi:10.3233/apc220028

Raab, C., Mayer, K. J., Kim, Y.-S., & Shoemaker, S. (2009). Price-Sensitivity Measurement: A Tool for Restaurant Menu Pricing. *Journal of Hospitality & Tourism Research, 33*(1), 93–105. doi:10.1177/1096348008329659

Rajkomar, A., Oren, E., Chen, K., Dai, A. M., Hajaj, N., Hardt, M., . . . Dean, J. M. (2018). Scalable and Accurate Deep Learning With Electronic Health Records. *NPJ Digital Medicine, 1*(1). doi:10.1038/s41746-018-0029-1

Sakib, S. M. N. (2023). Restaurant Sales Prediction Using Machine Learning. 202–226. doi:10.4018/978-1-6684-7105-0.ch011

Serag, R. H., Abdalzaher, M. S., Elsayed, H. A. E. A., Sobh, M., Krichen, M., & Salim, M. M. (2024). Machine-Learning-Based Traffic Classification in Software-Defined Networks. *Electronics, 13*(6), 1108. Retrieved from https://www.mdpi.com/2079-9292/13/6/1108

Şahinbaş, K. (2022). Price Prediction Model for Restaurants in Istanbul by Using Machine Learning Algorithms. *Ekonomi İşletme Ve Maliye Araştırmaları Dergisi, 4*(2), 159–171. doi:10.38009/ekimad.1148216

Wang, H., Chu, C. R., Xie, X., Wang, N., & Sun, J. (2018). Decision Tree Incremental Learning Algorithm Oriented Intelligence Data. doi:10.23940/ijpe.18.05.p3.849856

Yang, Y., Pan, B., & Song, H. (2013). Predicting Hotel Demand Using Destination Marketing Organization's Web Traffic Data. *Journal of Travel Research, 53*(4), 433–447. doi:10.1177/0047287513500391

Zhang, W., Chen, X., Liu, Y., & Xi, Q. (2020). A Distributed Storage and Computation k-Nearest Neighbor Algorithm Based Cloud-Edge Computing for Cyber-Physical-Social Systems. *Ieee Access, 8*, 50118–50130. doi:10.1109/ACCESS.2020.2974764

Zhou, G., & Wang, L. (2012). Co-Location Decision Tree for Enhancing Decision-Making of Pavement Maintenance and Rehabilitation. *Transportation Research Part C Emerging Technologies, 21*(1), 287–305. doi:10.1016/j.trc.2011.10.007

Samet Karahan[*] and Yüksel Öztürk[†]

Chapter 26 How Can Local Governments Deal with Overtourism?

Introduction

Tourism has become one of the most developing sectors in recent years and is expected to develop in the coming years. Factors such as changes in the understanding of leisure time, ease of travel and transportation, increasing general welfare, and the provision of unique experiences encourage people to participate in tourism activities more and more every day. Even while unforeseen circumstances like the pandemic have significantly impacted the industry, it can be argued that the tourist sector has gotten better at swiftly adjusting to the difficulties it encounters.

Overtourism, which has been seen as one of the most significant handicaps of tourism in recent years, refers to the deterioration in the quality of life and experience of both visitors and local people by exceeding capacity due to the excessive influx of tourists to destinations (Goodwin, 2017). Several destination stakeholders are actively searching for and delicately addressing solutions to this growing issue.

Local governments play a unique role in developing tourism in a region, notwithstanding the involvement of other stakeholders. Due to their roles in implementation, control, and decision-making, local governments are among the most potent entities in travel destinations (Godfrey, 1998). Local governments also have essential roles in sustainable tourism development and resource protection. Overtourism is a significant issue that local governments should address since it is the antithesis of sustainable tourism (Mihalic, 2020; Vagena, 2021). In this study, firstly, the concept and indicators of overtourism, which is seen as an essential threat to destinations, will be discussed. Then, the ways of dealing with overtourism from the perspective of local governments will be examined comprehensively.

* PhD., Bilecik Seyh Edebali University, Applied Science Faculty, Department of Tourism Management, samet.karahan@bilecik.edu.tr

† PhD., Ankara H.B.V. University, Tourism Faculty, Department of Tourism Management, yuksel.ozturk@hbv.edu.tr

The Concept of Overtourism

The phenomenal growth of the tourism sector in recent years has harmed many destinations. Despite the locals' tolerance, popular attractions are being overwhelmed by tourists, overwhelming the capacity of tourism zones (Volo, 2020). Especially in destinations with high tourism intensity, the number of media reports has recently increased, and a new term for this phenomenon, overtourism, has been developed (Goodwin, 2017; Milano, 2018; Atzori, 2020).

The phrase "overtourism" was first used as a search keyword in 2006, according to Google (Dodds & Butler, 2019). It was originally used in an academic study on coastal resource management in Vietnam in 2008 (An, Phung, & Chau, 2008). In 2012, it appeared on Twitter noticeably more frequently (Weissman, 2021). In 2016, Skift writer Rafat Ali highlighted the dangers of overtourism in his article "Exploring the Coming Perils of Overtourism" (Kuščer & Mihalič, 2019), which gained popularity within the online tourism community. In the late 2010s, it became the subject of academic studies as its adverse impacts on tourist destinations were felt increasingly, and anti-tourism movements increased.

One of the most common definitions of overtourism is defined by UNWTO (2018: 3) as "the impact of tourism on a destination, or parts thereof, that excessively influences the perceived quality of life of citizens and/or quality of visitors experiences negatively". The concept of overtourism refers to the flow of visitors much more than a destination can accommodate in a given period. Innerhofer, Erschbamer, and Pechlaner (2019) consider overtourism as a phenomenon that depends on the subjective perceptions of local people and state that the question of how much is too much is answered differently in different regions and destinations. As can be understood from the definitions of overtourism, many authors (Goodwin, 2017; Singh, 2018; UNWTO, 2018; Butler, 2018; Simancas & Peñarrubia, 2019) consider the focus of the concept to be an excessive influx of tourists to destinations.

In summary, overtourism can be defined as an unregulated and excessive influx of tourists to a destination, which can harm both locals and tourists. However, it is essential to note that the phenomenon is not limited to this. It is important to note that destinations have different signs and indicators. In the next section, we will discuss the indicators of overtourism.

Indicators of Overtourism

Although the causes and symptoms of overtourism in destinations have existed for many years in most popular destinations, it is relatively recent to perceive this phenomenon as a problem. As Karahan (2024) asserts, these symptoms are more fully elucidated by examining the case studies of destinations (Venice, Amsterdam, Brussels, Istanbul, Dubrovnik, and Barcelona) where overtourism is experienced.

A review of the literature on the subject of overtourism (Peeters et al., 2018; Mihalic, 2020; Blázquez-Salom et al., 2023) reveals that the indicators of overtourism are generally consistent with the reasons for tourism development. Consequently, the rationale behind tourism development in a region may also give rise to overtourism. For instance, Dodds and Butler (2019) posit that the primary indicator of overtourism is the substantial number of tourists, a notion that is frequently referenced in definitions. Other indicators of overtourism include more accessible travel opportunities, continuous growth-oriented destination management, short-term solutions, package tours that enable mass tourism, social media practices, and a power imbalance between stakeholders.

The causes of overtourism, as discussed by Avond et al. (2019) regarding supply and demand, can be understood as a complex set of interrelated factors. On the demand side, factors such as a high propensity for spending, tourists with more affluent economic conditions, the development of travel technology, globalization, sophistication, changing lifestyles, and women's emancipation have been identified as contributing to the phenomenon of overtourism. On the supply side, the following factors have been identified as contributing to the phenomenon of overtourism: a lack of resources, poor destination management, a sensitive and fragile structure of destinations, a lack of planning, a lack of technical expertise, and a lack of understanding of sustainability awareness.

In addition, inadequate management and incompatible stakeholders in destinations that do not aim for similar goals are also important indicators of overtourism (Atzori, 2020). Weber et al. (2017) posit that destination management organizations and destination stakeholders directly influence carrying capacity and overtourism. Among these stakeholders, local governments occupy a distinctive position. Jeffries (2007) posits that local governments are at the vanguard of tourism planning and policy formulation within their respective jurisdictions. From this perspective, local governments are responsible for preventing and addressing the consequences of overtourism in destinations. This topic will be explored in greater depth in the subsequent section of the study.

Overtourism and Coping Strategies in Terms of Local Governments

Tourism is an essential tool for development, providing local economies with income, employment, and infrastructure investments. Stakeholders in the macro and micro environments must collaborate and actively participate for this development to be realized entirely (Akhtar, Bukhari, & Najar, 2022). Among these stakeholders, local governments play a critical role in the sustainable management and development of tourism. Local governments in particular significantly impact the efficient use of destination resources, the adoption of sustainable practices, the direction of development, and the complete execution of legislation. Local governments, which are directly involved in many issues related to tourism development, assume much of this responsibility (Aronsson, 2000) and control much of the policy and development planning related to tourism (Dredge & Jenkins, 2007).

Holloway (2022: 528) lists the responsibilities of local governments about tourism as follows; providing recreational facilities (such as conference centers, theaters, parks, sports centers, and museums) for tourists and residents, planning urban and rural policies, controlling land use, licensing and inspecting accommodation facilities such as hotels, motels, hostels, pensions, camping and food and beverage facilities such as restaurants and cafes, providing daily visitor services, providing parking spaces for vehicles, producing tourism-related statistics, helping to promote the destination, maintaining historical buildings, ensuring the cleanliness and hygiene of garbage and public toilets. When all of these duties are taken into account, it becomes increasingly evident how crucial local governments are to the development, planning, and maintenance of tourism (Baş et al., 2007).

The negative effects of overtourism affect visitors as well as local people. Therefore, local governments should take measures to protect both sides. From this point of view, the next section will discuss the methods that local governments can use to deal with overtourism.

Focusing on Sustainability, Not the Number of Tourists

Overtourism is also defined as uncontrolled and unsustainable tourism (Mihalic, 2020), and it has been demonstrated to cause serious problems. When tourism is not managed in an appropriate manner, it has the potential to cause significant damage and disruption (Coldwell, 2017). Dodds and Butler (2019) liken overtourism to a volcano that has been about to erupt for years, emphasizing that sustainability in the tourism sector has still not been achieved.

Rather than attempting to maximize the number of tourists, destinations should prioritize maintaining or enhancing the quality of the tourism experience, the quality of life of local residents, and environmental sustainability. In this context, the most fundamental task falls to local governments. While central governments (such as ministries) are more preoccupied with statistical data, local governments should prioritize sustainability and the well-being of local residents.

A Rapid Response: Restriction

One of the most prominent indicators of overtourism is overcrowding (Peeters et al., 2018; Dodds & Butler, 2019; Avond et al., 2019). Local governments employ a variety of regulatory and control measures to manage the influx of tourists to specific destinations. To achieve this objective, tourists are frequently provided with a specified period within which they are permitted to remain in a particular region or area. In some instances, the entrances to these areas are entirely closed. At the Taj Mahal, a three-hour time limit is set for visitors. The Koh Khai Islands, a popular day-trip destination in Thailand, is off-limits to tourists. In the case of the Greek island of Santorini, the number of daily cruise passengers disembarking is subject to a limit (Gupana, 2018). In Cinque Terre, Italy, the number of annual visitors was subject to limitation. In addition to imposing a cap on the number of tourists permitted to visit, Peru's Machu Picchu also limits the time that can be spent at the site to four hours. The city of Venice has ceased to permit the use of its terminal by cruise ships (Jet, 2018). The prehistoric caves of Lascaux in France and Altamira in Spain are entirely inaccessible to tourists. As the aforementioned examples illustrate, local governments employ restrictions, bans, and blockades as a straightforward and expedient means of addressing overtourism.

Changing the Management Mindset

Tourtellot (2021) posits that traditional institutions are inadequate for destination management. Once mass tourism has been established in a given location, it can be exceedingly challenging to reverse the trend. Jamieson and Jamieson (2019) posit that this may occur when traditional destination management organizations (DMOs) are supplanted by destination development, marketing, and management organizations (DDPMOs). In contrast to traditional Destination Management Organizations (DMOs), which are primarily focused on marketing, Destination Governance, Planning, and Management Organizations (DGPMOs) encompass a broader range of responsibilities, including planning and management.

In their 2018 study, Koens et al. argue that local governments should now prioritize tourism potential in their urban planning and innovation efforts, making decisions that align with the interests of local stakeholders. Peeters et al. (2018) enumerate four measures that local governments most frequently adopt to control and manage overtourism: bureaucratic regulations, initiatives to better distribute tourist audiences across time and space, expanding the capacity of destinations, and increasing the cost of the visit.

Marketing or Demarketing

Demarketing is defined as a specific branch of marketing that aims to temporarily or permanently deter consumers or a class of consumers. In other words, demarketing aims to reduce the demand for a product from unwanted customers (Kotler & Levy, 1971). This strategy may be employed proactively to reduce the demand for and impacts of overtourism.

To effectively address the issue of overtourism, local governments should prioritize the implementation of strategies that aim to "reverse the marketing" of a destination in the short term (Frey, 2021). It is thus hoped that this approach will help to minimize the risks and tensions associated with overtourism and successfully improve the quality of the visitor experience (Weiler et al., 2019). For instance, tourists residing on the outskirts of Amsterdam are subject to a lower tax rate than those who choose to reside in the city center. In Bhutan, Indian tourists are subject to a different price structure than other tourists (Nyaupane & Timothy, 2010).

As part of the demarketing strategy, local governments may modify the marketing message to appeal to different market segments that are more likely to spend economically with fewer visitors. A comparable approach is employed in Cyprus and Egypt to target higher-income tourists. Additionally, they can inform potential visitors about experiences in neighboring destinations that are less crowded and offer a similar experience (Frey, 2021). Suh et al. (2009) propose that local governments should adopt a strategy of reducing rather than increasing advertising, reducing promotion costs, and increasing the availability of alternative products and services.

Fighting Back Against the Usual Suspect Sharing Economy

The Oxford Dictionary defines the sharing economy as "an economic system in which assets or services are shared between private individuals, either for free or for a fee, typically using the Internet" (Heo, 2016: 1). This economic model is frequently preferred in the fields of accommodation in tourism (Airbnb, Couchsurfing, 9Flats, etc.) and transportation (Uber, Zipcar, TaxiBeat, etc.). In numerous studies (Sans & Quaglieri, 2016; Postma & Schmuecker, 2017; Avond et al., 2019; Atzori, 2020; Celata & Romano, 2022), these peer-to-peer applications are identified as a significant contributor to overtourism due to their provision of affordable lodging. Peer-to-peer platforms eliminate intermediaries in the distribution channel, allowing tourists to enter destinations unsupervised and raising rental prices (Goodwin, 2017).

In the city of Las Vegas, those who wish to rent out their properties to tenants must obtain a license from the relevant authorities. Furthermore, in the case of short-term rentals, the landlord must be present at the property in question. In New York, landlords who rent for less than a month must be permanent residents, be present at the property during the rental period, and are prohibited from owning more than one property at a time. In Paris, all rented apartments must be registered with the relevant authorities, and short-term rentals are limited to a period of 120 days. In Amsterdam, all residential properties may be rented for a period of up to 30 days per year (Bustamante, 2020; Koçak & Ulema, 2020; Karahan & Öztürk, 2020; Genç, 2021).

As evidenced by the aforementioned examples, local governments can mitigate the adverse effects of excessive tourism in a destination by implementing measures such as auditing, restricting, and monitoring within the scope of the sharing economy. Local governments may choose to address the issue through the implementation of a policy of balance.

Making the City Smart

The "smart city" concept has emerged as a response to urban management's current challenges (Gretzel, 2018). Undoubtedly, the challenges faced by cities and urban areas include the phenomenon of overtourism. The primary objective of smart urbanism is to provide strategic solutions to specific travel needs (Khan et al., 2017). These solutions include using applications or technological tools to assist users in identifying their preferred destinations. In this manner, tourists can examine pre-designed virtual applications for destinations and gain insight (Tripathy et al., 2018).

Frey (2021) advocates using contemporary digital technologies to mitigate the adverse effects of excessive tourism in tourist destinations. It is possible to develop computer-generated multidimensional designs that are difficult to distinguish from the real thing, to create animation with hologram technology, and to travel to a destination with VR glasses. Consequently, local governments are better positioned to regulate access to cultural sites through online registration.

Managing Visitor Behaviors

In some destinations, the primary concern may not be the number of tourists, but rather their behavior (Postma & Schumacer, 2017). In response to the challenges posed by overtourism, authorities have initiated communication campaigns to foster mutual understanding between residents and tourists (Goodwin, 2019). For example, in Amsterdam, the construction of new hotels and souvenir shops was prohibited, and a public relations campaign entitled "Have Fun and Respect" was initiated.

Local governments have taken action to address visitor behavior in other locations. In the Canary Islands, the slogans "Tourist: respect my land!" and "Respect the City," as coined by the recently elected mayor of Dubrovnik, are visible throughout the city. In Spain's Balearic Islands, Amsterdam, and Munich, the implementation of alcohol restrictions has been necessitated by the occurrence of disruptive behavior by tourists. Iceland has implemented a pledge that requests tourists refrain from driving off-road or engaging in other inappropriate activities in natural areas. The Pacific islands of Palau have a version of their pledge, written by the country's children, which requests that visitors exercise caution and conservation.

It is possible that tourists may exhibit different behaviors in the locations they visit. This phenomenon can be attributed to the concept of "being a tourist," which entails engaging in activities that are not part of their daily routines. Although local governments lack the authority to alter visitor behavior entirely, they can exert influence through their messages and decisions. As Jamieson and Jamieson (2019) observe, these efforts may only sometimes be documented in writing, and there are no established rules. Nevertheless, positive outcomes can be attained through the use of so-called "soft" approaches.

Results

Overtourism results from a discrepancy between the desire to visit a destination and the destination's capacity to accommodate this demand (Cheer et al., 2019). On the demand side, several factors have contributed to the growth in tourism. These include improvements in the economic conditions for tourists, the development of travel technology, the ease of travel facilitated by globalization, changes in lifestyles and working conditions, and the impact of social media. On the supply side, several factors contribute to the problem of overtourism. These include inadequate infrastructure, a lack of resources, traditional destination management, unrealistic planning, and a lack of sustainability.

This section outlines the methods local governments should implement to address overtourism in their destinations. Local governments are the most influential tourism stakeholders at the regional level, given their powers and responsibilities.

A comprehensive approach to managing overtourism necessitates a shift in perspective, particularly among those employed in the public sector. Given the nature and objectives of the tourism industry, addressing the primary sources of overtourism necessitates significant policy changes. As a means of addressing overtourism through local governments, this chapter has underscored the necessity of a sustainable and collaborative management approach devoid of the traditional management mindset obsessed with tourist numbers.

Due to a shift in the approach of local governments, restrictions, and non-marketing strategies, exemplified in various destinations, are being discussed. It is important to consider reducing the number of tourists and targeting segments with a higher economic contribution. The following tactics have been proposed as potential solutions: increasing the price of the service, implementing a queuing system for visitors, limiting advertising to select market segments, promoting conservation needs through sustainable development, and diverting tourist traffic to alternative areas in the nearby region (Nared & Viscovic, 2014). Restrictions may be based on any of the following criteria: place, time, space, or people. In the short term, different limits are applied to achieve quick results. Although this approach may appear to contradict the discourse on the freedom of travel, it is an effective method for achieving sustainability.

Another contributing factor to the phenomenon of overtourism is the emergence of sharing economy practices that offer inexpensive and unregulated accommodation. Such practices permit a considerable number of individuals, who are challenging to monitor, to move freely within the destination, thereby damaging the sense of community in the city. Faced with a multitude of undesirable behaviors, including noise, environmental pollution, alcohol and drug consumption, and rioting, some residents have been compelled to leave their homes and relocate to other areas. The phenomenon of "touristification," whereby a destination becomes increasingly commercialized and geared towards tourism, is evident in numerous locations. By registering and inspecting rental housing in their respective jurisdictions, local governments can protect their local populations more effectively and ensure public order.

Another avenue for local governments to pursue in addressing overtourism is the design of smart and technological cities. Furthermore, replicas or holograms of historical sites or artifacts can be presented to tourists through augmented reality technologies, thereby enhancing their experience. In this manner, original artifacts can be preserved. Finally, local governments should take action to regulate visitor behavior. Before arrival at the destination, tourists should be provided with information regarding local culture and the quality of life of local people. This is particularly important for tourists who respect local culture and are sensitive to local people's quality of life. Information can be disseminated concisely and impactfully during the visit through various channels, including businesses, social media, vehicles, and even on the streets.

This chapter focuses on how local governments can combat overtourism. Local governments bear significant responsibility for addressing the adverse consequences of overtourism. It is assumed that the bleeding can be stopped if the mentioned issues are considered. Who knows, it could be cured completely.

References

Akhtar, S., Bukhari, S. A., & Najar, P. A. (2022). Key stakeholder's perspective towards sustainable tourism development. *Eco. Env. & Cons.* 28 (January Suppl. Issue), 388–392.

An, N. T., Phung, N. K., & Chau, T. B. (2008). Integrated coastal zone management in Vietnam: Pattern and perspectives. *Journal of Water Resources and Environmental Engineering*, 23, 297–304.

Aronsson, L. (2000). *The Development of Sustainable Tourism*. London: Continuum.

Atzori, R. (2020). Destination stakeholders' perceptions of overtourism impacts, causes, and responses: The case of Big Sur, California. *Journal of Destination Marketing & Management*, 17, 100440.

Avond, G., Bacari, C., Limea, I., Seraphin, H., Gowreesunkar, V., & Mhanna, R. (2019). Overtourism: A result of the Janus-faced character of the tourism industry. *Worldwide Hospitality and Tourism Themes*, 11(5), 552–565.

Baş, M., Kılıç, B., & Güçer, E. (2007). Türkiye'de yerel yönetimler ve turizm. *Mevzuat Dergisi*, 10(119), 1–12.

Blázquez-Salom, M., Cladera, M., & Sard, M. (2023). Identifying the sustainability indicators of overtourism and undertourism in Majorca. *Journal of Sustainable Tourism*, 31(7), 1694–1718.

Bustamante, J. (2020). *Airbnb statistics*. https://ipropertymanagement.com/research/airbnb-statistics. (Accessed: 15.05.2024).

Butler, R. (2018). Challenges and opportunities. *Worldwide Hospitality and Tourism Themes*, 10(6), 635–641.

Celata, F., & Romano, A. (2022). Overtourism and online short-term rental platforms in Italian cities. *Journal of Sustainable Tourism*, 30(5), 1020–1039.

Cheer, J. M., Milano, Claudio, & Novelli, Marina. (2019). Afterword: Over overtourism or just the beginning?. In *Overtourism: Excesses, Discontents and Measures in Travel and Tourism* (pp. 227–232). Wallingford UK: CAB International.

Coldwell, W. (2017). First Venice and Barcelona: Now anti-tourism marches spread across Europe. *Guardian*, https://www.theguardian.com/travel/2017/aug/10/anti-tourism-marches-spread-across-europe-venice-barcelona (Accessed: 17.05.2024).

Dodds, R., & Butler, R. (Eds.). (2019). *Overtourism: Issues, Realities and Solutions*. Boston: Walter de Gruyter GmbH & Co KG.

Dredge, D., & Jenkins, J. M. (2007). *Tourism Planning and Policy*. New Jersey: John Wiley & Sons.

Frey, B. S. (2021). Reactions to cultural overtourism. In *Overcoming Overtourism Creating Revived Originals*. (pp. 29–41). Basel: Springer.

Genç, K. (2021). Turizmde paylaşım ekonomisi tartışmaları ve İstanbul: Airbnb örneği. *Balıkesir Üniversitesi Sosyal Bilimler Enstitüsü Dergisi*, 24(46–1), 1343–1357.

Godfrey, K. B. (1998). Attitudes towards 'sustainable tourism'in the UK: A view from local government. *Tourism Management*, 19(3), 213–224.

Goodwin, H. (2017). The challenge of overtourism. *Responsible Tourism Partnership*, 4, 1–19.

Goodwin, H. (2019). Barcelona–crowding out the locals: A model for tourism management. In *Overtourism: Issues, Realities and Solutions*. (pp. 125–138). Boston: Walter de Gruyter GmbH & Co KG.

Gretzel, U. (2018). From smart destinations to smart tourism regions. *Investigaciones Regionales-Journal of Regional Research* (42), 171–184.

Gupana, S. (2018). 11 overrun destinations that are trying to curb tourism, Matador Network, 16 May. https://matadornetwork.com/read/11-overrun-destinations-trying-curb-tourism/ (Accessed: 17.05.2024).

Heo, Y. (2016). Sharing economy and prospects in tourism research. *Annals of Tourism Research*, 58, 166–170.

Holloway, J. C., & Humphreys, C. (2022). *The Business of Tourism*. London: Sage.

Innerhofer, E., Erschbamer, G., & Pechlaner, H. (2019). The challenge of managing the limits. In Pechlaner, H., Innerhofer, E., & Erschbamer, G. (Eds.) *Overtourism* (pp. 3–13). New York: Routledge.

Jamieson, W., & Jamieson, M. (2019). Managing overtourism at the municipal/destination level. In Dodds, R., & Butler, R. (Eds.). *Overtourism: Issues, Realities and Solutions*. (pp. 217–233). Boston: Walter de Gruyter GmbH & Co KG.

Jeffries, D. (2007). Governments and tourism. New York: Routledge.

Jet, J. (2018). How is overtourism impacting travel to popular destinations? Forbes, 20 August. https://www.forbes.com/sites/johnnyjet/2018/08/20/how-is-overtourism-impacting-travelto-popular-destinations/#77d39d335b84 (Accessed: 10.05.2024).

Karahan, S. (2024). *Aşırı turizm destinasyonlarında kalabalık algısı, destinasyon hizmet kalitesi ve turist deneyiminin memnuniyet ve davranışsal niyete etkisi: Kültür turizmi destinasyonu örneği*. Doktora Tezi. Ankara Hacı Bayram Veli Üniversitesi / Lisansüstü Eğitim Enstitüsü / Turizm İşletmeciliği Ana Bilim Dalı. Ankara.

Karahan, S., & Öztürk, Y. (2020). "Overtourısm" kavramına etimolojik bir yaklaşım. *Journal of Tourism & Gastronomy Studies*, 8(Special Issue 4), 191–203.

Khan, M. S., Woo, M., Nam, K., & Chathoth, P. K. (2017). Smart city and smart tourism: A case of Dubai. *Sustainability*, 9(12), 2279.

Koçak, Y. & Ulema, Ş. (2020). Paylaşım ekonomisinde yasal sorunlar ve çözüm önerileri: Airbnb Örneği, *Journal of Tourism and Gastronomy Studies*, 8(4), 3021–3038.

Koens, K., Postma, A., & Papp, B. (2018). Is overtourism overused? Understanding the impact of tourism in a city context. *Sustainability*, 10(12), 4384.

Kotler, P., & Levy, S. J. (1971). Demarketing, yes, demarketing. *Harvard Business Review*, 79, 74–80.

Kuščer, K., & Mihalič, T. (2019). Residents' attitudes towards overtourism from the perspective of tourism impacts and cooperation—The case of Ljubljana. *Sustainability*, 11(6), 1823.

Mihalic, T. (2020). Conceptualising overtourism: A sustainability approach. *Annals of Tourism Research*, 84, 103025.

Milano, C. (2018). Overtourism, social unrest and tourismphobia. A controversial debate. *PASOS: Revista de Turismo y Patrimonio Cultural*, 16(3), 551–564.

Nared, J., & Visković, N. R. (Eds.). (2014). *Managing Cultural Heritage Sites in Southeastern Europe*. Ljubjiana: Založba ZRC.

Nyaupane, G. P., & Timothy, D. J. (2010). Power, regionalism and tourism policy in Bhutan. *Annals of Tourism Research*, 37(4), 969–988.

Peeters, P., Gössling, S., Klijs, J., Milano, C., Novelli, M., Dijkmans, C., … & Postma, A. (2018). Research for TRAN Committee-Overtourism: impact and possible policy responses, European Parliament, Policy Department for Structural and Cohesion Policies, Brussels. chrome-extension://efaidnbmnnnibpcajpcglclefindmkaj/https:// researchdata.brighton.ac.uk/id/eprint/149/1/IPOL_STU(2018)629184_EN.pdf (Accessed: 12.05.2024).

Postma, A. & Schmuecker, D. (2017). Understanding and overcoming negative impacts of tourism in city destinations: Conceptual model and strategic framework, *Journal of Tourism Futures*, 3 (2): 144–156.

Sans, A. A., & Quaglieri, A. (2016). Unravelling Airbnb: Urban perspectives from Barcelona. In P. Russo, & G. Richards (Eds.), *Reinventing the Local in*

Tourism. Travel Communities and Peer-Produced Place Experiences (pp. 209–228). UK: Channel View Publications.

Simancas Cruz, M., & Peñarrubia Zaragoza, M. P. (2019). Analysis of the accommodation density in coastal tourism areas of insular destinations from the perspective of overtourism. *Sustainability*, 11(11), 3031.

Singh, T. (2018). Is over-tourism the downside of mass tourism?. *Tourism Recreation Research*, 43(4), 415–416.

Suh, M., Ahn, J., & Rho, T. (2009). A study on the relationship demarketing and customers' relationship behaviors. *Journal of Marketing Management Research* 15, 21–55.

Tourtellot, J. B. (2021). Overtourism and destination governance. In Honey, M., & Frenkiel, K. (Eds.). *Overtourism: Lessons for a Better Future*. Washington: Island Press.

Tripathy, A. K., Tripathy, P. K., Ray, N. K., & Mohanty, S. P. (2018). iTour: The future of smart tourism: An IoT framework for the independent mobility of tourists in smart cities. *IEEE Consumer Electronics Magazine*, 7(3), 32–37.

UNWTO (2018). *Overtourism? Understanding and Managing Urban Tourism Growth beyond Perceptions*, Madrid: UNWTO.

Vagena, A. (2021). Overtourism: Definition and impact. *Academia Letters*, 12(7).

Volo, S. (2020). Overtourism: Definitions, enablers, impacts and managerial challenges. In H. Séraphin, T. Gladkikh, & T. V. Thanh (Eds.), *Overtourism Causes, Implications and Solutions* (pp. 11–26). Cham: Springer.

Weber, F. (2017). Overtourism. An analysis of contextual factors contributing to negative developments in overcrowded tourism destinations. *BEST EN Think Tank XVII: Innovation and Progress in Sustainable Tourism*, 315.

Weiler, B., Moyle, B. D., Scherrer, P., & Hill, M. (2019). Demarketing an iconic national park experience: Receptiveness of past, current and potential visitors to selected strategies. *Journal of Outdoor Recreation and Tourism*, 25, 122–131.

Weissman, A. (2021). A growing problem. In Honey, M., & Frenkiel, K. (Eds.). *Overtourism: Lessons for a Better Future*. Washington: Island Press.

Serpil Kocaman*

Chapter 27 Exploring Digital Nomads in the Tourism Context

Introduction

The dynamics shaping today's global markets, coupled with advancements in transportation and communication technologies, have resulted in new social structures, trends, and expectations. Socio-economic changes, driven by wireless communication technologies and improvements in transportation systems that have made travel more affordable, have reshaped the labor market, giving rise to new forms of work and employment in the digital economy (Valenduc, 2019; Orel, 2019; Müller, 2016). The fusion of digitization and mobility, encompassing physical movement and technological interconnectedness, has enabled the widespread adoption of leisure and mobile lifestyles on national and international levels (Paris, 2011; Urry, 2007). Reflecting this new lifestyle, research on digital nomadism, which has recently emerged is gaining prominence and is particularly noteworthy for its potential impact on the tourism sector. The global population of digital nomads, which was 16.9 million in 2022, experienced a growth of over one hundred percent, reaching 35 million in 2023 (Statista, 2024). In a world increasingly characterized by digitalization, the expanding community of digital nomads presents tourism opportunities for the destinations they choose to visit.

Until recent years, digital nomads were primarily regarded as a lifestyle and working model. However, with their rapidly increasing numbers and the tourism potential they generate, they have recently started to attract attention as a new form of tourism. In the competitive environment of international tourism, leveraging the potential of digital nomads can provide a significant advantage. As this group grows, major tourist destinations worldwide have begun to claim their share of the digital nomad tourism market (Vagena et al., 2023). However, studies on the tourism potential created by digital nomads remain limited. This rapid technological change offers remote-working tourists the opportunity to continue working and producing while enjoying the

* Asst. Prof., Alanya Alaaddin Keykubat University, Tourism Faculty, Department of Tourism Management, serpil.kocaman@alanya.edu.tr

destination experience. This has led to the emergence of digital nomad tourists, who can work remotely thanks to digital technology infrastructure, creating a new tourism market for destinations (Kocaman, 2022b). Digital nomad tourists have different needs and expectations compared to traditional tourist profiles. Understanding the travel planning, booking habits, destination preferences, and selection criteria of digital nomads is crucial for developing appropriate marketing and service strategies and understanding how the tourism sector can adapt to these changing dynamics. This study will comprehensively examine digital nomads within the tourism literature from the perspective of a new tourism approach.

The Relationship Between Digital Nomadism and Tourism

Since the 1980s, there has been a global trend of increasing international mobility among individuals driven by changes in lifestyle, the desire for freedom of choice, and personal satisfaction (Olga, 2020). This trend, coupled with technological advancements in the 1990s, laid the groundwork for scientific studies on the concept of the 'digital nomad,' which has continued to grow in parallel with the rapid changes and developments in technology (Hermann & Paris, 2020). The term 'digital nomad' was coined by Makimoto and Manners in 1997 to describe the impact of technological advancements on people's lives (Makimoto & Manners, 1997). The emerging literature on digital nomadism initially began to define it primarily as a lifestyle, without sufficiently addressing the theoretical framework of digital nomadism (Olga, 2020). Although the term 'digital nomad' has started to be used in tourism literature in recent years, it has often been used interchangeably or ambiguously with the term 'remote worker,' without clearly defining the travel component that is a significant part of the digital nomad lifestyle. This has led to confusion between digital nomads and expatriates, business travelers, freelancers, and global nomads (Chevtaeva & Denizci-Guillet, 2021). Unlike digital nomads, other employees who travel to different destinations for work purposes usually travel only to locations designated by their employers, with travel expenses covered by the employer. However, digital nomads choose their travel destinations and durations and cover their travel expenses independently. In other words, digital nomads can travel according to their preferences, free from the directives of their workplaces (Kocaman, 2021a).

Mouratidis (2018) explains the term 'digital nomad' as a combination of 'digital,' involving the use of computer technology, and 'nomad,' which signifies continuous travel from one place to another. According to Reichenberger (2017), digital nomads are young professionals who work entirely online, are location-independent and lead a lifestyle that involves frequent travel. Thompson (2018) defines digital nomads as individuals who use technology to be independent of time and place constraints. In this lifestyle, the boundaries between work, leisure, and travel appear blurred (Reichenberger, 2017). Therefore, some studies have examined digital nomadism as a lifestyle (Bonneau et al., 2023; Green, 2020; Mouratidis, 2018; Bartosik-Purgat, 2018), while others have approached digital nomadism as a working model (Cook, 2023; Sánchez-Vergara et al., 2023; Schlagwein & Jarrahi, 2020; Orel, 2019; Kong et al., 2019; Nash et al., 2018; Reichenberger, 2017).

The blurred boundary between work and leisure for digital nomads, coupled with the need for travel and temporary accommodation inherent in digital nomadism, has inevitably attracted the interest of tourism literature. Research on the tourism potential of digital nomads has begun to explore their destination choices, accommodation types, travel motivations, and leisure activities (Handizar & Kusuma, 2023; Zerva et al., 2022; Floricic & Pavia, 2021; Kocaman, 2021a; Chevtaeva & Denizci-Guillet, 2021 Wiranatha et al., 2020; Chevtaeva, 2020; Hall et al., 2019; Putra & Agirachman, 2016). Research topics that examine digital nomads in the context of tourism can be grouped into four main themes: (i) the profile of digital nomad tourists, (ii) the travel motivations and destination selection criteria of digital nomads, (iii) the role of co-working spaces for digital nomads, and (iv) digital nomads' perceptions of work-leisure balance (Kocaman, 2022b).

The rise of digital nomadism has led to the emergence of 'digital nomad tourism,' with studies focusing on the perceptions and motivations of digital nomads in specific travel destinations (Sari & Suarmana, 2022). From a tourism sector perspective, digital nomadism, which allows individuals to work remotely while traveling, can be seen as a new and promising tourist typology (Düzgün & Akin, 2022). To satisfy digital nomads during their stay, the tourism market must consider various aspects such as digital nomad tourists' characteristics, perceptions, and satisfaction. Online resources such as Nomad List, which attracted over 3 million visitors in a year, evaluate various destinations around the world. They offer filtering and ranking options based on criteria such as internet speed, quality of life, and cost of living. These resources have a significant impact on choosing a destination (Nomad List, 2024; Vagena, 2021; Cook, 2020). They enable digital nomads to make well-informed decisions, share their experiences with others, and develop new concepts for work and travel.

As outlined above, the relationship between digital nomadism and tourism can be examined under specific prominent themes. In this study, based on research that addresses digital nomads in the context of tourism, we will focus on tourist profiles, destination selection criteria, and coliving and co-working spaces at destinations.

Digital Nomad Tourist Profile

Winarya & Petriwi (2020) observed that digital nomads distinguish themselves from other nomads by their eagerness to explore new destinations, their preference for working in areas with reliable internet connectivity, their affinity for technology, and their interest in weekend entertainment activities. These distinctions contribute to their inclination towards continuous travel, exploration of novel destinations, and engagement in leisure activities, positioning them as potential tourists.

In the realm of nomadic tourism, digital nomads are considered one of the types of nomadic tourists. Therefore, the emergence of digital nomads is not only related to the growth of living and working environments but also closely linked to tourism trends (Handizar & Kusuma, 2023). The tourism potential of digital nomads differs from traditional tourist profiles. As tourists, digital nomads tend to stay longer in destinations compared to traditional tourists; they engage more with the local community, contribute more to the local economy, prioritize the need for internet connectivity, and have a lower environmental impact (Mariati et al., 2023; Pecsek, 2018). This situation not only allows digital nomads to explore the country but also provides them with a long-term living experience, potentially increasing tourism revenue. Unlike traditional tourists, the cost of accommodation and living expenses in the chosen destination are significant criteria for digital nomads due to their longer-term stays (Mariati et al., 2023). According to Hall et al. (2019), while digital nomads are motivated to position themselves in places offering appealing leisure activities, they also prefer to settle in locations where they can connect with acquaintances, existing digital nomad communities, or markets and industries of interest.

According to NomadList (2024), the world's largest digital nomad platform, many digital nomads change locations only every 4 to 6 months, work from co-working spaces or cafes, establish close connections with the places they visit, and often revisit the same destinations. This lifestyle, characterized by long-term stays of digital nomad tourists, has led to the emergence of the concept of 'digital nomad tourism,' where accommodation plays a significant role and potentially increases revenue for the tourism sector (Astrama & Darsana, 2023).

Destination Selection Criteria of Digital Nomads

Digital nomads' destination preferences generally exhibit certain common characteristics. These prominent criteria are factors that digital nomads consider and that significantly influence their destination selection. Among the important factors that digital nomads take into account when deciding on their destination are the cost of living, climate, culture, internet connectivity, and visa requirements. These criteria are based on the valid assumption that digital nomads choose destinations based on leisure preferences rather than business requirements. It is understood that digital nomad tourists, who seek new experiences and cultural enrichment, are willing to immerse themselves in the local culture when they visit a destination (Tomeljenović et al., 2023).

The criteria effective in the destination selection of digital nomads have been identified by NomadList, the largest digital nomad platform in the world. These criteria include internet speed, availability of free Wi-Fi, proficiency in foreign languages, level of education, startup ecosystem, income level, cost-effectiveness, safety, traffic safety, walkability, population density, healthcare services, quality of life, family-friendly environment, level of happiness, peace, freedom of speech, entertainment options, nightlife, temperature, air quality (annual), humidity, heating and cooling systems, favorability among other digital nomads, tolerance towards women, LGBTQ+ friendliness, friendliness towards foreigners, and racial tolerance (NomadList, 2024).

The interest of workers in combining work and leisure to travel has increasingly become a subject of concern, leading many countries to regulate and amend residency permits and tax rules, and implement policies such as 'digital nomad visas' in many destinations (Vagena et al., 2023). Digital nomads have become a significant market globally, leading to increased efforts by tourism operators and government institutions to attract them (Hall et al., 2019). In just the year 2020 alone, high-ranking politicians from destinations such as Bermuda, the Cayman Islands, Georgia, Barbados, Antigua and Barbuda, Dubai, and Croatia made statements describing and inviting digital nomad tourism initiatives in the media. Some of the highlighted topics in these statements, which they emphasized to increase demand for digital nomads, include digital nomad visas, advanced infrastructure, global connectivity, fast internet, virtual working programs, multiculturalism and hospitality, historical and cultural attractions, social activities, security, and safety (Kocaman, 2021a).

Co-working and Co-living Spaces in Digital Nomad Destinations

Destinations and the accommodation industry must understand the role of collaboration and co-living in digital nomad tourism and research the attractions of this new type of tourism (Chevtaeva, 2020). The concept of co-living and co-working has facilitated the development of new services and products, creating a new concept called co-working spaces in popular digital nomad destinations. Co-working spaces have become popular establishments that offer digital nomads the opportunity to connect with like-minded individuals based on their lifestyles (Chevtaeva & Guillet, 2021; Pecsek, 2018). Co-working spaces in numerous destinations worldwide, such as Bali, Lisbon, Mexico City, and Barcelona, attract the interest of digital nomads (Chevtaeva, 2020). The tourism industry has begun to adapt to the increase in digital nomads and change its tourist facilities according to their needs (Orel, 2021). Therefore, creating co-working spaces in destinations and highlighting the tourism potential of digital nomads is essential (Kocaman, 2022a).

When the digital nomad experience is considered in isolation, it often involves loneliness. Digital nomads face the challenge of creating a living and working environment where they can collaborate with others or participate in networking events to find potential clients and collaborators if they choose to prefer this experience (Thompson, 2019). In this regard, co-living spaces constitute an important part of destinations that enhance the well-being of digital nomads and provide a solution to their loneliness (Chevtaeva & Guillet, 2021).

Co-living refers to a lifestyle in which individuals live and work together, sharing common areas. In other words, co-living is a form of shared accommodation where not only rent but also communal spaces are shared (Wong, 2020). Co-living plays a significant role in the accommodation experience of digital nomads. It focuses on the social value of living together to overcome urbanization and the isolation brought about by digital nomadism, while also providing an alternative to combat high rents in destinations (Von Zumbusc & Lalicic, 2020). Unlike other tourism accommodation establishments, communal living spaces are not homogeneous and are offered in various types targeting different subgroups of digital nomads. Community managers are a crucial feature that distinguishes these spaces from other types of accommodation. They facilitate networking, socializing, and community-building opportunities by organizing events, in addition to administrative tasks. In this way, communal living spaces help digital nomads overcome social isolation issues (Von Zumbusc & Lalicic, 2020). Communal living spaces can assist digital nomads in forming deeper and more sustainable relationships within limited periods in a destination. Thus, communal living spaces become an integral part of the digital nomad lifestyle (Wang et al., 2019).

The development and enhancement of the co-living concept in destinations can influence the preferences of digital nomads and become an attraction for the destination (Brown & White, 2019). One notable example is the Canggu region in Indonesia, which is a popular co-living destination among digital nomads. Another popular co-living destination is Bali, where 90 % of customers in co-living spaces identify themselves as digital nomads (Chevtaeva & Guillet, 2021). Therefore, destinations often seek to attract digital nomads by developing and enhancing the co-living concept.

Results

The digital age has ushered in rapid changes across various industries, including tourism (Mariati et al., 2023). As remote work evolves from being a trend to becoming mainstream, the emergence of digital nomads positions them as a market segment that every destination should cater to (Poulaki et al., 2023). At this juncture, there is a need to delve deeper into the travel patterns of digital nomads and assess their impact on tourism mobility through research (Reichenberger, 2017). Digital nomads, who have garnered attention in tourism literature in recent years due to their tourism potential, represent a valuable asset for destinations. Particularly with their allure, accessibility, tourism infrastructure, events, and climatic conditions, they offer a significant alternative for tourist destinations providing diverse types of tourism services in current circumstances.

The prominent destination selection criteria for digital nomad tourists include the cost of living, internet infrastructure, climate, entertainment, accommodation options, digital nomad visas, and shared living and working spaces. However, special consideration should be given to infrastructure for social networks such as shared living and working spaces, and other emerging needs that differ from traditional tourism. Local communities capable of providing professional and emotional support to help digital nomads integrate, perform, and succeed in their new environments constitute one aspect of developing social capital (networks and collaborations) among digital nomads and other institutions (Hall et al., 2019). On the other hand, considering digital nomads' activities on online platforms, it is crucial for potential digital nomad destinations to be visible on online digital nomad platforms and to engage in online promotional activities. Furthermore, unlike traditional tourism, digital nomad tourists may lack access to the healthcare services they need while living away from home. Understanding these challenges and incorporating features into the design that can help alleviate these issues make destinations more attractive to digital nomads (Hall et al., 2019).

Another significant criterion that warrants emphasis for digital nomad destinations is the provision of shared living and co-working spaces. This innovative approach to accommodation management, known as co-living, has emerged to cater to the lodging needs of digital nomads in tourism destinations. Co-living differs from traditional hotel management in terms of room design and the configuration of shared areas. These spaces offer personalized living and working zones, communal workspaces, shared or private kitchen facilities with food and beverage services, as well as various activities, sports, and cultural amenities alongside accommodation. The proliferation of co-living spaces, managed by community managers, which foster social interactions and cultivate environments conducive to creativity and inspiration, is on the rise (Kocaman, 2021b). An examination of the features and attributes of shared living spaces and co-living facilities preferred by digital nomads can assist destinations in understanding the prerequisites for developing such establishments.

After considering all these points, it becomes evident that there is a pressing need to enhance and expand research on digital nomads within the realm of tourism literature. Further exploration into topics such as destination selection criteria, motivations, work-leisure dynamics, and shared living-coworking spaces among digital nomads will not only enrich tourism literature but also provide valuable insights for the tourism industry. As a relatively novel research area in tourism, the majority of studies are descriptive and qualitative. Increasing the quantity of such studies and developing measurement tools for quantitative research could greatly benefit the field. Moreover, it is crucial to identify and categorize the characteristics of digital nomad destinations for effective tourism planning and marketing endeavors.

The suggestion to explore the inclination of digital nomad tourists towards adopting sharing economy-based lifestyles for sustainable tourism could assist destinations in formulating policies that prioritize sustainability to attract such tourists (Kocaman, 2022b). Finally, conducting a thorough analysis of the issues and obstacles encountered by digital nomad tourists during their journeys may enable destinations to enhance their readiness and refine their services to effectively tackle these challenges.

References

Astrama, I. and Darsana, I. (2023). Pengelolaan guest house di ubud bali sebagai pengembangan nomadic tourism. *Journal on Education*, 5(3), 8960–8970. https://doi.org/10.31004/joe.v5i3.1694

Bartosik-Purgat, M. (2018). Digital Nomads–Entrepreneurship or New Lifestyle?. *Przedsiębiorczość i Zarządzanie*, 19(2.1), 259–266.

Bonneau, C., Aroles, J., & Estagnasié, C. (2023). Romanticization and monetization of the digital nomad lifestyle: The role played by online narratives in shaping professional identity work. *Organization*, 30(1), 65–88.

Brown, K., & White, S. (2019). The Impact of Coliving on Destination Loyalty Among Digital Nomads. *Tourism Management*, 24(3), 187–201.

Chevtaeva, E., & Denizci-Guillet, B. (2021). Digital nomads' lifestyles and coworkation. *Journal of Destination Marketing and Management*, 21, 100633. https://doi.org/10.1016/J.JDMM.2021.100633.

Chevtaeva, E. (2020). Coworking and Coliving: The Attraction for Digital Nomad Tourists. *Information and Communication Technologies in Tourism* 2021, 202–209. https://doi.org/10.1007/978-3-030-65785-7_17.

Cook, D. (2023). What is a digital nomad? Definition and taxonomy in the era of mainstream remote work. *World Leisure Journal*, 65(2), 256–275.

Cook, D. (2020). The freedom trap: digital nomads and the use of disciplining practices to manage work/leisure boundaries. *Information Technology and Tourism*, 22, 355–390.

Düzgün, E., & Akin, A. (2022). A New Tourist Typology Proposal: Digital Nomad. Ahi Evran University Journal of Social Sciences Institute. https://doi.org/10.31592/aeusbed.1095735.

Floricic, T. & Pavia, N. (2021). Linkage of Leisure with Remote Work and Digital Nomadism in Tourist Accommodation Facilities. *Tourism in South East Europe*. 6, 247–268.

Green, P. (2020). Disruptions of Self, Place, and Mobility: Digital Nomads in Chiang Mai, Thailand. *Mobilities*, 15(3), 431–445.

Hall, G., Sigala, M., Rentschler, R. & Boyle, S. (2019). Motivations, Mobility, and Work Practices; The Conceptual Realities of Digital Nomads. In Information and Communication Technologies in Tourism (pp. 437–449). Springer, Cham.

Handizar, D. S., & Kusuma, S. (2023). Influence of The Experience of Digital Nomad Tourists On Destination Loyalty In Canggu, Badung District, Bali. International Journal of Sustainable Competitiveness on Tourism. https://doi.org/10.34013/ijscot.v2i02.1366.

Hermann, I. & Paris, C. M. (2020). Digital Nomadism: the nexus of remote working and travel mobility. *Information Technology & Tourism*. 22, 329–334.

Kocaman, S. (2022a). Co-working Spaces. Encycl. Tour. Manag. Mark. Edw. Elgar Publ, 25, 656–659.

Kocaman, S. (2022b). Digital Nomad Research in Tourism: A Systematic Literature Review. *Tourism Academic Journal*, 9(2), 447–462.

Kocaman, S. (2021a). Classification of the Most Preferred Destinations in 2020 By Digital Nomads in Nomad List According to Destination Selection Criteria. *Journal of Turkish Tourism Research,* 5(2), 799–815.

Kocaman, S. (2021b). “The New Accommodation Trend of Digital Nomad Tourists: CoLiving Hotels”. 2nd Managing Tourism Across Continents (MTCON21), April 1–3, 2021. ISBN 978–1-955833–02–8 Research Papers.

Kong, D., Schlagwein, D., & Cecez-Kecmanovic, D. (2019). “Issues In Digital Nomad-Corporatework: An Institutional Theory Perspective”. In Proceedings of the 27th European Conference on Information Systems (ECIS), Stockholm & Uppsala, Sweden, June 8–14, 2019. ISBN 978–1-7336325–0-8 Research Papers.

Makimoto, T. & Manners, D. (1997). Digital nomad. Wiley, Chichester.

Mariati, S., Gilitasha, A., Rahmanita, M., Djati, S., & Adawiyah, R. (2023). Analysis of Sustainable Tourism Destination Development for Digital Nomads (Comparative Study: Lisbon-Portugal and Canggu, Bali-Indonesia). *Technium Social Sciences Journal.* https://doi.org/10.47577/tssj.v43i1.8773.

Mouratidis, G. (2018). Digital Nomadism Travel, Remote Work and Alternative Lifestyles. Master of Applied Cultural Analysis. Lund University.

Müller, A. (2016). The digital nomad: Buzzword or research category?. *Transnational Social Review*, 6(3), 344–348.

Nash, C., Jarrahi, M. H., Sutherland, W., & Phillips, G. (2018, March). Digital nomads beyond the buzzword: Defining digital nomadic work and use of digital technologies. In International conference on information (pp. 207–217). Cham: Springer International Publishing.

NomadList (2024). About Nomad List. https://nomadlist.com/faq#traffic

Olga, H. (2020). In search of a digital nomad: defining the phenomenon. Information Technology & Tourism, 1–19. https://doi.org/10.1007/s40558-020-00177-z.

Orel, M. (2021). ‘Life is better in flip flops. Digital nomads and their transformational travels to Thailand’, *International Journal of Culture, Tourism and Hospitality Research*, 15(1), 3–9.

Orel, M. (2019). Coworking environments and digital nomadism: balancing work and leisure whilst on the move. World Leisure Journal, 61, 215 - 227. https://doi.org/10.1080/16078055.2019.1639275.

Paris C. (2011). *Affluence, mobility, and second home ownership.* Routledge: Abingdon.

Pecsek, B. (2018). ‘Working on holiday: the theory and practice of workcation’, *Balkans Journal of Emerging Trends in Social Sciences*, 1(1), 1–13.

Putra, G. B. & Agirachman, F. A. (2016). Urban Coworking Space: Creative Tourism in Digital Nomads Perspective. In Proceedings of Arte-Polis 6 international conference (pp. 169–178).

Poulaki, I., Mavragani, E., Kaziani, A., & Chatzimichali, E. (2023). Digital Nomads: Advances in Hospitality and Destination Attractiveness. *Tourism and Hospitality*. https://doi.org/10.3390/tourhosp4030030.

Reichenberger, I. (2017). Digital nomads–a quest for holistic freedom in work and leisure. *Annals of Leisure Research*, 21(3), 364–380.

Sánchez-Vergara, J. I., Orel, M., & Capdevila, I. (2023). "Home office is the here and now." Digital nomad visa systems and remote work-focused leisure policies. *World Leisure Journal*, 65(2), 236–255.

Sari, R. and Suarmana, I. (2022). Trends of nomadic tourism using campervan as a tourism attraction for millennials to travel in Bali. *TRJ Tourism Research Journal*, 6(2), 248. https://doi.org/10.30647/trj.v6i2.182

Schlagwein, D., & Jarrahi, M. H. (2020). *The Mobilities of Digital Work: The Case of Digital Nomadism*. In ECIS, Marrakesh, Morocco.

Statista (2024). Digital Nomad Statistics. https://www.statista.com/study/111371/digital-nomads/

Tomeljenović, R., Marušić, Z., & Gjurašić, M. (2023). Digital Nomads in the Mediterranean Tourism Hub. *Tourism in Southern and Eastern Europe*. 7, 425–439.

Thompson, B. Y. (2019). The digital nomad lifestyle: (remote) work/leisure balance, privilege, and constructed community. *International Journal of the Sociology of Leisure*, 2(1–2), 27–42. https://doi.org/10.1007/s41978-018-00030-y

Thompson, B. Y. (2018). Digital Nomad: Employment in The Online GIG Economy. *Journal of Culture, Politics and Innovation*, 1(1), 1–26.

Urry J. (2007). *Mobilities*. SAGE: London.

Vagena, A., Latyshev, O., & Papakonstantinidis, L. (2023). Greece as an Organic Space for the Development of Digital Nomadistry. Qeios. https://doi.org/10.32388/d7nx7l.

Vagena, A. (2021). Digital Nomads and Tourism Industry. *Academia Letters*, Article 765. https://www.academia.edu/45662229/Digital_Nomads_and_Tourism_Industry?auto=citations&from=cover_page

Valenduc, G. (2019). New forms of work and employment in the digital economy. In A. Serrano Pascual & M. Jepsen (Eds.), The deconstruction of employment as a political question (pp. 63–80). Cham: Palgrave Macmillan.

Von Zumbusch, J. S. H., & Lalicic, L. (2020). The role of co-living spaces in digital nomads' well-being. *Information Technology & Tourism*, 22(3), 439–453.

Wang, B., Schlagwein, D., Cecez-Kecmanovic, D., ve Cahalane, M. (2019). Digital nomadism and the market economy: resistance and compliance. International conference on information systems (ICIS), Munich.

Wiranatha, A., Antara, M., Wiranatha, A., Piartrini, P., Pujaastawa I. & Suryawardani, G. A. (2020). Dıgıtal Nomads Tourısm in Bali. *Journal of Development Economics and Finance*, 1(1), 1–16.

Winarya, S.; Raith Petriwi, P. The Digital Nomad Tourist Motivation in Bali: Exploratory Research Based on Push and Pull Theory. Athens J. Tour. 2020, 7, 161–174.

Wong, C. (2020, July 31). From old hotels to refreshed co-living. https://www.ppsintl.com/post/from-hotel-to-coliving.

Zerva, K., Huete, R., & Segovia-Pérez, M. (2022, May). Digital nomad tourism: The experience of living at the destination. In International Conference on Modern Trends in Business Hospitality and Tourism (pp. 15–26). Cham: Springer International Publishing.

Ahmet Erdem[†] and Mahmut Barakazi[‡]

Chapter 28 Earthquakes, Tourism, and Kahramanmaraş: Technological Solutions in the Reconstruction Process[*]

Introduction

The number and impact of natural disasters occurring on Earth have notably increased in recent times. One of the most prevalent disasters is earthquakes. 'An earthquake is a natural disaster caused by sudden movements in the earth's crust.' During these movements, a significant amount of energy is released, resulting in major tremors on the earth's surface (Disaster and Emergency Management Presidency, 2019). Despite advancements in technology, earthquakes occur without precise prediction of time and location, causing loss of life and property damage. Considering that the raw material of tourism consists of natural, cultural, and historical assets, it can be said that tourism and natural disasters are interconnected elements. Particularly due to the high degree of uncertainties associated with natural disasters, they are generally unpredictable and uncontrollable, necessitating careful consideration for the tourism sector (Xu, Zhu, Qu, & Ma, 2023).

After the occurrence of a high magnitude earthquake in any destination, tourist businesses suffer damage. Especially, the compromised factor of trust leads to a decrease in visitor numbers and a decline in tourism revenues (Tsai & Chen, 2010). Developing strategies for rapid and effective response after disasters plays a vital role in creating resilience in destinations. Natural disasters can affect tourism operations and development. Additionally, tourist attractions and even entire destinations may be permanently destroyed, leading to temporary disruptions in the tourism industry (Chan, Nozu, & Ting, 2020). Spatial impact

---

* The abstract of the study was presented at the 'I. International İstiklal Symposium' held in Kahramanmaraş in February 2024.

† PhD., Harran University, Tourism Faculty, Department of Tourism Management, ahmeterdem@harran.edu.tr

‡ PhD., Harran University, Tourism Faculty, Gastronomy and Culinary Arts, mahmutbarakazi@harran.edu.tr

is a significant factor affecting the transfer of tourism development to economic growth (Zhang, 2023).

The earthquakes referred to in the literature as the 'Kahramanmaraş earthquakes' or the 'February 6 Earthquakes' occurred in the districts of Pazarcık and Elbistan in Kahramanmaraş on February 6, 2023. These earthquakes, with approximately 9 hours apart (at 04:17 and 13:24 Turkish local time), had magnitudes of Mw 7.7 (Pazarcık) and Mw 7.6 (Elbistan). According to official sources, these earthquakes resulted in the loss of 50,783 lives, with 107,204 people injured, and a total of 311,000 buildings comprising 872,000 independent units rendered unusable (TMMOB, 2023). Institutions responsible for preserving the attractions in destinations were also affected by the losses incurred after these earthquakes. Many museums, libraries, historical buildings, and archaeological sites were destroyed (Arıcan, 2023). The question of how tourism destinations will recover after crises is a fundamental issue in tourism crisis management (Xu et al., 2023).

The restoration of tourism to its former state in the region, and making the city attractive to both domestic and international tourists, depends on the reconstruction of tourism service and infrastructure, and the seamless conduct of activities. In this regard, alongside addressing the damage to tourism supply resources in the destination, a comprehensive tourism perspective should be developed with the participation of all stakeholders during the region's redevelopment. With this vision in place, the destination can achieve better infrastructure than before (Presidency of Strategy and Budget, 2023). A series of policies, measurements, and programs are required to enhance the physical, social, economic, political, and ecological recovery and livelihood resilience of the region (Zhang, 2023).

In this section, the aim is to present technology-based solution proposals for the reconstruction process of tourism areas in Kahramanmaraş following the February 6 Earthquakes. Firstly, by examining the current situation of the destination, the framework for investments and planning to be made during the reconstruction process is attempted to be outlined within the context of technology. Integrating technology into tourism attractions and transportation infrastructure in Kahramanmaraş, particularly through the establishment of real-time data collection and analysis systems, enhancing customer experience, conducting risk assessments, preparing emergency disaster management plans, and increasing destination resilience, will provide significant benefits. When examining the existing national and international literature in terms of the subject and approach of the study, it is observed that studies focusing on earthquakes, tourism, and technology are quite limited, which enhances the originality of the study. Furthermore, the recommendations presented in this section also provide guidance to practitioners. Therefore, it is possible to say that the research yields significant results both in theory and in practice.

Literature Review

After the disaster referred to as the February 6 earthquakes, various studies have been conducted, with Kahramanmaraş as the epicenter. A literature review revealed that the majority of studies conducted after the earthquake were conceptual and qualitative in nature. The research generally focused on topics such as situational analysis of cities, recommendations for reconstruction processes, economic impacts, media reflections, and effects on tourism businesses.

In a study conducted by Göktaş (2023), the damage caused by the earthquakes centered in Kahramanmaraş in 11 provinces was examined. The study revealed that numerous historical structures were damaged, with particularly significant destruction observed in places of worship. Recommendations for short, medium, and long-term actions were proposed in the study regarding the topic. Karademir (2023) has suggested creative tourism as a means to mitigate the damage caused by the earthquake. In a study conducted specifically in the Andırın district of Kahramanmaraş, it was highlighted that there are options for creative tourism activities in the region, but there is a lack of infrastructure, investment, and incentives. In another study, it has been suggested that the earthquake-affected region could be evaluated within the scope of mourning tourism (Çolakoğlu, Temizel, & Temizel, 2023). The historical and cultural inventories within the Kahramanmaraş Archaeology Museum, Elbistan City Museum, and the Panorama Museum of the Independence Epic, all significant museums in the city, have not been damaged, and no losses have been incurred (According to data obtained from the Ministry of Culture and Tourism, General Directorate of Cultural Heritage and Museums, Protection Application and Inspection Offices (KUDEB)). However, the physical structures of the complexes have survived the earthquake with minor damage (Arıcan, 2023).

Following the earthquakes on February 6th, a study examining Southeastern Anatolia in terms of tourism highlighted a decrease in the number of tourists, indicating that, overall, there were no issues at museums and archaeological sites. It also pointed out that the duration of tours has shortened compared to previous years (Gök, 2023). Bayram and Çiftçi (2021), who examined the 1999 earthquake, highlighted that the cancellation of package tours for foreign tourists in Turkey after the earthquake had a negative impact on travel agencies, hotels, and airlines. They also stated that if there hadn't been an earthquake, the future number of tourists would have been approximately 8 % higher than the actual number of tourists visiting Turkey at that time. In both local and national media, effective communication strategies that particularly overcome security perceptions should be preferred (Aygün & Ünüvar, 2023). Regarding the social media reflections of the earthquake, a two-sided evaluation has been conducted considering both positive and negative effects. Especially, the transmission of incorrect and misleading information about the region has been mentioned as a deterrent factor for tourists in terms of making purchases (Korkmaz Orhan, Özdemir Akgül, & Cabi Bilge, 2023).

In another study evaluating the economic impacts of earthquakes, it was stated that the material damage of the earthquake amounted to 34.2 billion dollars, affecting 13.4 million people. Additionally, the total direct and indirect cost of the earthquake could be 103.6 billion dollars according to calculations by the Presidency of the Republic of Turkey Strategy and Budget Directorate. Furthermore, it was anticipated that various macroeconomic conditions such as inflation, budget deficits, and disruptions in the balance of payments could arise.

Lastly, in terms of tourism supply in the region, 23.8 % of the total tourism supply of Turkey and 1 % of Turkey's total tourism supply encountered issues. Although it constitutes a small portion on a national scale, it represents a significant part of the region that has been affected (Kara & Sezgin, 2023). In a study conducted regarding the effects of the earthquake on Manavgat, an important holiday destination, interviews were conducted with managers, and content analysis was performed. The managers stated that visitors had security concerns, there were reservation cancellations, and the season started late (İlbay Vatan & Vatan, 2023). However, especially after the initial shock was overcome, tourist mobility returned to normal. Turkey welcomed 56.7 million visitors in 2023, marking a 10 % increase compared to the previous year, and achieved a tourism revenue of 54.3 billion dollars, indicating a 17 % increase compared to 2022. At this point, international news agencies reported that the earthquake was particularly distant from popular destinations and would not pose any problems for tourists choosing Turkey, thus minimizing the impact of the earthquake. However, various agencies highlighted the existence of various restrictions and travel bans in the earthquake-affected region (Aygün & Ünüvar, 2023).

Research Area: Kahramanmaraş

Kahramanmaraş, with a total area of 14,346 km^2 (59.7 % mountains, 24 % plateaus, and 16.3 % covered by plains), is the 11th largest province in Turkey

(Kahramanmaraş Governorship, 2024). As of 2022, Kahramanmaraş has a total of 561 immovable cultural assets requiring protection and 313 conservation areas. In the UNESCO World Heritage Temporary List, Eshab-ı Kehf (2015) from Kahramanmaraş is included. The Cave and Complex of Eshab-ı Kehf, located in the Afşin district, holds great importance for religious tourism due to its widespread recognition among different religions and cultures (General Directorate of Cultural Heritage and Museums, 2023c). Table 1 lists the immovable cultural assets and conservation areas of Kahramanmaraş.

Table 1. Immovable Cultural Assets and Conservation Areas of Kahramanmaraş

Cultural Assets	**Number**
Monuments and Memorials	9
Administrative Buildings	5
Cultural Structures	58
Martyrdoms	1
Military Buildings	25
Industrial and Commercial Buildings	41
Religious Structures	46
Cemeteries	32
Examples of Civil Architecture	327
Remains	17
Total	**561**
Conservation Areas	**Number**
Urban Conservation Area	3
Historical Conservation Area	1
Archaeological Conservation Area	309
Total	**313**

Source: General Directorate of Cultural Heritage and Museums, 2023a.

When examining the Museums and Affiliated Units Temporarily Closed, it is observed that Kahramanmaraş Museum and its affiliated units were closed as of February 6, 2023, with an estimated reopening date stated as "until the completion of works." Additionally, another important tourism attraction, the Germanicia Mosaic Area in Kahramanmaraş, has been closed as of May 5, 2021. The latest update for museum-related information is dated January 12, 2024 (General Directorate of Cultural Heritage and Museums, 2023b). The mosaics located in the Germanicia Mosaic Area possess unique characteristics with their quality craftsmanship, depictions, and diverse themes and iconography (Küçükdağlı, 2012). However, it has been noted that among the 56 museums offered on the Ministry of Culture and Tourism's website as virtual museums, there is no museum from Kahramanmaraş.

The 2020–2024 Kahramanmaraş Tourism Strategy and Action Plan, conducted by the Eastern Mediterranean Development Agency, includes an assessment of the city's current status, potential, identification of priority areas, and recommendations. Upon reviewing the report, it is noted that tourist safety and the city's tourism attractions have the highest average scores, whereas guidance services for visitors, promotion, entertainment facilities, and information services have the lowest average scores (Eastern Mediterranean Development Agency, 2020). Especially in the conclusion section of the report, necessary recommendations have been provided, particularly regarding guidance and information services. News related to the tourism supply resources, safety, accessibility, and tourism infrastructure in cities located in earthquake zones is crucial in both national and international channels (Aygün & Ünüvar, 2023).

Lastly, when examining the technological applications in Kahramanmaraş, it is observed that the website provided by Kahramanmaraş Metropolitan Municipality serves as a digital inventory of the city's culture and tourism. However, it can be said that information regarding tourism supply resources in Kahramanmaraş has not been updated following the earthquake. For instance, when clicking on the "One Day in Kahramanmaraş" tab under the "My Routes" section, destinations such as Kahramanmaraş Museum, Kahramanmaraş Castle, and the Museum of Seven Beauties of Literature are listed. However, as of February 6, 2023, these places have been closed to visitors, and there has been no official announcement regarding when they will reopen. Internet websites, due to their dynamic structure, offer significant advantages in terms of updating information instantly. From a mobile application perspective, the Digital Culture Inventory mobile application also has some shortcomings in sharing up-to-date information about attractions, as mentioned above. Similarly, the Gezede mobile application does not provide current information about tourism attractions in the destination (as of January 30, 2024). It has been reported that the Kahramanmaraş Metropolitan Municipality provides WiFi service (Erdem, 2022); however, the locations where this service is available have not been determined. Considering that technology has entered all aspects of life, providing free WiFi service at tourism attractions is crucial for enhancing visitor experience.

Results

Earthquakes can result in damage to tourism supply resources and businesses. Additionally, employees and local residents who earn income from the sector can be negatively affected. To mitigate all the adverse effects of earthquakes, proper plans and programs should be implemented. Particularly, post-disaster recovery, identification of the needs of major tourist attractions that have been damaged or destroyed, planning restoration efforts, and re-marketing initiatives are crucial for the destination's image (Göktaş, 2023). During the reconstruction process, technology plays a significant role. The use of smart technologies in shaping tourism infrastructure and superstructure will help alleviate the damage caused by the earthquake and lead to the destination's improved position. Investments made will not only focus on a single sector or economic approach but will also have an impact on residents' living conditions, social welfare, and employment (Zhang, 2023).

Smart infrastructure investments should not only focus on technological innovations but also on governance, economy, social factors, and environmental sustainability. For example, smart transportation systems should be considered as part of sustainable transportation policies. Specifically, measures such as promoting public transportation, building bicycle lanes, reducing vehicle traffic, and expanding pedestrian walkways should be supported (Nazlı & Gök, 2023). The establishment of a smart ecosystem in the city will facilitate planning and decision-making processes. As a result of the evaluation conducted in the study, recommendations have been provided under five dimensions: tourism-travel, energy-waste, advertising-promotion, infrastructure-superstructure, and others. Especially, with the participation of all local tourism stakeholders, a Tourism Sector Roadmap for Kahramanmaraş should be created, and all activities should be carried out within a comprehensive planning framework. Within this framework, the roadmap should be divided into short, medium, and long-term goals as part of the strategic plan. The responsibilities of the involved activities should be allocated to relevant institutions, and a monitoring mechanism (e.g., a commission established under the leadership of the Governorship) should oversee the process. This way, the transition from traditional urban planning to smart city investments (in areas such as living, transportation, economy, health, and governance) should be initiated. The recommendations presented in the scope of the research are as follows:

Tourism and Travel

- Free WiFi infrastructure should be installed in all tourist areas in Kahramanmaraş (including Kahramanmaraş Castle, Grand Bazaar, Kahramanmaraş Archaeology Museum, etc.).
- When planning tourism areas, access to all tourist attractions in the city should be provided under the scope of accessible travel.
- Navigational information tailored to the destination should be provided to tourists through mobile travel applications supported by artificial intelligence (such as ChatGPT).
- Travel routes should be updated according to the new city plan of Kahramanmaraş, and 3D modeling of these routes should be created.
- Virtual museums should be established for museums in Kahramanmaraş (Elbistan City Museum, Germanicia Archaeological Site, Ethnography Museum, Liberation Epic Panorama Museum, Museum of Seven Beauties of Literature, Ice Cream Museum, Thematic Kitchen Museum, Intangible Cultural Heritage Museum).
- Kiosks capable of providing services in various languages about tourist areas should be placed in the city center.
- Effective planning and establishment of intra–city transportation network should be carried out for tourist attractions in the destination (especially increasing connectivity options between the airport and bus terminal).
- To create a focal point for tourists, a strategic street should be designated as a tourism/prestige street where shopping and cultural products can be showcased (For example, an area including Trabzon Street, Cyprus Square, and Kahramanmaraş Castle). When planning the tourism/prestige street, special attention should be given to providing an area where tourist buses can park without entering the city center, moving walkways to facilitate the transportation of disabled and elderly visitors within the street, and shelters provided for the convenience of tourists.
- Historical events can be showcased or information about attractions can be diversified and enriched using hologram, virtual reality, and augmented reality applications, especially in attractions like Kahramanmaraş Castle and Germanicia Archaeological Site.
- A Kahramanmaraş virtual guiding application should be prepared to assist tourists during their travels and provide services in various languages.
- Retro concerts (such as a concert by Barış MANÇO) can be organized in the amphitheater area at the Expo site using hologram technology.

Energy and Waste

- Sustainable energy sources should be utilized when planning tourism areas.
- Sensor-controlled lighting should be used in tourist areas and city centers.
- An intelligent sidewalk system that converts energy should be installed in areas heavily visited by tourists and locals. The energy generated can be used for street lighting.
- Smart charging stations should be installed in areas heavily visited by tourists.
- Recycling of waste, especially food waste, generated in tourist areas should be ensured.
- Walkways should be expanded and bicycle lanes should be constructed for environmental sustainability.
- Buildings such as museums, archaeological sites, and accommodation facilities should have automatic fire alarms, automatic water cutoff systems, smart lighting systems, and smart heating and cooling systems.

Advertising and Promotion

- Advertising, promotion, and public relations activities should be implemented systematically, and an urgent action plan should be prepared through stakeholders (the perception that Maraş is destroyed should be eliminated).
- During the reconstruction process, the "Smart Tourism Destination Kahramanmaraş" label should be utilized in the product differentiation process thanks to the technological investments to be made.

Infrastructure and Superstructure

- Emergency warning systems should be developed through sensors and smart cameras installed especially in tourism facilities (For example, in the event of a possible flood, flood danger can be identified in advance through water level sensors).
- A Earthquake Museum equipped with technology reflecting the traces of earthquakes should be established for visitors coming to Kahramanmaraş.
- Emergency buttons should be available in tourist areas.
- In the city center, smart signaling application should be implemented to ensure rapid arrival of especially ambulances, fire brigades, and security personnel to the scene.
- Tourism resources should be planned in a resilient manner against natural disasters, and in case of possible disaster situations, they should exhibit a feature where disaster victims can take refuge.

- Considering the issues experienced globally, especially in the face of chemical attacks, special fortified shelters should be built at certain points in the city.

Other Suggestions

1. A Tourism Sector Roadmap for Kahramanmaraş should be developed.
2. A destination management organization (such as the Kahramanmaraş Tourism Management Center) involving all tourism stakeholders should be established.
3. Emergency training should be provided to tourism employees.
4. Evacuation plans for tourist areas in case of emergencies should be prepared for all tourist attractions in Kahramanmaraş.

References

Arıcan, B. (2023). Deprem Sonrasında Bölgede Bulunan Müzelerin Durumu: 2023 Kahramanmaraş Depremi, Gaziantep ve Kahramanmaraş Değerlendirmesi. Deprem ve Turizm (Maraş Depremi Sonrası Turizm Sektörü Değerlendirmeleri) Inside, Ed. Prof. Dr. Çağatay Ünüsan, Prof. Dr. Mete Sezgin, Öğr. Gör. Abdullah Köseoğlu. Eğitim Yayınevi

Aygün, Y. & Ünüvar, Ş. (2023). Yabancı Basında Deprem Haberleri Yansımaları ve Turizme Etkileri. İçinde Deprem & Turizm. Ed. Ünüsan, Ç., Sezgin, M. & Köseoğlu, A. Eğitim Yayınevi. 119–128.

Bayram, S. & Çiftçi, G. (2021). The Effects of Earthquakes on Tourism: Evidence from Turkey. Journal of Tourism Leisure and Hospitality, 3(2), 82–94.

Chan, C. S., Nozu, K., & Ting, L. C. (2020). Tourism and natural disaster management process: perception of tourism stakeholders in the case of Kumamoto earthquake in Japan. Current Issues in Tourism, 23(15), 1864–1885. https://doi.org/10.1080/13683500.2019.1666809

Çolakoğlu, M. A., Temizel, G. & Temizel, M. (2023). 6 Şubat 2023 Kahramanmaraş Depremlerinin Hüzün Turizmi Kapsamında Değerlendirilmesi. Inside Deprem ve Turizm. Ed. Ünüsan, Ç., Sezgin, M. Ve Köseoğlu, A. Eğitim Yayınevi.147–162.

Disaster and Emergency Management Presidency. (2019). What is an Earthquake? Access address: https://www.afad.gov.tr/deprem-nedir (Accessed: 30.01.2024).

Eastern Mediterranean Development Agency. (2020). Kahramanmaraş Tourism Strategies and Action Plan (2020–2024). Access address: https://www.dogaka.gov.tr/assets/upload/dosyalar/dogaka_kahramanmaras_tsep_web.pdf (23.01.2024).

Erdem, A. (2022). Akıllı Şehirler Gerçekten Akıllı Turizm Destinasyonları Mı?. Safran Kültür ve Turizm Araştırmaları Dergisi, 5(1), 5–26.

Gök, B. (2023). Asrın Felaketi Sonrası Güneydoğu'da Turizm. İçinde Asrın Felaketi Sosyoekonomik Etkileri. Ed. Erdoğan Tarakçı, İ. & Aslan R. Efe Akademi Yayınları.

Göktaş, L. S. (2023). Kahramanmaraş Merkezli Depremlerin Turizm Sektörüne Etkisi Ve Deprem Sonrası Atılması Gereken Adımlar Hakkında Öneriler. Journal of Gastronomy, Hospitality and Travel, 2023, 6(2), 624–635 Doi: 10.33083/joghat.2023.288

İlbay Vatan, B., & Vatan, R. O. (2023). Kahramanmaraş Merkezli Depremlerin Otel İşletmelerine Potansiyel Etkilerinin Otel Yöneticileri Perspektifinden Değerlendirilmesi: Manavgat İlçesi Örneği. Journal of Tourism & Gastronomy Studies, 11(3), 2031–2048.

Kahramanmaraş Governorship, 2024. İlimiz Hakkında. Access address: https://kahramanmaras.csb.gov.tr/ilimiz-hakkinda-i-824#:~:text=%C4%B0lin%20Co%C4%9Frafi%20Konumu%3A,ilin%20kuzey%20kesimleri%20olduk%C3%A7a%20da%C4%9Fl%C4%B1kt%C4%B1r. (30.01.2024)

Kara, G. & Sezgin, M. (2023). Kahramanmaraş Merkezli Depremlerin Türkiye'nin Turizm Ekonomisi Üzerindeki Etkisi. inside Deprem ve Turizm. Ed. Ünüsan, Ç., Sezgin, M. ve Köseoğlu, A. Eğitim Yayınevi. 21–36.

Karademir, N. (2023). Yaratıcı Turizm Uygulamaları ile Andırın'da Deprem Yaralarını Sarabiliriz. Kahramanmaraş Sütçü İmam Üniversitesi Sosyal Bilimler Dergisi, 20(3), 1006–1020.

Korkmaz Orhan, B., Özdemir Akgül, S. & Cabi Bilge, A. (2023). Turizm Sektöründe Deprem Haberlerinin Sosyal Medyada Yansıması. inside Deprem ve Turizm. Ed. Ünüsan, Ç., Sezgin, M. & Köseoğlu, A. Eğitim Yayınevi. 37–52.

Küçükdağlı, S. (2012). Kahramanmaraş Germanicia Mozaikleri. Journal of Mosaic Research (5), 97–101.

General Directorate of Cultural Heritage and Museums, 2023a. Statistics of Immovable Cultural Assets to be Protected by Provinces. Access address: https://kvmgm.ktb.gov.tr/TR-44799/illere-gore-korunmasi-gerekli-tasinmaz-kultur-varligi-istatistigi.html (11.01.2024).

General Directorate of Cultural Heritage and Museums, 2023b. Temporarily Closed Museums and Affiliated Units. Access address: https://kvmgm.ktb.gov.tr/TR-46782/gecici-sureyle-kapali-olan-muzeler-ve-bagli-birimler.html (16.01.2024).

General Directorate of Cultural Heritage and Museums, 2023c. Access address: World Heritage Tentative List. https://kvmgm.ktb.gov.tr/TR-44395/dunya-miras-gecici-listesi.html (16.01.2024).

Nazlı, T. & Gök, T (2023). Akıllı Şehirler ve Deprem. Deprem ve Turizm (Maraş Depremi Sonrası Turizm Sektörü Değerlendirmeleri) inside, Ed. Prof. Dr. Çağatay Ünüsan, Prof. Dr. Mete Sezgin & Öğr. Gör. Abdullah Köseoğlu. Eğitim Yayınevi

Presidency of Strategy and Budget (2023). 2023 Kahramanmaraş ve Hatay depremleri raporu. Access address: https://www.sbb.gov.tr/wp-content/uploads/2023/03/2023-Kahramanmaras-ve-Hatay-Depremleri-Raporu.pdf (11.01.2024).

TMMOB. 2023. TMMOB 6 Şubat Depremleri 8. Ay Değerlendirme Raporu. Access address: https://www.tmmob.org.tr/icerik/tmmob-6-subat-depremleri-8-ay-degerlendirme-raporu-yayimlandi (01.02.2024).

Tsai, C. H., & Chen, C. W. (2010). An earthquake disaster management mechanism based on risk assessment information for the tourism industry-a case study from the island of Taiwan. Tourism Management, 31(4), 470–481.

UNESCO World Heritage Centre. (2015). Eshab-ı Kehf Kulliye (Islamic-Ottoman Social Complex). Retrieved from https://whc.unesco.org/en/tentativelists/?action=listtentative&state=tr&order=states.

Xu, Q., Zhu, G., Qu, Z., & Ma, G. (2023). Earthquake and Tourism Destination Resilience from the Perspective of Regional Economic Resilience. Sustainability, 15(10), 7766.

Zhang, J. (2023). Spatial effects of tourism development on economic resilience: an empirical study of Wenchuan earthquake based on dynamic spatial Durbin model. Natural Hazards, 115(1), 309–329.

Halil Ibrahim Karakan*

Chapter 29 Analyzing Switzerland's Policy as an Eco-Tourism Destination Concerning the United Nations' Sustainable Development Goals

Introduction

Sustainable development, as defined by the World Commission on Environment and Development (WCED, 1987), is a development approach that meets the needs of the present without compromising the ability of future generations to meet their own needs. This approach is centered on maintaining the integrity of natural systems and ensuring that human activities do not harm the climate or ecosystems. It also strongly emphasizes human development, ensuring the equitable distribution of resources necessary for well-being within and between generations (Wuelser et al., 2020).

Globalization in tourism has heightened the importance of competitiveness, leading to various definitions and approaches in this field. Many authors attribute the success of a tourism destination to its competitiveness (Dwyer and Kim, 2003; Otto and Ritchie, 1996; Ritchie and Crouch, 2003). There is a consensus within the scientific community that competitiveness is a critical element in ensuring the success of a tourism destination (Cronje and Plessis, 2020). Ritchie and Crouch (2003) argue that sustainable tourism competitiveness is only possible. They assert that for a destination to be truly competitive, it must be economically, environmentally, socially, culturally, and politically sustainable.

Similarly, Cronje and Plessis (2020) emphasize in their literature review on the competitiveness of tourism destinations that sustainability has recently gained more importance and should be researched in this context. Ritchie and Crouch identify six main aspects of a destination's competitiveness: increasing tourist spending, increasing the number of tourists, providing satisfying and memorable experiences, operating profitably, enhancing the well-being of the local population, and preserving natural resources for future generations. In this

* Dr., Gaziantep University, Associate Professor., Gaziantep University, Vocational School Of Tourism and Hotel Management, Department of Tourism Management, hikarakan@gmail.com

way, sustainability and competitiveness can coexist, creating a successful model for the tourism industry.

Eco-tourism, a sub-component of sustainable tourism, holds immense potential as a powerful tool for sustainable development. This potential serves as a beacon of hope, driving the adoption and integration of eco-tourism into the economic development and conservation strategies of developing countries (Kiper, 2013). The combined concepts of eco-tourism and sustainable development offer a promising alternative to current challenges and a means to prevent the negative impacts of mass tourism, particularly in small countries with high population density (Cidalia, 2011).

Switzerland, despite its limited surface area, boasts a robust eco-tourism market segmentation (Utama et al., 2023). This study delves into Switzerland's policies as an eco-tourism destination in relation to the United Nations' Sustainable Development Goals. Specifically, it examines the key messages in Switzerland's sustainable development policies and their alignment with the United Nations' Sustainable Development Goals.

Literature Review

Switzerland is among the leading countries aiming to balance environmental, economic, and social sustainability through its sustainable development policies. Numerous studies have been conducted in this area, and below is a literature review covering the past five years, sourced from international publications.

Huber and Schlegelmilch (2020) aimed to examine the impacts of Switzerland's agricultural policies on sustainability and evaluate their contributions to environmental and economic sustainability goals. Using case studies and economic analyses, they found that Switzerland's agricultural policies promote sustainable farming practices and significantly contribute to environmental protection. However, some farmers face financial difficulties, indicating that the policies need to balance economic and environmental sustainability goals better.

Moser and Schwager (2021) conducted an analysis of Switzerland's energy policies, focusing on their role in sustainability and climate change mitigation. Through energy policy analysis and modeling studies, they found that Switzerland's energy policies are geared towards increasing the use of renewable energy sources and have made significant strides in this direction. They also underscored Switzerland's solid commitments to reducing carbon emissions and effectively combating climate change. However, they stressed the need for a more accelerated energy transition process.

Breu et al. (2020) propose an eight-step process for creating knowledge to implement the 2030 Agenda at the national level, specifically using Switzerland as a case study. The first four empirically tested steps are: translating the 2030 Agenda to national contexts, identifying priority targets, systematically assessing SDG target interactions, and evaluating the systemic relevance of priority targets. The subsequent steps, inspired by current debates, include establishing collaboration structures, developing implementation tools, setting up monitoring and evaluation mechanisms, and adapting policies based on feedback.

Buser and Koch (2022) undertook an evaluation of Switzerland's urban planning and sustainable urban development policies, with a focus on their social and environmental impacts. Using urban planning analysis and qualitative research methods, they found that Switzerland's urban development policies are making significant contributions to sustainable urbanization goals and improving social welfare. The study particularly highlighted the importance of green space conservation and energy efficiency. However, they also pointed out the pressing need for more innovative solutions to address the challenges posed by rapid urbanization and population growth.

Meier and Baumgartner (2023) analyzed Switzerland's sustainable tourism policies and their impacts on the tourism sector. Using tourism policy analysis and case studies, they found that Switzerland's sustainable tourism policies effectively protect natural resources and support local economies. They observed that environmental sustainability practices in the tourism sector are widespread and positively received by tourists. However, they concluded that more infrastructure and resource management improvements are needed in regions with high tourism density.

Bornemann and Christen (2020) analyzed the potential of sustainability governance in implementing the 2030 Agenda for Sustainable Development in Swiss cantons. The study used policy analysis and qualitative research methods at the cantonal level. They examined four different sustainability governance arrangements representing Switzerland's first generation of sustainability governance. The study assessed how well these arrangements align with the changes envisioned by the 2030 Agenda. The results showed that Swiss cantons have innovative approaches to sustainability policies and that local governments play a crucial role in achieving the 2030 Agenda goals. However, the study also highlighted the need for increased cooperation and information sharing among cantons and the necessity of context-specific reforms for first-generation sustainability governance in line with the new Agenda.

These studies represent the recent literature on Switzerland's sustainability policies over the past five years. Overall, Switzerland's sustainability policies are supported by robust legal frameworks and practical implementations, but there are areas where additional efforts are required.

Methodology

Using a qualitative research design, this study analyses Switzerland's policies as an ecotourism destination regarding the United Nations Sustainable Development Goals (SDGs). The main focus is on examining the key messages and policies outlined in Switzerland's document, "Implementing the 2030 Agenda for Sustainable Development: Voluntary National Review of Switzerland 2022 (Priority issues for Switzerland—key messages)." In addition, relevant literature and policy documents related to Switzerland's sustainable tourism and ecotourism strategies will be reviewed.

Data analysis uses thematic analysis with MAXQDA 2020 Analytics Pro software. This process includes familiarizing oneself with the data, generating initial codes, identifying themes, reviewing themes, defining and naming themes, and reporting findings.

The study employs several methodological rigor practices to ensure the reliability and validity of the findings. Triangulation involves using multiple sources of data to cross-check findings. Peer debriefing allows for discussing findings with peers and experts to validate interpretations. An audit trail provides a detailed record of the data collection and analysis process, ensuring transparency and replicability.

Findings

The United Nations has established 17 Sustainable Development Goals (SDGs) (UN, 2015). Switzerland has grouped these development goals under three main categories. Switzerland's sustainability policies, shaped by the United Nations Sustainable Development Goals (SDGs) framework, are not just a collection of measures, but a comprehensive and holistic approach to sustainability. As depicted in Visual 1, these policies are structured into three main areas: Equal Opportunities and Social Cohesion, Sustainable Consumption and Production, and Climate, Energy, and Biodiversity (Swiss Confederation, 2022). This comprehensive and holistic approach serves as a reassuring testament to Switzerland's unwavering commitment to addressing all aspects of sustainability.

Under the category of Equal Opportunities and Social Cohesion, Switzerland's commitment to the UN's SDGs is not a solitary endeavor. It is a collaborative effort that involves ending poverty in all its forms everywhere (SDG 1: No Poverty), ensuring healthy lives and promoting well-being for all at all ages (SDG 3: Good Health and Well-being), providing inclusive and equitable quality education (SDG 4: Quality Education), achieving gender equality (SDG 5: Gender Equality), promoting decent work and economic growth (SDG 8: Decent Work and Economic Growth), reducing inequality within and among countries (SDG 10: Reduced Inequality), making cities and human settlements inclusive, safe, resilient, and sustainable (SDG 11: Sustainable Cities and Communities), and promoting peaceful and inclusive societies (SDG 16: Peace, Justice, and Strong Institutions). Switzerland's commitment to partnerships (SDG 17: Partnership for the Goals) emphasizes the inclusivity of these global sustainability efforts.

The category of Sustainable Consumption and Production focuses on strategies such as ending hunger, achieving food security, promoting sustainable agriculture (SDG 2: Zero Hunger), promoting sustained, inclusive, and sustainable economic growth and decent work for all (SDG 8: Decent Work and Economic Growth), building resilient infrastructure, promoting inclusive and sustainable industrialization, and fostering innovation (SDG 9: Industry, Innovation, and Infrastructure), ensuring sustainable consumption and production patterns (SDG 12: Responsible Consumption and Production), conserving and sustainably using the oceans, seas, and marine resources (SDG 14: Life Below Water), and partnerships for achieving these goals (SDG 17: Partnership for the Goals).

Under the category of Climate, Energy, and Biodiversity, Switzerland focuses on ensuring availability and sustainable management of water and sanitation for all (SDG 6: Clean Water and Sanitation), ensuring access to affordable, reliable, sustainable, and modern energy for all (SDG 7: Affordable and Clean Energy), making cities and human settlements inclusive, safe, resilient, and sustainable (SDG 11: Sustainable Cities and Communities), taking urgent action to combat climate change and its impacts (SDG 13: Climate Action), protecting, restoring, and promoting sustainable use of terrestrial ecosystems (SDG 15: Life on Land), and partnerships for achieving these goals (SDG 17: Partnership for the Goals).

As part of the UN's SDGs framework, Switzerland's sustainability policies are not just a collection of isolated efforts. They are a comprehensive strategy that balances social cohesion, economic development, and environmental protection. This holistic approach instills confidence in the effectiveness of Switzerland's sustainable development goals, both at national and international levels.

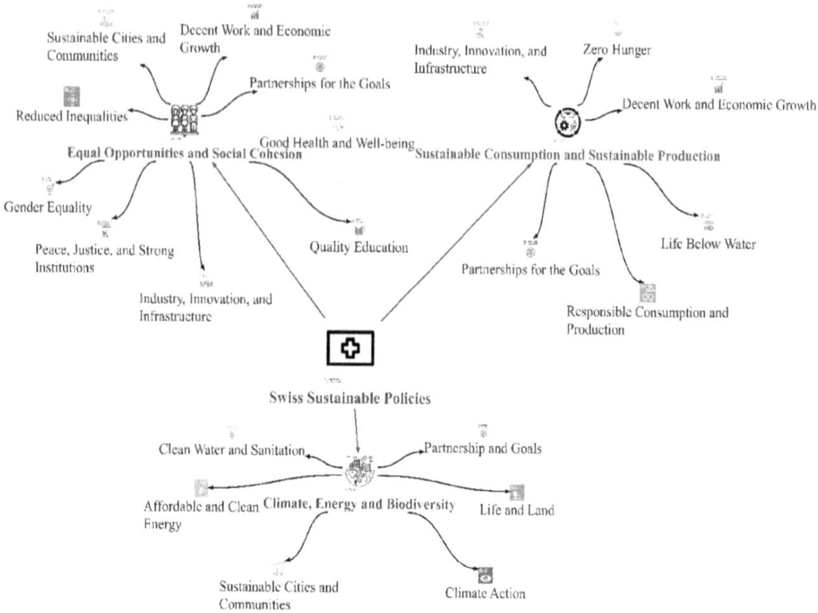

Figure 1. Results of Thematic Analysis of Swiss Sustainable Policies

Switzerland's Sustainable Consumption and Production Policy analysis delves into various themes. The theme of Sustainable Food Systems is underpinned by key codes such as the Swiss Nutrition Policy, the National Roadmap for Food Systems Transformation, and Healthy Eating and Environment. These codes are not mere guidelines but robust strategies to foster healthy and balanced eating habits, reduce environmental impacts, curb food waste, and promote sustainable agricultural practices.

The theme of Sustainable Development Strategy consists of codes like interconnected goals and policy coherence. These codes focus on reducing hunger, protecting biodiversity, and contributing to climate change mitigation through sustainable agricultural practices.

The theme of Civil Society Participation is not just a cornerstone but the driving force of Switzerland's Sustainable Consumption and Production Policy. It encompasses codes such as individual and institutional roles and effective collaboration. The Individual and Institutional Roles Code underscores individuals and organizations' pivotal and empowering contributions to sustainable development, emphasizing that this participation can be realized through awareness-raising, education, and changes in individual behavior. The Effective Collaboration Code further underscores the importance of cooperation and coordination with civil society, emphasizing its crucial and integral role in achieving sustainability goals.

The theme of Environmental Assessments includes codes such as Global Impact and Resource Efficiency. The Global Impact Code demonstrates the positive effects of Switzerland's local measures on the global environment and emphasizes the global responsibility we all share in environmental sustainability. It evaluates the global impacts of environmental sustainability, underlining the importance of our local actions in a global context. The Resource Efficiency Code encompasses efforts to minimize environmental impacts, use resources efficiently and sustainably, and ensure long-term economic sustainability, further reinforcing the global significance of our local actions.

The theme of Economic and Financial Systems is a comprehensive pillar of Switzerland's Sustainable Consumption and Production Policy. It focuses on sustainable economic activities and the role of the financial sector. The Sustainable Economic Activities Code underscores the importance of structuring economic activities to support environmental sustainability. The role of the financial sector is not just about profit but about developing financial tools and policies that support sustainable development and promoting investments aligned with environmental sustainability criteria.

Lastly, the theme of Collaborative Governance, a testament to the policy's inclusive nature, addresses public-private partnerships and multi-level collaboration. The Public-Private Partnerships Code highlights the importance of cooperation between the public and private sectors to achieve sustainable development goals, facilitating efficient resource use and innovative solutions.

Switzerland's strategy, a comprehensive and multi-faceted approach, has successfully focused on sustainable food systems, integrated development strategies, civil society participation, environmental considerations, economic and financial systems transformation, and collaborative governance. This strategy, which is not just about setting goals but also about implementing them to address their interconnectedness and the importance of a multi-faceted approach to sustainable development, has been a beacon of success. This holistic and integrated strategy has inspired and set a high standard for sustainable development, focusing on sustainable food systems, integrated development strategies, civil society participation, environmental considerations, transformation of economic and financial systems, and collaborative governance.

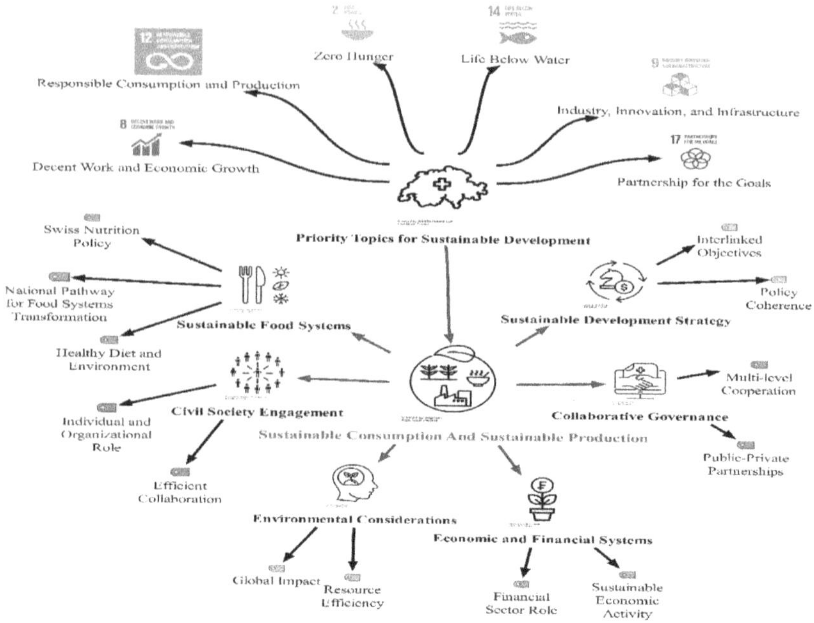

Figure 2. Results of Thematic Analysis of Sustainable Consumption and Sustainable Production Policies of Switzerland

Visual 2 shows several themes and codes under the heading of Climate, Energy, and Biodiversity. The theme of resilience to natural hazards includes the codes "Policy Measures" and "Impact of Climate Change." These codes focus on implementing policy measures to enhance resilience against natural hazards and addressing the impacts of climate change, respectively. The sustainable resource management theme encompasses the codes "Resource Efficiency" and "Food Systems." This theme aims to ensure the efficient use of resources to minimize environmental impacts and to develop sustainable food systems that reduce the environmental footprint and promote food security.

The theme of climate action and mitigation includes the codes "Process and Challenges," "Goals and Strategies," and "Regional Initiatives." These codes emphasize the importance of setting and implementing clear targets for effective climate action, addressing the processes and challenges involved, and promoting regional initiatives to combat climate change. The public engagement and education theme is represented by the codes "Identical Initiatives," "Civil Society Role," and "Educational Initiatives." This theme highlights the need to increase public awareness and education on sustainability issues, emphasizing the crucial role of civil society and outlining the next steps in public engagement.

The energy transition and efficiency theme involves the codes "Long-term Objectives," "Energy Strategy 2050," and "Next Steps." This theme is critical for guiding the transition towards sustainable energy use, establishing long-term objectives, and implementing the Energy Strategy 2050. The biodiversity conservation theme includes the codes "Current State," "Societal Impact," and "Industry, Innovation, and Infrastructure." This theme underscores the positive societal impacts of biodiversity conservation, assessing the current state of biodiversity efforts and integrating conservation with industry, innovation, and infrastructure.

Switzerland's approach to equal opportunities and social cohesion is comprehensive and integrated. Switzerland aims to create a more equitable and cohesive society while advancing its sustainability goals by focusing on policy coherence, equal opportunities, and sustainable supply and demand. This policy provides a clear roadmap for policymakers and stakeholders to understand and implement effective, sustainable development initiatives.

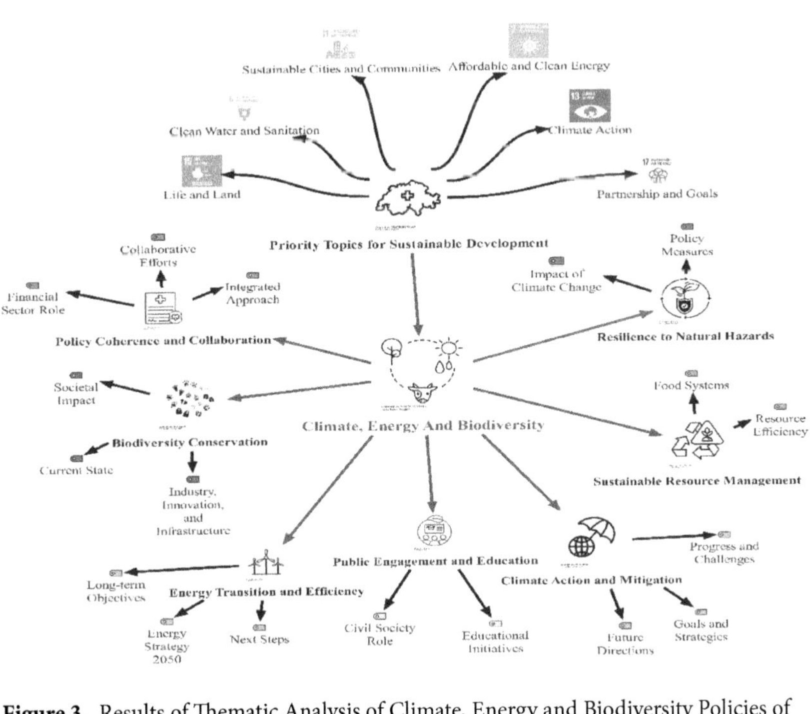

Figure 3. Results of Thematic Analysis of Climate, Energy and Biodiversity Policies of Switzerland

Visual 4 shows Switzerland's themes and codes of the "Equal Opportunities and Social Cohesion" policy. The policy is structured around three main themes, each with specific focus areas or codes. In this context, 'codes' refer to specific policy measures, strategies, or actions designed to address the focus areas under each theme.

The first theme, Policy Coherence and Collaboration, ensures that policies across sectors are harmonized and supported to promote equal opportunities and social cohesion. The "Integrated Policy Approach" emphasizes aligning transparency and accountability in financial operations. "Collaboration Across Sectors" also encourages cooperation between government, private industry, and civil society to pool resources and efforts for common goals.

The second theme, Equal Opportunities, catalyzes inclusivity and equity, ensuring access to opportunities for all individuals. The "Education System" is a key driver, focusing on providing quality education and vocational training to reduce disparities and foster social mobility. "Gender Equality" is a significant goal, aiming to eliminate gender disparities and promote equal opportunities in all life aspects, including education and employment. Furthermore, "Inclusion and Equality" is a transformative approach, fostering social inclusion and reducing inequalities through targeted policies and programs, ensuring marginalized groups are included.

The third theme, Sustainable Supply and Demand, is a strategic move towards creating sustainable economic systems that balance supply and demand while minimizing environmental impact. The "Environmental Impact of Consumption Patterns" is a proactive measure, assessing and mitigating the environmental impacts of consumption by promoting sustainable habits and reducing waste. "Sustainable Food Systems" is a solution that promotes environmentally sustainable and economically viable food systems, ensures food security, and reduces waste. Lastly, "Economic Prosperity and Resource Efficiency" is a key strategy, supporting inclusive and sustainable economic growth by emphasizing efficient resource use.

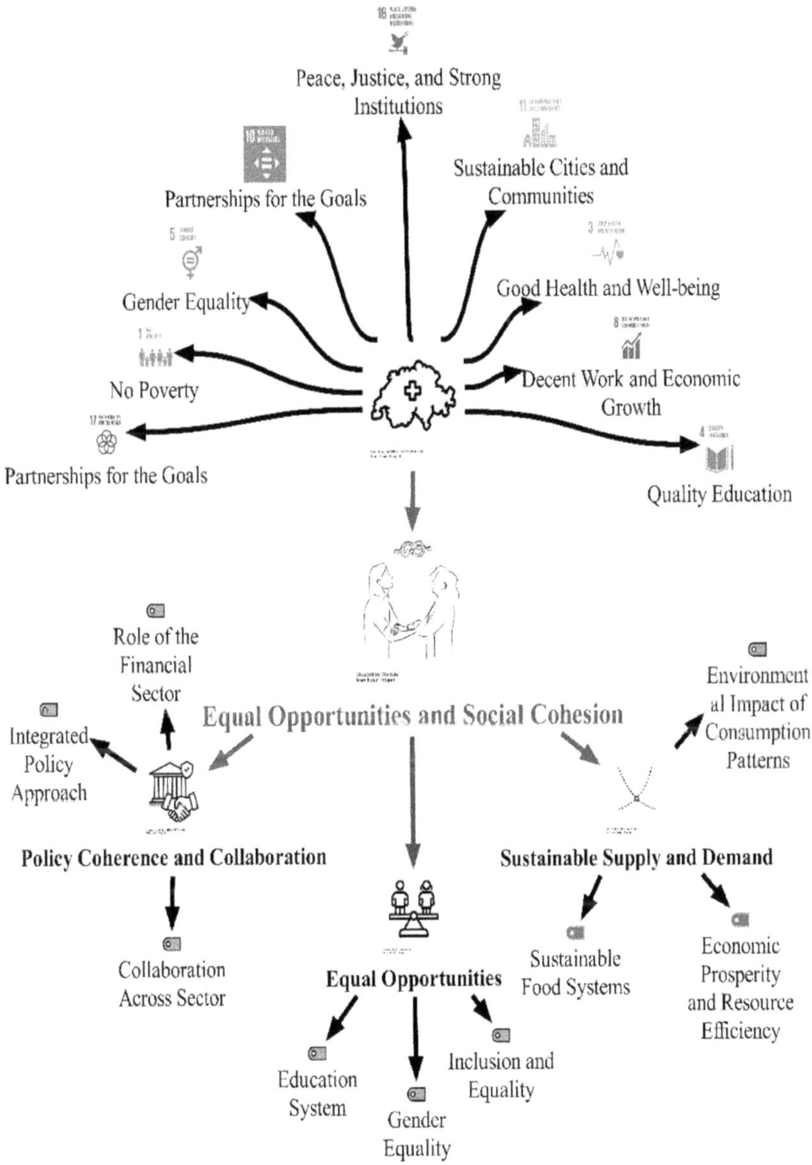

Figure 4. Results of Thematic Analysis of Equal Opportunities and Social Cohesion Policies of Switzerland

Conclusion

This research delves into the unique aspects of Switzerland's policies as an ecotourism destination within the United Nations Sustainable Development Goals (SDGs) framework. Switzerland's sustainable development policies stand out for their series of measures and a comprehensive and holistic approach. These policies are distinctively structured under three main headings: Equal Opportunities and Social Cohesion, Sustainable Consumption and Production, and Climate, Energy, and Biodiversity.

Switzerland's commitment to the UN's Sustainable Development Goals (SDGs) under the category of Equal Opportunities and Social Cohesion is a comprehensive and collaborative effort. It addresses a wide range of issues including poverty (SDG 1), health (SDG 3), education (SDG 4), gender equality (SDG 5), economic growth (SDG 8), inequality (SDG 10), sustainable cities (SDG 11), and peace (SDG 16), all while emphasizing partnerships (SDG 17).

In the category of Sustainable Consumption and Production, Switzerland's strategies are designed for maximum effectiveness. They focus on ending hunger (SDG 2), promoting economic growth (SDG 8), fostering innovation (SDG 9), ensuring sustainable consumption (SDG 12), conserving marine resources (SDG 14), and emphasizing partnerships (SDG 17).

Under Climate, Energy, and Biodiversity, Switzerland's approach is built on the crucial role of partnerships. It addresses water and sanitation (SDG 6), clean energy (SDG 7), sustainable cities (SDG 11), climate action (SDG 13), land ecosystems (SDG 15), and emphasizes the importance of partnerships (SDG 17).

Switzerland's policy on equal opportunities and social cohesion is shaped around three main themes: policy coherence, equal opportunities, and sustainable supply and demand. The Policy Coherence and Collaboration theme ensures that sectoral policies are harmonious and supportive. The integrated policy approach and intersectoral cooperation enhance transparency and accountability in this context. The Equal Opportunities theme includes providing quality education and vocational training, promoting gender equality, and enhancing social inclusion. The Sustainable Supply and Demand theme aims to reduce the environmental impacts of consumption patterns, promote sustainable food systems, and increase resource efficiency to achieve economic prosperity. This holistic approach simultaneously targets social justice and economic sustainability, addressing all societal segments.

The Sustainable Consumption and Production policy is shaped around various themes and codes. The Sustainable Food Systems theme is supported by key codes such as the Swiss Nutrition Policy, the National Food Systems

Transformation Roadmap, and Healthy Eating and Environment. These strategies promote balanced eating habits, reduce environmental impacts, prevent food waste, and support sustainable agricultural practices. The Sustainable Development Strategy theme focuses on reducing hunger, protecting biodiversity, and combating climate change through interconnected goals and policy coherence. The Civil Society Participation theme includes codes for individual and institutional roles and effective collaboration, aiming to contribute to sustainable development through awareness-raising, education, and behavioral changes. The Environmental Assessments theme includes global impact and resource efficiency codes, emphasizing the positive effects of local measures on the global environment. The Economic and Financial Systems theme focuses on sustainable economic activities and the role of the financial sector. This policy set aims to achieve the necessary economic, environmental, and social balance for sustainable development.

The Climate, Energy, and Biodiversity policy is shaped around various themes, including resilience to natural hazards, sustainable resource management, climate action and mitigation, public participation and education, energy transition and efficiency, and biodiversity conservation. The Resilience to Natural Hazards theme aims to address policy measures and the impacts of climate change. The Sustainable Resource Management theme includes resource efficiency and sustainable food systems codes. The Climate Action and Mitigation theme includes processes and challenges, goals and strategies, and regional initiatives codes. The Public Participation and Education theme aims to increase public awareness and emphasize the role of civil society. The Energy Transition and Efficiency theme includes long-term goals, Energy Strategy 2050, and following steps codes. The Biodiversity Conservation theme includes codes for the current state, societal impact, industry, innovation, and infrastructure. These themes reflect Switzerland's commitment to combating climate change and protecting biodiversity.

This research highlights the remarkable success of Switzerland's sustainable development strategy, which is shaped around various themes, such as sustainable food systems, integrated development strategies, civil society participation, environmental assessments, economic and financial systems transformation, and collaborative governance. Switzerland's comprehensive and holistic approach has not only set a high standard for sustainable development but has also become an inspiring model. This approach, which can be emulated by other countries, is a testament to Switzerland's commitment to achieving sustainable development goals.

Through this research, it becomes evident that Switzerland's sustainable development policies have tangible and significant implications not only for the country but also for the global community. Switzerland's policies, deeply rooted in environmental protection, economic sustainability, social participation, and international appeal, actively contribute to the conservation of natural resources, long-term economic growth, social cohesion, and cultural richness. The active civil society participation and educational efforts not only raise awareness of sustainability but also enhance the country's international reputation. These policies firmly establish Switzerland as a global leader in sustainability and create a compelling model in the ecotourism sector, promoting sustainable development both locally and globally.

References

Bornemann, B., & Christen, M. (2020). A new generation of sustainability governance: Potentials for 2030 agenda implementation in Swiss cantons. Sustainability, 12(3), 1234–1250.

Breu, T., Bergöö, M., Ebneter, L., Pham-Truffert, M., Bieri, S., Messerli, P., Ott, C., & Bader, C. (2020). "Defining National Strategies for Implementing the 2030 Agenda: The Case of Switzerland."

Buser, M., & Koch, P. (2022). Sustainable urban development in Switzerland: Evaluating policies and social impacts. Journal of Urban Planning and Development, 148(2), 180–192.

Cidalia, C. (2011). Eco-tourism as a means to sustainable development in developing countries. Tourism Management Perspectives, 1(1), 85–92.

Cronje, P., & Plessis, E. (2020). The competitiveness of tourism destinations: An overview. Tourism Review, 75(1), 1–15.

Dwyer, L., & Kim, C. (2003). Destination competitiveness: Determinants and indicators. Current Issues in Tourism, 6(5), 369–414.

Huber, R., & Schlegelmilch, M. (2020). Evaluating the environmental and economic impacts of agricultural policies in Switzerland. Environmental Policy and Governance, 30(4), 245–258.

Kiper, T. (2013). Role of eco-tourism in sustainable development. In Özyavuz, M. (Ed.), Advances in Landscape Architecture (pp. 773–802). InTech.

Meier, R., & Baumgartner, D. (2023). Assessing the impacts of sustainable tourism policies in Switzerland. Tourism Management Perspectives, 37, 100819.

Moser, S. C., & Schwager, P. (2021). The role of energy policies in combating climate change: The Swiss experience. Renewable Energy, 169, 1015–1027.

Otto, J. E., & Ritchie, J. R. B. (1996). The service experience in tourism. Tourism Management, 17(3), 165–174.

Ritchie, J. R. B., & Crouch, G. I. (2003). The competitive destination: A sustainable tourism perspective. CABI Publishing.

Swiss Confederation. (2022). Implementing the 2030 Agenda for Sustainable Development: Voluntary National Review of Switzerland 2022. Retrieved Retrieved June 7, 2024, from [https://www.sdgital2030.ch/docs/CountryReport/country-report-switzerland-2022_en.pdf]

United Nations. (2015). *Sustainable Development Goals*. United Nations. Retrieved June 7, 2024, from https://www.un.org/sustainabledevelopment/sustainable-development-goals/

Utama, M., Kunz, F., & Wilhelm, S. (2023). Eco-tourism in Switzerland: Aligning policies with the United Nations' Sustainable Development Goals. Tourism Planning & Development, 20(2), 154–168.

WCED. (1987). Our common future. Oxford University Press.

Wuelser, G., Pohl, C., & Hirsch Hadorn, G. (2020). Structuring complexity for sustainability: Criteria for success. Sustainability Science, 15(1), 123–134.

Emel Cengiz Kaynakci[*], Aleyna Kavalci[†], and Meryem Ece Mutlu[‡]

Chapter 30 Vegetarianism: A Terminology in Food Service and Menu Planning

Introduction

Nutrition is the process of taking **in**, digesting, absorbing, and metabolizing the nutrients necessary for growth, maintenance of life, and protection of health. Organizations that serve to meet the nutritional needs of people outside the home are called **"mass nutrition systems."** Organizations providing mass nutrition services, such as hotels, restaurants, and catering companies, which are used by individuals of all ages and walks of life, should meet consumer needs and prevent public health problems. **In this context,** it is important for accommodation establishments, hospitals, nursing homes, nurseries, **and other entities** that provide food and beverage services to plan menus **according to** the consumer profile and needs (Sezgin & Özkaya, 2014). In the tourism sector, which has a **significant** economic **impact** on a global scale, meeting the nutritional needs of guests coming for various purposes in terms of physiological, social, and psychological aspects, and **integrating** the traditional food and beverages of our country's destinations **into** touristic products, should be one of the most important goals of mass nutrition systems.

The tourism sector consists of a large number of businesses operating around the world and the travel of masses, and **it is** on its way to becoming the world's largest industry in the 21st century. The phenomenon of globalization, especially strengthened by regional integration movements, brings new approaches and changes in the field of tourism, as in all other fields (Eren, Çetin, Eren & Çetin, 2021; Çeken & Ateşoğlu, 2008). The world tourism industry has developed rapidly, especially after World War II, and has spread to wider masses and distances. Turkey, on the other hand, became acquainted with the phenomenon of tourism, which is framed by laws, after the 1980s. Considering that the number of tourists visiting Turkey in those years was 1 million 300 thousand, and today

---

* Assistant Professor, Akdeniz University, Faculty of Health Sciences, Department of Nutrition and Dietetics, ekaynakci@akdeniz.edu.tr

† Akdeniz University, Institute of Health Sciences, Department of Medical Biotechnology

‡ Akdeniz University, Institute of Health Sciences, Department of Medical Biotechnology

this number has been exceeded by tens of times, the **progress** Turkey has **made** can be clearly seen. Today, tourism is seen as an industry of hope for the whole country, especially in the Mediterranean and Aegean regions, and therefore plays a very important role in Turkey's economic development strategies (Bilgiçli & Altınkaynak, 2016). In the European Union (EU), the tourism sector is growing steadily, becoming an industry with **significant** potential, together with the sub-sectors it affects. The EU is emerging as the most important tourism destination in the world and is one of the **leading regions** in terms of tourism revenues (Çeken & Ateşoğlu, 2008). However, nowadays, people do not limit their expectations to sea, sand, and sun; instead, they engage in different touristic activities that cater to their interests. The fact that food and beverage activities, which are among these, are a compulsory need, and tourists' desire to enjoy every moment of their travels makes the gastronomic value of the **chosen vacation destination** an integral part of the tourism experience (Küçükkömürler, Şırvan & Sezgin, 2019).

A destination's food and beverage services can be part of its tourist products or can be the main attraction on their own. Either way, they play a critical role in tourist experiences and are an important factor affecting visitors' overall satisfaction. Individuals travel with dietary habits, culinary cultures, and food allergies. Food allergies and different dietary approaches affect tourists' destination choices and vacation types (Eren et al., 2021). However, vegetarian and vegan diets stand out as dietary approaches that are becoming increasingly recognized today (Vatan & Türkbaş, 2018). Vegetarianism/veganism, which is gradually increasing worldwide, is accepted not only as a diet but also as a lifestyle, philosophy of life, stance, and ethical approach (Kaytez & Tunçay, 2020).

The majority of the world's vegetarian and vegan population is found in Asia, with India topping the list with 375 million people (31 % of the population). The reason for such a high proportion of the vegan/vegetarian population in India is attributed to religious choices and what is considered sacred (e.g., the cow is considered sacred). Although the number of qualified studies to determine the vegan/vegetarian population in Europe is small, it is estimated that between 10 and 50 million people (between 2 % and 10 % of the total population) are vegetarian or vegan. In the USA, an average of 15 million people (2 % of the total population) are known to adopt this diet (Meat Atlas, 2014; Dilek, 2018). In addition, many studies on human nutrition have reported that this number has been increasing from the past to the present and will continue to increase globally in the future (Fox & Ward, 2008; Tantamango-Bartley, Jaceldo-Siegl, Fan & Fraser, 2013).

Food allergies and dietary behaviors such as vegan and vegetarian **diets** are important factors that restrict individuals' food intake and affect their quality of life. Lee, Scott, and Packer (2014) emphasized that some tourists' food preferences and problems with food culture are an important obstacle **to** their travel motivation. In this context, veganism and vegetarianism appear **as concepts** that can be considered **travel barriers** (Altaş, 2017; Eren, 2020; Huang, Chen, Lu & Shen, 2020). The trend of veganism and vegetarianism has been a topic that has been emphasized recently **for its** consumption dimension and the market and economic impact it has started to create. Despite these developments, the services and infrastructure related to veganism and vegetarianism are still open to debate (Altaş, 2017; Sünnetçioğlu, Mercan, Yıldırım & Türkmen, 2017; Huang et al., 2020). Considering the health and food preferences of these individuals, it is important to meet their energy and nutrient requirements and to create menus **consisting of** adequate and balanced nutrients using appropriate food products. In many studies, it has been reported that menu design has a great influence on food choice. At this point, it is important to influence and guide consumer choice in the right direction (Baiomy, Jones & Goode, 2019; Kızıldemir & Hülağa Kaderoğlu, 2021; Şahin & Yazıcıoğlu, 2018).

A menu is a carefully curated list that presents the names, prices, and descriptions of the food and beverages offered to guests in an establishment. This list can be illustrated or unillustrated and is typically organized in a specific manner to facilitate customers' decision-making. The design and layout of a menu can significantly impact the customers' perception and choice. For instance, a well-structured menu can guide customers through their options, enhancing their overall dining experience (Altınel, 2011; Bulduk, 2013; Yılmaz, 2006).

Moreover, the menu serves as one of the most important tools for facilitating communication between the business and the customer. It conveys not only the offerings but also the identity and values of the establishment. Bowen and Morris (1995) highlighted that a menu can be a powerful communication tool that captures attention and leaves a lasting impression. They further noted that the language used in menu descriptions can evoke emotions and stimulate appetite, thereby influencing customers' dining decisions (Şahin & Yazıcıoğlu, 2018).

Knowing the boundaries and types of vegetarianism and veganism is of great importance for customer satisfaction in tourism industries that produce goods and services. It is thought that businesses that want to meet the needs of vegetarian and vegan tourists and differentiate themselves from their competitors can ensure the satisfaction of customers with vegetarian and vegan diets by offering vegan/vegetarian food options in their menus, enriching and labeling these options (Vatan & Türkbaş, 2018). Labeling vegan/vegetarian food alternatives on menus will help customers make the right choices (Eryücel, Yalçınkaya & Küçükaltan, 2020).

The "V-Label" symbol, created by the European Vegetarian Union (EVU) in 1996, is one of the most well-known and trusted labels in this field. As Uçan and Bozok (2019) point out, the V-Label targets conscious food consumers, individuals who want to reduce meat consumption, vegans and vegetarians, allergy sufferers and religious groups with special diets. The V-Label is an important reference point for vegans to recognize vegan products and for **them** to make restaurant or food choices. This label increases the participation of vegans in gastronomy activities and ensures that the products of food companies producing vegan products are recognized. Thus, vegan tourists can make choices with confidence, and businesses can gain a competitive advantage by appealing to a wide target audience (Uçan & Bozok; 2019).

Turkish cuisine, with Turkey's geographical location, the variety of products **available,** and the new expansions brought to the tradition of eating and drinking thanks to the civilizations with which relations have been established in the historical process, **includes many options suitable for vegetarian diets.** Turkish cuisine, which provides rich examples in every branch of culinary art, is one of the three richest cuisines in the world with more than 2500 types of dishes. It is possible to classify the dishes suitable for a vegetarian diet in Turkish cuisine as soups, vegetable dishes with olive oil, legume dishes with olive oil, pilafs, pastries, salads, **and** desserts (Sezgin & Ayyıldız, 2014). It would be a strategic step for accommodation establishments that provide food and beverage services, especially in the field of tourism, to meet the nutritional needs of customers who follow a vegan/vegetarian diet with traditional Turkish dishes in terms of both the promotion of Turkish cuisine and the advancement of gastronomy tourism.

For this purpose, this chapter aims to provide information about traditional Turkish dishes that can be consumed by vegan and vegetarian individuals who have specifically chosen Turkey as a holiday destination and to provide information to food and beverage staff in public catering systems to guide the needs of this group. In addition, it is also aimed to provide useful information to food service businesses, students studying in the field of gastronomy and gastronomy employees in order to understand the guests and develop menu items and services that will best meet their needs by addressing the difficulties experienced by individuals with special diets while meeting their nutritional needs outdoors.

Vegetarian

The vegetarian diet is a sustainable dietary model **dating back to the** 6th century BC for ethical, cultural, and religious reasons and has been proven to have positive effects on health. The concept of "vegetarian," which originates from the

Latin word "vegetus" meaning full of life, healthy, and lively, was first defined in 1842. According to the definition, **a** vegetarian diet is a dietary model in which red meat, chicken, fish, and other seafood are not consumed, and eggs and dairy products are consumed on demand (Tabu, 2020). As of the 20th century, it is known that the vegetarian population has increased due to ethical and bioethical approaches, ecological concerns, economic situations, taste, culture, religion, and positive effects on health (Balcı, 2018). According to the 2014 Meat Atlas Report, approximately 2 % of the US population, 2 % to 10 % of the European population, and 31 % of the Indian population are vegetarian (Vatan & Türkbaş, 2018). In another report prepared by Meticulous Research, it was reported that, as of 2023, 2.6 million people in Europe are vegan, which corresponds to 3.2 % of the European population (The Vegan Society, 2024). It is **evident** that this huge market of more than one billion consumers worldwide who prefer vegetarian and vegan diets and avoid the consumption of animal foods will make a great contribution to the tourism industry (Vatan & Türkbaş, 2018).

An increasing number of vegetarian and vegan individuals face various challenges in food and beverage establishments. Vegetarians and vegans, whose most important criteria in business preferences are that the business is ethical and environmentally conscious, often have to **resort to** fast food due to the lack of menu alternatives. In addition, high prices, lack of trust in establishments, and insufficient knowledge of the staff about the menu content are among the most common problems experienced by individuals. Accordingly, it is necessary to prepare a reliable and transparent menu, increase the variety of menus, and ensure sustainable service quality (Ayyıldız & Sezgin, 2021).

Many studies on plant-based nutrition have found that vegetarian diets rich in whole grains, legumes, vegetables, fruits, and oilseeds have prophylactic and therapeutic effects on blood pressure, blood glucose, insulin sensitivity, serum lipid profile, and inflammation (Key, Fraser, Thorogood, Appleby, Beral, Reeves, Burr, Chang-Claude, Frentzel-Beyme, Kuzma, Mann & McPherson, 1999; Ferdowsian et al., 2010; Rizzo, Sabaté, Jaceldo-Siegl & Fraser, 2011; Crowe, Appleby, Travis & Key, 2013; Orlich & Fraser, 2014; Yokoyama, Nishimura, Barnard, Takegami, Watanabe, Sekikawa, Okamura & Miyamoto, 2014; Wang, Zheng, Yang, Jiang, Fu & Li, 2015; Papier, Appleby, Fensom, Knuppel, Perez-Cornago, Schmidt, Tong & Key, 2019; Craig, Mangels, Fresán, Marsh, Miles, Saunders, Haddad, Heskey, Johnston, Larson-Meyer & Orlich, 2021; Barnard, Alwarith, Rembert, Brandon, Nguyen, Goergen, Horne, do Nascimento, Lakkadi, Tura, Holubkov & Kahleova, 2022). A meta-analysis published in 2020 showed that **a** vegetarian dietary pattern **has** positive effects on total cholesterol, LDL, and body weight; vegetarians **have a** lower risk of diabetes, ischemic heart disease, and cancer (Oussalah, Levy,

Berthezène, Alpers & Guéant, 2020). However, there are also studies showing that it reduces body weight, body fat mass, and body mass index (BMI) compared to omnivorous diets (Wright, Wilson, Smith, Duncan & McHugh, 2017; Kahleova, Fleeman, Hlozkova, Holubkov & Barnard, 2018; Kahleova, Hlozkova, Fleeman, Fletcher, Holubkov & Barnard, 2019). Current research suggests that a vegetarian dietary pattern regulates gut microbiota due to its high dietary fiber and phytochemical content, which helps promote optimal energy balance and can be used to combat chronic diseases (Djekic, Shi, Brolin, Brolin, Carlsson, Särnqvist, Savolainen, Cao, Bäckhed, Tremaroli, Landberg & Frøbert, 2020; Beam, Clinger & Hao, 2021; Craig et al., 2021).

The vegetarian diet is sub-categorized as lacto-ovo vegetarian, lacto vegetarian, ovo vegetarian, semi-vegetarian, polo vegetarian, pesco vegetarian, and vegan. Traditional Turkish dishes representing these categories are given in table form (Kutluay Merdol, 2018).

Lacto-Ovo Vegetarian

The lacto-ovo vegetarian diet, which is formed from the words "lacto" meaning "milk" and "ovo" meaning "egg", originates from Buddhism and Jainism, which are based on the philosophy that life is sacred and killing animals is against moral values. This dietary model excludes red meat, poultry, fish and seafood. The diet includes dairy products and eggs in addition to plant foods. (Karabudak, 2012; Vatan & Türkbaş, 2018). In 2021, in a study conducted on overweight and obese non-alcoholic fatty liver (NAFLD) patients, it was observed that patients who were on a lacto-ovo vegetarian diet for three months showed improvement in NAFLD findings, glycemic indices and serum lipid profile, and a decrease in body weight and waist circumference (Garousi, Tamizifar, Pourmasoumi, Feizi, Askari, Clark & Entezari, 2023). Table 1 shows lacto-ovo vegetarian dishes from traditional Turkish cuisine. However, the dishes shown in Tables 2, 3 and 6 can also be consumed by lacto-ovo vegetarians.

Table 1. Lacto-ovo vegetarian dishes of traditional Turkish cuisine.

Main course	**Oven hash browns**	Sunflower oil, olive oil, wheat flour, dill, zucchini, black pepper, dried garlic, spring onion, salt, yogurt, egg
	Zucchini hash browns	sunflower oil, wheat flour, dill, zucchini, baking powder, black pepper, paprika, dried onion, parsley, spring onion, salt, yogurt, egg
	Frying	sunflower oil, wheat flour, carrot, cauliflower, red pepper, dried garlic, potato, salt, egg, yogurt
	Roasted cauliflower with yogurt	Sunflower oil, wheat flour, tomato paste, cauliflower, dried garlic, lemon, parsley, salt, egg, yogurt
	Poached eggs with yoghurt (cılbır)	Sunflower oil, olive oil, paprika, parsley, vinegar, salt, yogurt, egg
Soup	Yayla soup (rice, yoghurt and mint soup)	Sunflower oil, olive oil, wheat flour, dried mint, rice, salt, yogurt, egg
	Toyga soup (soup made with yoghurt, hazelnut, rice, egg and mint)	Sunflower oil, olive oil, wheat, wheat flour, red pepper, dried mint, chickpeas, salt, strained yogurt, egg
Rice/pastry	Baked pasta	sunflower oil, wheat flour, cheddar cheese, black pepper, pasta, parsley, milk, salt, butter, egg
	Erişte with kashar and walnuts	sunflower oil, butter, cheddar cheese, noodles, walnuts, salt
	Börek with cheese (savoury pastry filled with cheese)	Sunflower oil, feta cheese, parsley, milk, phyllo dough, egg
	Börek with spinach (savoury pastry filled with spinach)	Sunflower oil, spinach, black pepper, onion, milk, salt, phyllo dough, egg
	Börek with cheddar (savoury pastry filled with cheddar)	Wheat flour, fresh yeast, salt, cheese, egg

(*continued*)

Table 1. Continued

Dessert	Baklava	Wheat flour, walnut, lemon, starch, granulated sugar, salt, egg, butter
	Sütlaç (milk pudding with rice)	Starch, rice, rice flour, milk, granulated sugar, egg
	Kalburabastı (riddled egg shaped pastries)	Wheat flour, yeast, lemon, semolina, powdered sugar, granulated sugar, walnut, butter, egg
	Kazandibi (pudding with a caramel base)	Hazelnut, starch, milk, cinnamon, granulated sugar, egg
	Keşkül (milk and almond pudding)	Pistachios, almonds, coconut, starch, milk, granulated sugar, egg
	Şekerpare (baked soft pastry dipped in thick syrup)	wheat flour, hazelnut, baking powder, lemon, semolina, powdered sugar, granulated sugar, vanilla, butter, egg
	Tulumba dessert (fluted deep fried fritters in thick syrup)	Sunflower oil, wheat flour, starch, granulated sugar, butter, egg

Lacto Vegetarian

In the lacto-vegetarian diet model, which refuses red meat, poultry, fish and seafood as well as eggs, only milk and dairy products are consumed in addition to plant foods (Ayyıldız & Sezgin, 2021). Table 2 shows lacto-vegetarian dishes from traditional Turkish cuisine. However, the dishes shown in table 6 can also be consumed by lacto vegetarians.

Table 2. Lacto-vegetarian dishes from traditional Turkish cuisine.

Main course	Cauliflower gratin	sunflower oil, olive oil, wheat flour, black pepper, cauliflower, cheddar cheese, paprika, milk, salt, butter
	Spinach gratin	sunflower oil, olive oil, wheat flour, spinach, black pepper, cheddar cheese, dried onion, milk, dried garlic, capia pepper, thyme, salt, butter
	Fava (mashed broad beans with dill)	wheat flour, dill, tomato, tomato paste, carrot, dried onion, lemon, fresh broad beans, fresh onion, granulated sugar, salt, yoghurt, olive oil

Table 2. Continued

Soup	Creamy broccoli soup	sunflower oil, olive oil, wheat flour, black pepper, dried garlic, dried onion, broccoli, cream, milk, salt
	Creamy mushroom soup	sunflower oil, olive oil, wheat flour, black pepper, dried garlic, dried onion, mushrooms, cream, milk, salt
	Capia pepper soup	Sunflower oil, olive oil, wheat flour, capia pepper, black pepper, milk, salt
	Creamy corn soup	sunflower oil, olive oil, wheat flour, parsley, onion, cream, milk, salt, corn
	Tomato soup	Sunflower oil, olive oil, wheat flour, tomato, tomato paste, black pepper, cheese, milk, salt
	Gemici soup	Wheat, chickpeas, dried beans, yoghurt, flour, oil, mint, salt, lentils
	Spinach soup	Sunflower oil, olive oil, wheat flour, salt, milk, spinach
	Köylü soup (peasant soup of mixed vegetables)	sunflower oil, olive oil, wheat flour, carrot, black pepper, celery, onion, mushroom, leek, milk, salt
	Tarhana soup	Sunflower oil, olive oil, wheat flour, tomato paste, red chilli, dried garlic, tarhana, salt
	Carrot soup	Sunflower oil, olive oil, wheat flour, carrot, black pepper, milk, salt
Rice/pastry	Rice pilaf with tomatoes	Sunflower oil, tomatoes, rice, salt, butter
	İç pilaf (rice with onions, nuts, diced liver and currants)	Sunflower oil, pine nuts, black pepper, currants, rice, salt, butter, dried mint, new spring
	Rice pilaf	Sunflower oil, butter, rice, salt
	Rice pilaf with eggplant	sunflower oil, butter, tomato, black pepper, eggplant, rice, salt
	Rice with vegetables	Sunflower oil, butter, carrot, peas, rice, salt
Salad	Tsatsiki (cucumber-yoghourt)	Olive oil, dill, dried mint, dried garlic, cucumber, strained yogurt, salt
	Haydari (thick yoghurt with garlic and dill)	Dried garlic, dried mint, salt, strained yogurt, olive oil
	Carrot salad with yogurt	Carrot, dried garlic, parsley, salt, dill, yogurt
	Purslane salad with yogurt	Dried garlic, purslane, salt, yogurt, olive oil
	Green salad with yogurt	Curly, fresh mint, spring onion, salt, yogurt

(*continued*)

Table 2. Continued

Dessert	Turkish Semolina Halva	Pine nuts, semolina, milk, granulated sugar, butter
	Quince dessert	Quince, cream, granulated sugar, walnuts, cloves
	Aşure (noah's own pudding)	Wheat, walnut kernels, hazelnut, rose water, dried beans, dried figs, dried apricots, raisins, pomegranate, chickpeas, cinnamon, starch, milk, granulated sugar
	Supangle with cocoa	Cocoa, starch, granulated sugar, milk, coconut
	Milk dessert with fruit	Milk, starch, granulated sugar, cherry/ strawberry, vanilla
	Güllaç (rice wafers stuffed with nuts cooked)	Walnut, rose water, rose pudding leaves (flour, starch, water), pomegranate, milk, granulated sugar, vanilla

Ovo Vegetarian

In the ovo-vegetarian diet model, which refuses the consumption of red meat, poultry, fish and seafood as well as milk and dairy products, eggs are consumed in addition to plant foods (Ayyıldız & Sezgin, 2021). In a 2017 study conducted on Taiwanese women, it was found that ovo-vegetarian diet showed an anti-hypertensive effect (Ho, Yu & Lee, 2017). Table 3 shows ovo-vegetarian dishes from traditional Turkish cuisine. However, the dishes shown in Table 6 can also be consumed by ovo vegetarians.

Table 3. Ovo vegetarian dishes of traditional Turkish cuisine.

Main course	Spinach with egg	sunflower oil, olive oil, spinach, black pepper, red pepper, onion, capia pepper, salt, egg
	Menemen (spicy turkish omlette)	sunflower oil, olive oil, chilli, tomato, black pepper, onion, parsley, salt, egg
	Potato omelette (Spanish omelette)	Sunflower oil, olive oil, black pepper, red pepper, parsley, potato, salt, egg
	Scrambled eggs with leeks	Sunflower oil, olive oil, leek, salt, egg
Soup	Vermicelli soup	Vermicelli, tomato paste, oil, lemon, egg, salt, flour

Table 3. Continued

Salad	Potato salad	Potatoes, olive oil, lemon, parsley, onion, salt, olives, tomatoes, chilli, egg, dried mint, chilli flakes
Dessert	Hanım göbeği	Sunflower oil, wheat flour, granulated sugar, wet yeast, egg

Semi Vegetarian

In the semi vegetarian diet model, which is called "semi-vegetarian" and is not recognized as vegetarian by the Vegetarian Society, red meat consumption is rejected. The diet includes limited consumption of poultry, fish and seafood in addition to plant foods and secondary animal foods such as dairy products, eggs and honey (Vatan & Türkbaş, 2018). The dishes shown in Tables 4, 5 and 6 can also be consumed by semi vegetarians.

Polo Vegetarian

The Polo vegetarian dietary model includes the consumption of poultry in addition to plant foods and secondary animal foods such as dairy products, eggs and honey. This dietary model is not recognized as vegetarian by the Vegetarian Society because it includes meat consumption (Vatan & Türkbaş, 2018). Table 4 shows polo vegetarian dishes from traditional Turkish cuisine. However, the dishes shown in table 6 can also be consumed by polo vegetarians.

Table 4. Polo vegetarian dishes of traditional Turkish cuisine.

Main course	Grilled chicken	Chicken meat (with bones), tomato paste, salt, oil, thyme, black pepper
	Chicken curry	Chicken meat (boneless), potato, carrot, onion, tomato paste, salt, oil, black pepper
	Chicken casserole	Sunflower oil, tomato, tomato paste, carrot, black pepper, oregano, red pepper, peas, dried garlic, dried onion, potato, aubergine, chicken thigh (boneless), salt
	Boiled chicken	Sunflower oil, wheat flour, carrot, potato, chicken thigh (with bone), peas, lemon, salt
	Chicken shish	Sunflower oil, tomato, tomato paste, black pepper, red pepper, dried garlic, green pepper, chicken breast, salt
	Baked chicken	Sunflower oil, tomato, tomato paste, black pepper, onion, potato, green pepper, chicken thigh (with bone), salt
	Chicken sautéed with mushrooms	Sunflower oil, tomato, tomato paste, mushrooms, black pepper, oregano, red pepper, dried onion, chicken thigh (boneless), salt
	Pasha kebab	Chicken meat (boneless), potato, tomato, chilli, onion, tomato paste, salt, oil, black pepper
	Baked turkey	sunflower oil, tomato, tomato paste, black pepper, onion, potato, green pepper, turkey thigh, salt
	Circassian chicken	Sunflower oil, chicken breast, dried garlic, dried bread, walnut kernels, black pepper, paprika, salt
	Chicken schnitzel	sunflower oil, wheat flour, breadcrumbs, black pepper, chicken breast, potatoes, milk, butter, salt, egg
Soup	Chicken vermicelli soup	Sunflower oil, olive oil, wheat flour, tomato, tomato paste, black pepper, lemon, parsley, vermicelli, chicken breast, salt
Rice/pastry	Rice pilaf with chicken	Sunflower oil, butter, chicken breast, black pepper, rice, salt

Pesco Vegetarian

The pesco vegetarian or pescetarian diet originates from the Italian word "pesce", which means "fish". In this dietary pattern, freshwater fish such as trout and sea bass, saltwater fish such as salmon and tuna, shrimp, oysters and other shellfish are consumed, along with secondary animal foods such as plant foods, dairy products, eggs and honey. In contrast, red meat and poultry are avoided (Tabu, 2020). Pesco vegetarian diet is not recognized as vegetarian by the Vegetarian Society due to meat consumption (Vatan & Türkbaş, 2018).

Fish and other seafood are important sources of omega-3 fatty acids, B vitamins, zinc, iodine, selenium, calcium, magnesium and phosphorus, and provide the body with high-quality protein (Estruch, Sacanella & Ros, 2021; O'Keefe, Torres-Acosta, O'Keefe, Saeed, Lavie, Smith & Ros, 2020). Studies show that pesco vegetarians have a lower risk of vision problems, bone diseases and Alzheimer's disease due to the consumption of fish and seafood (Tabu, 2020). In a meta-analysis of 5 prospective dietary studies evaluating coronary artery disease mortality rates, it was found that pesco vegetarians had 34 % lower mortality rates due to coronary artery disease than omnivores (Key et al., 1999). Table 5 shows pesco vegetarian dishes from traditional Turkish cuisine. However, the dishes shown in table 6 can also be consumed by pesco vegetarians.

Table 5: Pesko vegetarian dishes of traditional Turkish cuisine.

Main course	Steamed fish	Sunflower oil, tomato paste, onion, lemon, parsley, fresh fish, salt
	Grilled fish	Sunflower oil, rocket, red onion, cress, lemon, radish, fresh fish, salt
	Baked fish	Sunflower oil, tomato, onion, lemon, parsley, potato, fresh fish, salt
	Fried fish	Sunflower oil, wheat flour, cornflour, rocket, red onion, cress, lemon, radish, fresh fish, salt
	Breaded fish	Sunflower oil, wheat flour, cornflour, tomato paste, lemon, soda, fish fillet, salt

Vegan

Although it is a subcategory of vegetarianism, veganism, which is handled separately, is a lifestyle in which animal foods are not consumed and animal products such as wool, silk and leather are not used (Tabu, 2020). The Vegan Society's definition of veganism in 2024; *"It is a philosophy and way of life that aims to avoid all forms of exploitation and cruelty to which animals are exposed for food, clothing or any other purpose, to the extent possible and feasible. It supports the development and use of alternatives that do not involve the use of animals for the benefit of people, animals and the environment"* (The Vegan Society, 2024).

In the vegan diet model, consumption of red meat, poultry, fish and other seafood, as well as products derived from animals such as eggs, honey, milk, yogurt, yogurt and kefir is avoided. The diet includes the consumption of plant foods such as vegetables, fruits, legumes, grains and oilseeds (Tabu, 2020). A vegan diet, which is a more restricted dietary model compared to other vegetarian dietary models, can cause nutrient deficiencies such as vitamin B12, iron, zinc, calcium, high quality protein, omega-3 fatty acids. Accordingly, the risk of bone fractures, sarcopenia, neurocognitive disorders, anemia and immunodeficiency increases (Estruch, Sacanella & Ros, 2021).

Existing research on vegan diet shows that vegan diet causes significant reductions in LDL, fasting insulin levels and body weight (Landry, Ward, Cunanan, Durand, Perelman, Robinson, Hennings, Koh, Dant, Zeitlin, Ebel, Sonnenburg, Sonnenburg & Gardner, 2023; Turner-McGrievy, Davidson, Wingard, Wilcox & Frongillo, 2015). In a study conducted by Kahleova and colleagues (2020), it was shown that a low-fat vegan diet caused a decrease in body weight and body fat mass, increased insulin sensitivity, and improved gut microbiota in overweight adults (Kahleova, Rembert, Alwarith, Yonas, Tura, Holubkov, Agnello, Chutkan & Barnard, 2020). In another study, it was determined that vegans had higher diet quality compared to vegetarians and semi-vegetarians (Peruvemba, Gieng, Chen & Pignotti, 2022). When veganism and vegetarianism are compared, it is concluded that veganism is a strategic dietary model that prioritizes animal rights and promotes better health (Zamir, 2004). Table 6 shows the vegan dishes of traditional Turkish cuisine and their ingredients.

Table 6: Vegan dishes of traditional Turkish cuisine

Main course	Imam bayildi (eggplant cooked in olive oil with onions, garlic and tomatoes)	sunflower oil, tomato, dried garlic, dried onion, parsley, aubergine, sharp pepper, salt, olive oil
	Stuffed Zucchini boat	sunflower oil, tomato, tomato paste, courgette, onion, parsley, sharp pepper, salt, olive oil
	Sour zucchini	Sunflower oil, olive oil, tomato paste, onion, courgette, green lentils, chickpeas, lemon juice, salt
	Shakshouka (dried fried aubergines in tomato garlic sauce)	Aubergine, tomato, chilli, potato, salt, oil, tomato paste

Table 6: Continued

	Babagannus	Olive oil, tomato, dried garlic, dried garlic, dried mint, red pepper, parsley, aubergine, green pepper, lemon juice, salt
	Okra with olive oil	Tomato, tomato paste, dried onion, parsley, green pepper, okra, granulated sugar, salt, lemon, olive oil
	Stuffedbell peppers with olive oil	pine nuts, dill, bell pepper, tomato, dried onion, currants, lemon, parsley, dried mint, rice, cinnamon, granulated sugar, salt, new spring, olive oil
	Cabbage rolls with olive oil	white cabbage, pine nuts, dill, black pepper, dried onion, parsley, rice, granulated sugar, salt, lemon, cinnamon, new spring, olive oil
	Stuffed aubergine with olive oil	pine nuts, dill, tomatoes, dried onion, currants, lemon, aubergine, parsley, rice, cinnamon, granulated sugar, salt, new spring, olive oil
	Stuffed dried eggplant with olive oil	pine nuts, dill, tomatoes, dried onion, currants, parsley, dried aubergine, rice, cinnamon, granulated sugar, salt, new spring, lemon, olive oil
	Artichokes with olive oil	Tomato paste, artichoke, carrot, peas, onion, parsley, dill, potato, lemon, granulated sugar, salt, olive oil
	Celery with olive oil	Wheat flour, dill, tomato paste, carrot, celery, peas, onion, lemon, potato, granulated sugar, orange, salt, olive oil
	Turkish barbunya pilaki (borlotti beans)	kidney beans/dry beans, tomatoes, tomato paste, carrots, dried garlic, dried onion, parsley, potatoes, green pepper, granulated sugar, salt, olive oil
	Aubergine with olive oil	Tomato, tomato paste, dried onion, parsley, green pepper, green beans, granulated sugar, salt, lemon, olive oil
	Leek with olive oil	Tomato, tomato paste, carrot, leek, rice, granulated sugar, salt, lemon, olive oil
	Green beans with olive oil	Beans, onion, tomato paste, salt, olive oil, tomato
	Stuffed grape leavs with olive oil	Pickled leaves, rice, dried onion, parsley, dill, currants, pistachio, allspice, black pepper, dried mint, salt, lemon
	Piyaz (white bean onion salad vinaigrette)	Tomato, tomato paste, carrot, dried beans, dried garlic, dried onion, parsley, potato, green pepper, granulated sugar, salt, olive oil

(*continued*)

Table 6: Continued

	Ezogelin soup	Sunflower oil, olive oil, wheat flour, tomato paste, carrot, red lentil, dried mint, black pepper, dried onion, potato, bulgur, garlic, lemon, rice, salt
	Lentil soup	Sunflower oil, olive oil, wheat flour, tomato paste, carrot, red pepper, dried mint, dried onion, potato, lemon, salt
	Bahcevan soup	Sunflower oil, olive oil, wheat flour, tomato, tomato paste, black pepper, peas, carrot, potato, salt
	Rice soup	Sunflower oil, olive oil, wheat flour, tomato, tomato paste, rice, salt
	Vegetable soup	Sunflower oil, olive oil, white cabbage, dill, carrot, black pepper, celery, lemon, potato, leek, onion, rice, salt
	Potato soup	sunflower oil, olive oil, wheat flour, tomato, tomato paste, carrot, black pepper, red pepper, onion, potato, salt
	Flour soup	Wheat flour, oil, salt
Rice/pastry	Bulgur pilaf (rice of wheat grains)	Sunflower oil, bulgur, chilli, red (capia) pepper, tomato, tomato paste, chilli paste, dried mint, dried onion, salt
	Bulgur pilaf with lentils	Sunflower oil, bulgur, black pepper, red pepper, onion, salt, green lentil
	Bulgur pilaf with mushrooms	Sunflower oil, bulgur, chilli, red (capia) pepper, tomato, tomato paste, chilli paste, mushroom, dried onion, salt
	Bulgur pilaf with chickpeas	Sunflower oil, bulgur, tomato paste, chickpeas, salt
	Rice pilaf with chickpeas	Sunflower oil, rice, tomato paste, chickpeas, salt
	Spinach pita	Wheat flour, wet yeast, salt, spinach, onion, sunflower oil, black pepper, paprika
Salad	Turkish shepherd's salad	Tomato, cucumber, onion, parsley, green pepper, lemon, salt, olive oil
	Tomato salad	Tomato, lemon, olive oil, salt
	Broccoli salad	Broccoli, red (capia) pepper, vinegar, lemon, garlic, parsley, olive oil, salt
	Carrot salad	Carrot, lemon, olive oil, salt
	Mixed salad	Cucumber, lettuce, tomato, parsley, lemon, olive oil, salt
	Mediterranean salad	Lemon, parsley, vinegar, carrot, lettuce, corn, tomato, salt, olive oil

Table 6: Continued

	Aubergine salad	aubergine, dried onion, olive oil, dried garlic, tomato, chilli, parsley, black pepper, red pepper, lemon salt
	Red cabbage salad	Cabbage, lemon, olive oil, salt
	Leaf lettuce salad	Leaf lettuce, lemon, olive oil, salt
	Lettuce salad	Lettuce, lemon, olive oil, salt
	Turkish kisir salad	Olive oil, tomato, tomato paste, chilli paste, pomegranate molasses, red pepper, black pepper, cumin, bulgur, dried garlic, dried onion, lettuce, red (capia) pepper, green pepper, lemon, parsley, mint, spring onion, salt
Dessert	Stuffed figs	Cinnamon, walnut, dried figs, granulated sugar
	Pumpkin dessert	Pumpkin, granulated sugar, walnuts
	Compote	Fresh fruit, sugar
	Dried Fruit Compote (Hoşaf)	Dried fruit, sugar
	Lokma dessert (fried sweet dough balls)	Sunflower oil, wheat flour, yeast, granulated sugar

An Overview of Nutritional Diversity in Congregate Nutrition Systems

Although **the** vegan lifestyle is a type of diet that restricts travel decisions and travel experiences, vegan tourists are determined to travel. Vegan tourists plan their trips by researching food options and environmentally friendly accommodations in advance. This suggests the need for more understanding and education about the vegan lifestyle in the tourism industry (Ottenbacher, Busam, Harrington & Allhoff, 2023). As mentioned above, many Turkish dishes have the majority and quality to meet the nutritional needs of individuals who prefer vegan diets. The most important issue here is menu planning in accordance with dietary types, sustainable purchasing, controlled processing, and serving meals to vegan/vegetarian guests with staff who are informed about the menu. The prominence of veganism and nutrition understanding in public nutrition services and the quality to be achieved in this regard are compiled from various studies and discussed below (Sezgin & Özkaya, 2014; Nebioğlu, 2023; Özdemir & Altıner, 2019).

1. The vegetarian and vegan meals served must be able to meet the energy and nutritional requirements of the target group in adequate amounts according to the

standards in line with their age, gender, body weight, physical activity, special conditions, etc. The most important point here is to obtain information about the food allergies and food preferences of the guests upon arrival and to work in coordination with the kitchen department, food engineer and/or dietician in charge.

2. Labeling of vegan products plays a critical role in protecting consumer health, promoting ethical and environmental values, enhancing market credibility and ensuring legal compliance. Therefore, accurate and transparent labeling of vegan products is of great importance for both consumers and producers (Erdoğan & Gürbüz, 2023). Therefore, it is important to check that the V-Label emblem is on the package for sustainable purchases of packaged products that can be consumed by vegan and vegetarian individuals in public nutrition services. This label symbol indicates that the food is free from animal products and by-products.
3. In menus planned in accordance with the needs of consumers, menu planning should be done in an appropriate number within the principle of diversity and balanced between meals within the principle of sufficiency. When making vegan and vegetarian menu lists, menus should be organized periodically according to the season in which the products are grown so that the carbon footprint is the lowest depending on the change in the supply season of fruits and vegetables.
4. One of the critical points in menu planning is the calculation of the nutritional value of meals per portion. These calculations should be made by working with the dietitian working in the institution and the guest should be informed that the information will be transferred by the executive dietitian when entering the accommodation establishment. This transfer will create a sense of trust in guests who have a different dietary approach compared to the traditional one and increase satisfaction.
5. One of the most important issues in accommodation establishments, catering establishments and all other institutions/organizations providing collective nutrition services is the preparation of all menu meals by fully implementing food safety and quality management systems. This process is a journey that starts in the receiving area and continues until it is presented to the consumer. Therefore, it would be the best decision to work with personnel trained in food safety (ISO 22000), quality management systems and vegan/vegetarian food.
6. Being aware of Turkey's traditional food and cultural diversity, it is important to identify strengths, weaknesses, opportunities and threats in order to create strategies to increase the number of guests eating vegan and vegetarian diets and to create targets to increase market power. On Veggie-hotels.com, one of the websites used by vegans to create their vacation plans, there is only one hotel from Turkey, revealing the gap in the country's sector.
7. Studies should be carried out to re-prescribe traditional Turkish dishes by taking into account their taste, smell, texture, color, etc. to cover the vegan diet and standardization should be implemented. In this regard, it is important that gastronomy departments of universities work on the subject and raise awareness by organizing trainings.

Vegan diets have attracted increasing attention in recent years for health, environmental and ethical reasons. Research in the field of gastronomy addresses various aspects of this diet. For example, a study conducted at Istanbul Gelisim University examined restaurant employees' knowledge and attitudes towards vegan and vegetarian diets. This study revealed that the knowledge level of the employees was partially sufficient and that improvements could be made in this regard with further training (Yılmaz & Albayrak; 2023).

The spread of vegan nutrition in the field of gastronomy affects and changes many factors from menu planning to customer satisfaction. Studies such as examining the cuisine of Giresun region within the scope of vegetarian nutrition show how local cuisines can be adapted to vegetarian diets (Sezgin & Ayyıldız; 2019). Such studies will contribute to the integration of local cuisines into sustainable and healthy eating habits.

The effects of a vegan diet on gastronomy tourism, which is one of the special interest tourism types, are also important. The increasing popularity of vegetarian and vegan tourism makes it imperative for destinations to offer services suitable for these diets, and it is important to create various systems that serve the purpose. It would be appropriate to plan vegan and vegetarian food and beverages with various colors or shapes on menus or in the buffets of accommodation establishments as a guide. Developing standardized recipes for meals in mass nutrition systems according to vegan/vegetarian diets and presenting them on menu cards with appropriate coding will guide consumers in food selection. Informing consumers with a colored sign to be placed next to the dishes that can be consumed by ovo-vegetarians in menus is one of the methods that can be applied.

In conclusion, research on vegetarian nutrition in the field of gastronomy reveals the different dimensions of this diet and increases the level of knowledge of those working in the sector while contributing to the spread of sustainable healthy eating habits. In this context, the development of more research and training programs in the future is important for a better understanding and implementation of vegetarian nutrition in the gastronomy sector.

Conclusion

Veganism and vegetarianism are globally widespread dietary patterns. Health, environmental sustainability, animal welfare and cultural factors are important factors in the adoption of these diets. With the increasing popularity of vegetarian and vegan tourism, the importance of vegan and vegetarian menus in the tourism sector is increasing. Accordingly, it is important for food and beverage service providers to develop standardized recipes and menu planning according to vegan/vegetarian diets. In this way, guest satisfaction will be increased and the business will appeal to a wider audience.

References

Altaş, A. (2017). Vegetarianism and Veganism: Current Situation in Turkey in the Light of Examples in the World. Journal of Tourism and Gastronomy Studies, 5(4), 403–421.

Altınel, H. (2011). Menü Yönetimi ve Menü Planlama, Ankara: Detay Yayıncılık.

Aytaç, G. (2022). Menülerinde vegan ürünlere yer veren yiyecek içecek işletmelerinin sürdürülebilirlik açısından incelenmesi: İstanbul'daki mutfak şefleri üzerine bir araştırma. Yüksek Lisans Tezi, İstanbul Gelişim Üniversitesi

Ayyıldız, S., & Sezgin, A. C. (2021). Vejetaryen/vegan bireylerin yiyecek içecek işletmeleri tercihleri ve sorunları. Journal of Tourism Research Institute, 2(1), 1–16.

Baiomy, A. E., Jones, E., & Goode, M. M. H. (2019). The influence of menu design, menu item descriptions and menu variety on customer satisfaction. A case study of Egypt. Tourism and Hospitality Research, 19(2), 213–224. https://doi.org/10.1177/1467358417708228

Balcı, T. N. (2018). Türkiye'de yaşayan vegan ve vejetaryen bireylere özgü besin tüketim sıklığı anketi geliştirilmesi. Yüksek lisans tezi, Hacettepe Üniversitesi, Ankara.

Barnard, N. D., Alwarith, J., Rembert, E., Brandon, L., Nguyen, M., Goergen, A., Horne, T., do Nascimento, G. F., Lakkadi, K., Tura, A., Holubkov, R., & Kahleova, H. (2022). A Mediterranean Diet and Low-Fat Vegan Diet to Improve Body Weight and Cardiometabolic Risk Factors: A Randomized, Cross-over Trial. *Journal of the American Nutrition Association*, *41*(2), 127–139. https://doi.org/10.1080/07315724.2020.1869625

Beam, A., Clinger, E., & Hao, L. (2021). Effect of Diet and Dietary Components on the Composition of the Gut Microbiota. *Nutrients*, *13*(8), 2795. https://doi.org/10.3390/nu13082795

Bilgiçli, İ. & Altınkaynak, F. (2016). Turizm endüstrisinin Türkiye ekonomi içindeki yeri ve önemi; Ekonomi paradigmasıyla yaklaşım. Uluslararası Yönetim iktisat ve İşletme Dergisi, 12(12), 560–580.

Bulduk, S. (2013). Beslenme İlkeleri ve Mönü Planlama, Ankara: Detay Yayıncılık.

Bowen, J.T. & Morris, A.J. (1995). Menu Design: Can Menus Sell? *International Journal of Contemporary Hospitality Management*, 7 (4): 4-9.

Craig, W. J., Mangels, A. R., Fresán, U., Marsh, K., Miles, F. L., Saunders, A. V., Haddad, E. H., Heskey, C. E., Johnston, P., Larson-Meyer, E., & Orlich, M. (2021). The Safe and Effective Use of Plant-Based Diets with Guidelines for Health Professionals. *Nutrients, 13*(11), 4144. https://doi.org/10.3390/nu13114144

Crowe, F. L., Appleby, P. N., Travis, R. C., & Key, T. J. (2013). Risk of hospitalization or death from ischemic heart disease among British vegetarians and non-vegetarians: results from the EPIC-Oxford cohort study. *The American journal of clinical nutrition, 97*(3), 597–603. https://doi.org/10.3945/ajcn.112.044073

Çeken, H. & Ateşoğlu, L. (2008). Küreselleşme sürecinde turizm endüstrisinin Avrupa Birliği ve Türkiye ekonomisindeki yeri ve önemi. Gazi Üniversitesi Ticaret ve Turizm Eğitim Fakültesi Dergisi (1), 136–151.

Dilek, S. E. (2018). Türkiye'de Vejetaryen/Vegan Oteller Mümkün Mü? Kavramsal Bir Tartışma. Dokuz Eylül Üniversitesi İşletme Fakültesi Dergisi, 19(1), 1–18.

Djekic, D., Shi, L., Brolin, H., Carlsson, F., Särnqvist, C., Savolainen, O., Cao, Y., Bäckhed, F., Tremaroli, V., Landberg, R., & Frøbert, O. (2020). Effects of a Vegetarian Diet on Cardiometabolic Risk Factors, Gut Microbiota, and Plasma Metabolome in Subjects With Ischemic Heart Disease: A Randomized, Crossover Study. *Journal of the American Heart Association, 9*(18), e016518. https://doi.org/10.1161/JAHA.120.016518.

Erdoğan, Z., & Gürbüz, F. (2023). Vegan ve Vejetaryen Tüketim Anlayışının Sürdürülebilirlik Rolünde Vegan Markaların Analizi. Journal Of Marketing & Marketing Research/Pazarlama & Pazarlama Arastirmalari Dernegi, 16(3).

Eren, R., Çetin, M., Eren, A., & Çetin, K. (2021). Food allergy knowledge, attitude, and practices of chefs in resort hotels in Turkey. *International Journal of Gastronomy and Food Science, 24*, 100345.

Eren, S. (2020). Food-Related Barriers to Gastronomy and Tourism: The Role of Food Allergies and Diet Preferences. MANAS Journal of Social Studies, 9(2), 1117–1129.

Eryücel, Ö., Yalçınkaya, T. ve Küçükaltan, D. (2020). Restoran Çalışanlarının Bilgi ve Bilinç Düzeylerinin Vegan/Vejetaryenler Tarafından Değerlendirilmesi. Içinde Ş. Aydın, Y. Karakuş, Ö. Çoban ve N. Çalışkan (Ed.), Gastronomi Turizminde Güncel Araştırmalar (ss. 167–180). Detay Yayıncılık.

Estruch, R., Sacanella, E., Ros, E. (2021). Should we all go pesco-vegetarian? *European Heart Journal*, Volume 42, Issue 12(21), 1144–1146, https://doi.org/10.1093/eurheartj/ehaa1088

Ferdowsian, H. R., Barnard, N. D., Hoover, V. J., Katcher, H. I., Levin, S. M., Green, A. A., & Cohen, J. L. (2010). A multicomponent intervention reduces body weight and cardiovascular risk at a GEICO corporate site. *American journal of health promotion: AJHP*, *24*(6), 384–387. https://doi.org/10.4278/ajhp.081027-QUAN-255

Fox, N. & Ward, K. (2008). Health, Ethics and Environment: A Qualitative Study of Vegetarian Motivations. Appetite, 50, 422429.

Garousi, N., Tamizifar, B., Pourmasoumi, M., Feizi, A., Askari, G., Clark, C. C. T., & Entezari, M. H. (2023). Effects of lacto-ovo-vegetarian diet *vs.* standard-weight-loss diet on obese and overweight adults with non-alcoholic fatty liver disease: a randomised clinical trial. *Archives of Physiology and Biochemistry*, *129*(4), 975–983. https://doi.org/10.1080/13813455.2021.1890128

Ho, C. P., Yu, J. H., & Lee, T. J. F. (2017). Ovo-vegetarian diet is associated with lower systemic blood pressure in Taiwanese women. *Public health*, *153*, 70–77. https://doi.org/10.1016/j.puhe.2017.07.032

Huang, Y. C., Chen, L. H., Lu, C. W. & Shen, J. L. (2020). Being A Vegetarian Traveller is not Easy. British Food Journal, 122(6) 1983–1998.

Kahleova, H., Fleeman, R., Hlozkova, A., Holubkov, R., & Barnard, N. D. (2018). A plant-based diet in overweight individuals in a 16-week randomized clinical trial: metabolic benefits of plant protein. *Nutrition & diabetes*, *8*(1), 58. https://doi.org/10.1038/s41387-018-0067-4

Kahleova, H., Hlozkova, A., Fleeman, R., Fletcher, K., Holubkov, R., & Barnard, N. D. (2019). Fat Quantity and Quality, as Part of a Low-Fat, Vegan Diet, Are Associated with Changes in Body Composition, Insulin Resistance, and Insulin Secretion. A 16-Week Randomized Controlled Trial. *Nutrients*, *11*(3), 615. https://doi.org/10.3390/nu11030615

Kahleova, H., Rembert, E., Alwarith, J., Yonas, W. N., Tura, A., Holubkov, R., Agnello, M., Chutkan, R., & Barnard, N. D. (2020). Effects of a Low-Fat Vegan Diet on Gut Microbiota in Overweight Individuals and Relationships with Body Weight, Body Composition, and Insulin Sensitivity. A Randomized Clinical Trial. *Nutrients*, *12*(10), 2917. https://doi.org/10.3390/nu12102917

Karabudak, E. (2012). Vejetaryen Beslenmesi, Sağlık Bakanlığı Yayını, No: 7, s. 8

Kaytez, N. & Tunçay, G. Y. (2020). Effect of being vegan/ vegetarian on social relations (University student samples). Çankırı Karatekin Üniversitesi Sosyal Bilimler Enstitüsü Dergisi, 11(1), 210–231.

Key, T. J., Fraser, G. E., Thorogood, M., Appleby, P. N., Beral, V., Reeves, G., Burr, M. L., Chang-Claude, J., Frentzel-Beyme, R., Kuzma, J. W., Mann, J., & McPherson, K. (1999). Mortality in vegetarians and nonvegetarians: detailed findings from a collaborative analysis of 5 prospective studies. *The American journal of clinical nutrition*, *70*(3 Suppl), 516S–524S. https://doi.org/10.1093/ajcn/70.3.516s

Kızıldemir, Ö. & Hülağa Kaderoğlu, G. (2021). Yiyecek içecek işletmelerindeki menü tasarılarının sürdürülebilirlik kapsamında değerlendirilmesi. Journal of Tourism Intelligence and Smatness, 4(2), 296–322.

Kutluay Merdol, T. (2018). Standart yemek tarifeleri. (7. Baskı). Ankara: Hatiboğlu Yayınları.

Küçükkömürler, S., Şırvan, N. B. & Sezgin, A. C. (2019). Dünya'da ve Türkiye'de Gastronomi Turizmi. Uluslararası Turizm Ekonomi ve İşletme Bilimleri Dergisi, 2(2), 78–85.

Landry, M. J., Ward, C. P., Cunanan, K. M., Durand, L. R., Perelman, D., Robinson, J. L., Hennings, T., Koh, L., Dant, C., Zeitlin, A., Ebel, E. R., Sonnenburg, E. D., Sonnenburg, J. L., & Gardner, C. D. (2023). Cardiometabolic Effects of Omnivorous vs Vegan Diets in Identical Twins: A Randomized Clinical Trial. *JAMA network open*, *6*(11), e2344457. https://doi.org/10.1001/jamanetworkopen.2023.44457

Lee, K. H., Scott, N., & Packer, J. (2014). Where does food fit in tourism? Tourism Recreation Research, 39(2), 269–274.

Meat Atlas-Facts and Figures About the Animals We Eat. http://www.foeeurope.org/sites/default/files/publications/foee_hbf_ meatatlas_jan2014 .pdf, E.T.: 06.11.2022.

Nebioğlu, O. (2023). Yiyecek içecek işletmelerinde menü araştırmaları: Türkiye'de yazılmış lisansüstü tezler üzerine sistematik bir literatür taraması. *Nevşehir Hacı Bektaş Veli Üniversitesi Sbe Dergisi, 13*(2), 982 1001.

O'Keefe, J. H., Torres-Acosta, N., O'Keefe, E. L., Saeed, I. M., Lavie, C. J., Smith, S. E., & Ros, E. (2020). A Pesco-Mediterranean Diet With Intermittent Fasting: JACC Review Topic of the Week. *Journal of the American College of Cardiology*, *76*(12), 1484–1493. https://doi.org/10.1016/j.jacc.2020.07.049

Orlich, M. J., & Fraser, G. E. (2014). Vegetarian diets in the Adventist Health Study 2: a review of initial published findings. *The American journal of clinical nutrition, 100 Suppl 1*(1), 353S–8S. https://doi.org/10.3945/ajcn.113.071233

Ottenbacher, M., Busam, S., Harrington, R. J., & Allhoff, J. (2023). The factors impacting on the travel experience of a vegan lifestyle. *International Hospitality Review.*

Oussalah, A., Levy, J., Berthezène, C., Alpers, D. H., & Guéant, J. L. (2020). Health outcomes associated with vegetarian diets: An umbrella review of systematic reviews and meta-analyses. *Clinical Nutrition*, *39*(11), 3283–3307. https://doi.org/10.1016/j.clnu.2020.02.037

Özdemir, G., & Altıner, D. D. (2019). Gastronomi kavramlari ve gastronomi turizmi üzerine bir inceleme. *Erzincan Üniversitesi Sosyal Bilimler Enstitüsü Dergisi*, *12*(1), 1–14.

Papier, K., Appleby, P. N., Fensom, G. K., Knuppel, A., Perez-Cornago, A., Schmidt, J. A., Tong, T. Y. N., & Key, T. J. (2019). Vegetarian diets and risk of hospitalisation or death with diabetes in British adults: results from the EPIC-Oxford study. *Nutrition & diabetes*, *9*(1), 7. https://doi.org/10.1038/s41387-019-0074-0

Peruvemba, S., Gieng, J., Chen, S. & Pignotti, G. (2022). Diet Motives and Dietary Quality in Vegans, Vegetarians, and Semi-Vegetarians, Journal of Nutrition Education and Behavior, Volume 54, Issue 7, Supplement, Page S22, ISSN 1499-4046, https://doi.org/10.1016/j.jneb.2022.04.049.

Eren, R., Çetin, M., Eren, A., & Çetin, K. (2021). Food allergy knowledge, attitude, and practices of chefs in resort hotels in Turkey. International Journal of Gastronomy and Food Science, 24, 100345.

Rizzo, N. S., Sabaté, J., Jaceldo-Siegl, K., & Fraser, G. E. (2011). Vegetarian dietary patterns are associated with a lower risk of metabolic syndrome: the adventist health study 2. *Diabetes care*, *34*(5), 1225–1227. https://doi.org/10.2337/dc10-1221

Sezgin, A. C., & Ayyıldız, S. (2019). Gastronomi alanında vejetaryen/vegan yaklaşımı; Giresun yöre mutfağı'nın vejetaryen mutfak kapsamında incelenmesi. *Siyasi Sosyal ve Kültürel Yönleriyle Türkiye ve Rusya*, *1*(1), 505–53.

Sezgin, A. C., & Özkaya, F. D. (2014). Toplu Beslenme Sistemlerine Genel Bir Bakış. Akademik Gıda, 12(1), 124–128.

Sünnetçioğlu, S., Mercan, Ş. O., Yıldırım, H. M. & Türkmen, S. (2017). Veganların Restoranlarda Karşılaştıkları Sorunlar Üzerine Bir Araştırma. Journal of Tourism and Gastronomy Studies, 5(Special Issue 2), 241–252.

Şahin, E., Yazıcıoğlu, İ. (2018). Menü Tasarımının Yemek Seçim Kararına Etkisi Üzerine Bir Değerlendirme. Journal of Tourism and Gastronomy Studies, 6(4), 900–913. 10.21325/jotags.2018.340

Tabu, Ö. (2020). Gastronomi Turizmi Kapsamında Türkiye'nin Vegan, Vejetaryen, Pesketaryen Yöresel Yemekleri. Yüksek Lisans Tezi, Gümüşhane Üniversitesi, Gümşhane.

Tantamango-Bartley, Y., Jaceldo-Siegl, K., Fan, J., & Fraser, G. (2013). Vegetarian diets and the incidence of cancer in a low-risk population. *Cancer epidemiology, biomarkers & prevention*, *22*(2), 286–294.

The Vegan Society. (2024). Vegan tanımı. 22 Mayıs 2024 tarihinde https://www.vegansociety.com/go-vegan/definition-veganism adresinden erişildi.

TOPLU BESLENME SİSTEMLERİ (TOPLU TÜKETİM YERLERİ) İÇİN ULUSAL MENÜ PLANLAMA VE UYGULAMA REHBERİ, Beyhan Yasemin, Bilici Saniye, Kızıl Mevlüde, Editör, T.C. Sağlık Bakanlığı Halk Sağlığı Genel

Müdürlüğü Sağlıklı Beslenme ve Hareketli Hayat Dairesi Başkanlığı, Ankara, ss. 1–48, 2020

Turner-McGrievy, G. M., Davidson, C. R., Wingard, E. E., Wilcox, S., & Frongillo, E. A. (2015). Comparative effectiveness of plant-based diets for weight loss: a randomized controlled trial of five different diets. *Nutrition (Burbank, Los Angeles County, Calif.), 31*(2), 350–358. https://doi.org/10.1016/j.nut.2014.09.002

Uçan, B. Z. ve Bozok, D. (2019). Veganizm ve Gastronomi. Içinde Ş. Aydın, Ö. Çoban, Y. Karakuş ve N. Çalışkan (Ed.), 4. Uluslararası Gastronomi Turizmi Araştırmaları Kongresi Bildiri Kitabı (ss. 32–39).

Vatan, A. & Türkbaş, S. (2018). Vejetaryen Turist ve Vegan Turist Kimdir? Journal of Tourism and Gastronomy Studies, 6(3), 24–39. 10.21325/jotags.2018.270

Wang, F., Zheng, J., Yang, B., Jiang, J., Fu, Y., & Li, D. (2015). Effects of Vegetarian Diets on Blood Lipids: A Systematic Review and Meta-Analysis of Randomized Controlled Trials. *Journal of the American Heart Association, 4*(10), e002408. https://doi.org/10.1161/JAHA.115.002408

Wright, N., Wilson, L., Smith, M., Duncan, B., & McHugh, P. (2017). The BROAD study: A randomised controlled trial using a whole food plant-based diet in the community for obesity, ischaemic heart disease or diabetes. *Nutrition & diabetes, 7*(3), e256. https://doi.org/10.1038/nutd.2017.3

Yılmaz, Y. (2006). Konaklama& Ağırlama İşletmelerinde Servis Tekniği ve Yönetimi, Ankara: Detay Yayıncılık.

Yılmaz, Y., & Albayrak, A. (2023). Restoran Çalışanlarının Vegan ve Vejetaryen Beslenmeyle İlgili Bilgi Düzeyleri ve Tutumlarının Ölçülmesi. İstanbul Gelişim Üniversitesi Sosyal Bilimler Dergisi, 10(1), 214–233. https://doi.org/10.17336/igusbd.995215

Yokoyama, Y., Nishimura, K., Barnard, N. D., Takegami, M., Watanabe, M., Sekikawa, A., Okamura, T., & Miyamoto, Y. (2014). Vegetarian diets and blood pressure: a meta-analysis. *JAMA internal medicine, 174*(4), 577–587. https://doi.org/10.1001/jamainternmed.2013.14547

Zamir, T. (2004). Veganism, Journal of Social Philosophy, Vol:35, No:3, pp. 368

Necibe Şen* and Ayşe Çelik Yetim†

Chapter 31 Digital Transformation and Environmental Sustainability in Tourism Businesses

Introduction

The tourism sector has witnessed a surge in digital transformation, particularly accelerated by the pandemic. There's a growing belief that leveraging technological advancements can serve as an asset for promoting sustainable tourism. This section unveils the interplay between digital transformation and environmental sustainability through a review of relevant literature. By examining the current state of affairs among businesses in the tourism sector, potential advancements are explored. Technological breakthroughs like mobile apps, online booking systems, and digital marketing strategies are reshaping tourists' preferences and behaviors, making travel experiences more accessible and personalized. Concurrently, challenges such as global climate change, dwindling natural resources, and environmental pollution profoundly impact the tourism sector. In response, businesses must pioneer innovative solutions in areas like renewable energy utilization, waste management, and water conservation. The integration of digital technologies into these solutions amplifies businesses' capacity to mitigate environmental footprints and enhance resource efficiency. With the use of technologies such as artificial intelligence, big data analytics and the internet of things, businesses are expected to become more competent in measuring, improving and reporting their environmental performance. The adoption of sustainable digital transformation practices among businesses not only fosters sustainability but also elevates customer satisfaction. This section aims to unveil existing practices and potential advancements while contributing to the literature by linking digital transformation with sustainability concerns. Additionally,

* PhD., Erzincan Binali Yıldırım University, Tourism and Hotel Management Vocational School, Department of Travel Tourism and Entertainment Services, nemin@erzincan.edu.tr

† PhD., Muğla Sıtkı Koçman University, Fethiye Faculty of Business Administration, Department of Tourism Management, aysecelik@mu.edu.tr

businesses will receive recommendations regarding the technologies essential for shaping a sustainable future.

Digital Transformation in Tourism

Digitalization involves using digital technologies to improve existing business processes like distribution and communication to save costs and enhance efficiency, potentially improving customer experiences. Otherwise, digital transformation is a comprehensive shift that leads to creative business models, extending beyond the mere digitalization of individual procedures or assignments (Verhoef et al., 2021). In order to achieve the strategic business goals, digitalization and transformation are essential (Bekele ve Raj, 2024).

Digitalization has a major influence on the present phase of global economic and social development (Andriushchenko et al. 2020). Innovation and digital transformation have become important factors impacting the sector's businesses' competitiveness and sustainability (Abeba, 2024). Businesses should think about how to provide value for customers while also taking sustainability into account (Chiang et al. 2019; Kandampully et al., 2023). The digital terrain of the world has significantly improved with the rise in globalisation because of the high-speed advancements in technology, including the internet, big data, blockchain, and artificial intelligence (AI)(Buhalis and Sinarta, 2019). It is possible to add all the topics brought by industry 4.0 such as cloud technologies, augmented reality (AR), virtual reality (VR), extended reality (ER), mixed reality (MR), internet of things (IoT), robotic technologies, social media, smart destinations and smart cities to the developments in technology.

During the Covid-19 pandemic, digitalization in the tourism industry increased (Wen et al., 2021). According to the OECD Tourism Trends and Policies report, countries have taken many precautions to escape the effects of the Pandemic. Chief among these are sustainability studies and competition (OECD, 2022). The nexus between digital technology and the tourism sector has been the subject of numerous study contributions in the past few years (Bekele and Raj, 2024). As global businesses and societies embrace digitalization, hotel firms are leveraging advanced technologies to enhance their operations, services, and overall performance. A long-lasting competitive advantage is guaranteed by these quick advancements (Centobelli et al. 2020). Digital transformation is required in many areas of the hotel industry's operations by reason of the mismatch between IT operations and business strategy (Wynn and Jones, 2022). Because of this, integrating digital solutions is not only advantageous but also required to maintain competitiveness in the ever-changing tourism industry.

Businesses in the tourism and travel industry are implementing new technologies to solve also operational problems (Queiroz and Wamba, 2019). Supply chain digitalization enables companies to enhance their operations through sustainable methods (Gusmão Caiado et al. 2018). Effective resource coordination, energy conservation, and the reduction of environmental hazards, data technology (Babu et al. 2018), paperless check-ins and electronic invoicing enhances operational efficiency and promotes environmental sustainability (Xiang, 2017).

Digital technology is essential for efficiency, communication, and promotion in tourism businesses, and enable online reservations, marketing, and cost-saving measures for both businesses and tourists (Kurniawati, Kohar, Meiji, Handayati & Ilies, 2022). Digital tools expand market reach, and reduce infrastructure costs. Online platforms like social media, search engines, and websites facilitate digitalization in tourism businesses (Kurniawati et al. 2022). Implementing mobile apps and platforms for online reservations streamlines the procedure for making reservations, enhances guest experiences, improves efficiency, and boosts revenue by expanding online market reach (Buhalis & Leung, 2018). Integrating data analytics and AI helps businesses understand consumer behavior and preferences, facilitating customised goods and targeted marketing (Sigala, 2020). The literature on sustainability has given blockchain technology a great deal of attention (Prados-Castillo et al. 2023). This technology, which allows reliable digital currency exchange using the blockchain technical infrastructure, has found its counterpart in the industry in a short time. UAE Wallet, Lufthansa, Tui Group Tripio, Snapcity are some usage examples.

As examined in the literature, there are studies on digitalization and digital transformation in tourism businesses. Studies on this subject show how comprehensive the subject is. In the following section, Madzik et al. (2023) and the topics covered in Bekele and Raj's (2024) work are briefly mentioned. According to Madzík, et al. (2023) subjects are cultural heritage, digital destination marketing, social media, data analytics, sustainable and economic development, technology-based experience and interaction, city and urban planning, and intelligent tourist management. According to Bekele and Raj (2024) studies in the field are grouped under the topics of digital innovation, ecosystem for smart tourism, e-Tourism and smart tourism experience. Future research in the multidisciplinary topic of digitization and transformation is expected to be very important and some of them are given below.

Smart Tourism: Smart tourism includes the smart technology application in the travel industry. Understanding the effects of smart tourism policies and tactics, the veracity of smart tourism offers, smart visitors' behaviour, and the contribution of smartness to the achievement of sustainable development goals (SDGs) is crucial (Bekele and Raj, 2024).

İmmersive Technologies: According to Flavian et al. (2019), immersive technologies improve user experience. Through the seamless integration of virtual entities into the physical world, immersive technologies provide users with an interactive, visibly stimulating, and immersive experience by overlaping artificial visuals onto actual ones (Pratisto et al., 2022). Virtual, augmented, mixed realities (Flavian et al. 2019), extended reality are reshaping customer experiences.

Generative Artificial İnteligence (GAI): GAI is a subset of AI that can produce language and creative content that resembles that of a human being and combine data from many sources for analytical reasons (Dasborough 2023). GAI can improve the productivity of tourism businesses by simplifying processes, automating regular jobs, enhancing procedures for making decisions, and enhancing the businesses' capacity to provide personalised data and goods based on previous searches and preferences (Carvalho and Ivanov 2023).

Digital Transformation: Businesses must start a digital transformation journey in order to survive and thrive in the current disruptive market environment (Nadkarni and Prügl, 2021). Digital transformation is driven by strategy and in this situation workforce is important. The most effective digital revolutions begin with a mentality change among staff members, executives, and organisations. The company's culture changes as a result of this, making it more flexible, risk-taking, experimental, and cooperative (Kane, 2019). According to the digital economy report, 8/10 businesses are in the digital transformation process(Cançelik, 2021).

Sustainable Tourism Development (STD): Understanding how the tourist sector is undergoing digital transformation is essential to accomplishing the agenda for STD and the Sustainable Development Goals (SDGs) (Bekele and Raj, 2024). Social, economic, especially environmental sustainability, has always been important for the continuity of the tourism sector. Organisations can lessen their dependency on physical resources, cut down on carbon emissions, use less energy, and produce less waste by integrating digitalization with sustainable business practices (Moghrabi et al., 2023).

Environmental Sustainability in Tourism

Global businesses are adopting sustainable strategies nowadays to accomplish sustainable objectives (Luthra and Mangla, 2018), particularly in the travel and tourism industry (Riaz et al. 2020). Sustainable tourism aims to benefit local communities, raise awareness, and support conservation (Kurniawati et al. 2022). The tourist industry is essential to improving the natural environment on Earth (Han et al. 2019). Environmental sustainability can be explained as without harming the ecosystem balance of destinations while meeting the needs of people (Morelli, 2011: 2). Moldan et al. (2012: 6) Environmental sustainability can be achieved through prioritizing nature and biodiversity, efficiently planning, using, and managing natural resources, ensuring ecosystem integrity, utilizing indicators in decision-making, improving quality of life (especially for local communities), fostering global environmental commitment, emphasizing recycling, preventing hazardous substances from entering or being released into the environment, efficiently utilizing non-renewable resources, and promoting stakeholder participation. It is the most utilized dimension in maintaining destination continuity, aiming to reduce resource use and pollution while ensuring optimal use of resources, including renewable and natural resources, and protecting ecological processes (Mathew ve Sreejesh, 2017). Tourism businesses must adhere to environmental standards in light of the growing sustainability consciousness (Sarfraz et al. 2023). Feroz et al. (2021) categories include pollution control, waste management, sustainable production and urban sustainability;

Pollution control: Pollution control contains Air pollution, C02 emission, Water treatment and climate change. Destinations should control pollution for the sake of improving quality of life. Enhancing air quality will also boost tourism demand. Measures need to be implemented to mitigate adverse conditions that may arise due to the growing tourism demand, which correlates with improved air quality. Local governments should spearhead initiatives to foster the development of tourism both within and between cities and regions, ensuring its sustainability (Dong, Xu, Yu ve Zhao, 2019). Digital technologies transform the ways pollution is measured, controlled, and managed (Feroz et. al. 2021).

Waste management: Feroz et al. (2021) categorizes waste management into solid waste, e-waste, food waste anda gri waste. Domestic and industrial waste harm the environment. Digital technologies are employed to find ways to manage such waste, including waste from electronic devices. Additionally, industrial and domestic food waste can be disposed of using digital technologies such as IoT and big data. These technologies are also utilized to manage the waste generated in the process of quality agricultural production (Feroz et al. 2021: 9). AI has applications in waste management, geoscience, waste management, wildlife care, environmental risk assessment, and land use planning. (Sharma, Yadav, and Chopra, 2020). Digital technologies are used to transform waste management. It is possible to encourage recycling habits among the general public by leveraging the impact of social media and the involvement of non-governmental organizations. (Sujata et al. 2019).

Sustainable production: Feroz et al. (2021) categorize sustainable production into sustainable manufacturing and sustainable supply chain. Sustainable Manufacturing: With the rise of the circular economy, clean industrial techniques driven by digital technology, aiming to avoid environmental harm, are sought. These methods can cut costs while lessening ecological impacts. In recent years, academia and industry have prioritized smart manufacturing for its competitive advantages and improvements in sustainability and efficiency (Ren et al. 2019). Sustainable Supply Chain: Businesses can eliminate waste across entire value chains, promote sustainable consumption, and mitigate environmentally harmful waste residues through the utilization of digital technologies. (Feroz et al. 2021: 9). The tourism industry is adopting effective supply chain tactics today to gain a long-term competitive edge. (Hatamifar et al. 2018). Manavalan et al. (2021) state that current supply chain tactics fail to mitigate environmental degradation's impact on nature. Experts suggest tourism businesses adopt a Sustainable Supply Chain Strategy (SSCS) for optimization, innovation, and growth (Sarfraz et al. 2023).

Urban sustainability: In the perspective of digital transformation, cities can be divided into smart cities and sustainable cities Feroz et al. (2021). By balancing the three dimensions of sustainability, minimizing resource consumption and environmental damage, enhancing resource use efficiency, and guaranteeing equality and democracy, the definitions aim to improve long-term human well-being (Huang et al. 2015). Smart Cities: Urban centers are becoming "smart cities" due to technology advancements that improve sustainability, increase well-being of inhabitants, increase the amount of energy that can be used efficiently, and create an environment that encourages good habits. Sustainable cities: With the use of digital technologies, smart cities may become even more sustainable by implementing clean energy usage, zero-waste creation methods, and sustainable consumption habits. (Feroz et al. 2021: 10).

Industry 4.0 technologies contribute significantly to achieving sustainability goals, primarily through addressing the environmental and economic aspects of sustainability. Industry 4.0 holds great promise for promoting sustainable industrial value creation across social dimensions, in addition to economic and environmental sustainability (Kamble et al. 2018).Kamble et al. (2018: 419) define

Sustainable 4.0 technology as incorporating process integration and delivering sustainable outcomes. They highlight how rapid industrialization has resulted in declining health and safety standards for workers and environmental degradation. The proposed framework by Kamble et al. (2018) acknowledges that Industry 4.0 enables business process integration and activities through cyber-physical interactions and process innovations, making production systems more flexible, cost-effective, and environmentally friendly.

Digital transformation enables businesses to access information more easily, leading to increased energy efficiency, expense control, and management efficiency. It contributes to the success of the target region in areas such as demand forecasting, crisis management, and tourism planning (Yıldız, 2019:93). Digital transformation can also be employed to achieve the environmental sustainability goals of businesses. It can improve infrastructure and transportation systems in smart cities, alleviate traffic congestion, facilitate waste management, and enhance the quality of human life (Mehmood et al. 2017).

Accor Hotels has sustainability strategy and has implemented the Planet 21 program, leveraging digital tools to monitor and reduce environmental impacts (reducing food waste, optimize energy consumption) (Accor.com, 2021). TUI Group uses digital platforms to enhance sustainable tourism. Their environmental management system tracks energy, water use and offer virtual tours to reduce the need for physical travel (Tuigroup.com, 2024).

Conclusion

Digitalization is a process that requires continuous adaptation. Businesses must stay abreast of technological developments to compete effectively and remain in the market. Incorporating ever-evolving current technologies into business processes is essential. Digitalization positively impacts customer satisfaction, loyalty, and perceived service quality. Customers actively follow technological innovations and share their experiences on social media. Through word-of-mouth marketing and mobile word-of-mouth marketing, these comments and ideas can reach a wide audience quickly. Customer experience is crucial in building customer satisfaction. Therefore, businesses aiming to enhance the customer experience should adopt and integrate new technologies. Moreover, during the digital transformation process, businesses should demonstrate their commitment to the environment and society by adopting sustainable practices.

It can be said that transferring the trust we received from our ancestors to future generations of resources and destinations is one of the most used expressions for sustainability. Improving the quality of life of the community while ensuring this transition should be one of the most important goals. Digital transformation is a technology that is used today to ensure sustainability and its impact is predicted to increase rapidly. In this context, pollution control, waste management, sustainable production and urban sustainability should benefit from areas called digital transformation (IoT, cyber-physical systems, cloud computing, AI, big data analytics, and digital twin). While the use of technology ensures that the footprint we will leave in nature is minimized, it also ensures that the activities are spread to large masses.

References

Abeba , L. (2024). Role of Digital Transformation and Innovation in Enhancing the Competitiveness and Sustainability of Hospitality and Tourism Businesses in South Africa. International Journal of Modern Hospitality and Tourism, 4(1), 50–62. https://doi.org/10.47604/ijmht.2375.

Andriushchenko, K., et al. (2020). Peculiarities of sustainable development of enterprises in the context of digital transformation. Entrepren. and Sustain. Issues, 7(3), 2255–2270. http://doi.org/10.9770/jesi.2020.7.3(53)

Buhalis, D., & Leung, D. (2018). Smart hospitality—Interconnectivity and interoperability towards an ecosystem. International Journal of Hospitality Management, 71, 41–50.

Buhalis, D., & Sinarta, Y. (2019). Real-time co-creation and nowness service: lessons from tourism and hospitality. J Travel Tourism Marketing 36(5):563–582. https://doi.org/10.1080/10548408.2019.1592059

Cançelik, M. (2021). "Dijital Dönüşüm ve Turizm" içinde Dijital Dönemde İşletmecilik, Ed. Ülkü, Abdullah; Nobel Akademik Yayıncılık

Carvalho, I. and Ivanov, S. (2023). "ChatGPT for tourism: applications, benefits and risks", Tourism Review, Vol. 79 No. 2, doi: 10.1108/TR-02-2023-0088.

Centobelli, P., Cerchione, R., Esposito, E. (2020). Pursuing supply chain sustainable development goals through the adoption of green practices and enabling technologies: A cross-country analysis of LSPs. Tech. Forec Soc Change 153:119920 https://doi.org/10.1016/j.techfore.2020.119920

Chiang, C. -F., Chen, W. -Y., Hsu, C. -Y. (2019). Classifying technological innovation attributes for hotels: an application of the Kano model. J Travel Tourism Marketing 36(7):796–807. https://doi.org/10.1080/10548408.2019.1575786

Dasborough, M. T. (2023). "Awe-inspiring advancements in AI: the impact of ChatGPT on the field of organizational behavior", Journal of Organizational Behavior, Vol. 44 No. 2, pp. 177–179. https://doi.org/10.1002/job.2695

Dong, D.; Xu, X.; Yu, H.; Zhao, Y. (2019) The Impact of Air Pollution on Domestic Tourism in China: A Spatial Econometric Analysis. Sustainability, 11,4148. https://doi.org/10.3390/su11154148

Feroz, A.K.; Zo, H.; Chiravuri, A. Digital Transformation and Environmental Sustainability: A Review and Research Agenda. Sustainability 2021, 13, 1530. https:// doi.org/10.3390/su13031530.

Flavian, C., Ibanez- Sanchez, S. and Orus, C. (2019). "The impact of virtual, augmented and mixed reality technologies on the customer experience", Journal of Business Research, Vol. 100, pp. 547–560, doi:10.1016/j.jbusres.2018.10.050.

Gusmão C. R. G., Leal F. W, Quelhas OLG, Luiz de Mattos Nascimento D, Ávila LV (2018). A literature-based review on potentials and constraints in the implementation of the sustainable development goals. J Cleaner Prod 198:1276–1288. https://doi.org/10.1016/j.jclepro.2018.07.102

Han, H., Yu, J., Kim, W. (2019). Environmental corporate social responsibility and the strategy to boost the airline's image and customer loyalty intentions. J Travel Tourism Marketing 36(3):371–383

Hatamifar, P., Darban Astane, A. & Rezvani, M. R. (2018). Analyzing Quality of Supply Chain Management in Hotels of Isfahan Using the Partial Least Squares (PLS). J Quality Assurance Hospitality Tourism 19(2):172–191. https://doi.org/10.1080/1528008X.2017.1331152

Huang, L., Wu, J., & Yan, L. (2015). Defining and measuring urban sustainability: a review of indicators. Landscape ecology, 30, 1175-1193.

Kandampully, J., et al. (2023). "Toward holistic experience-oriented service innovation: co-creating sustainable value with customers and society", Cornell Hospitality Quarterly, Vol. 64 No. 2, pp. 161–183, doi: 10.1177/19389655221108334.

Kamble, S. S., Gunasekaran, A., & Gawankar, S. A. (2018). Sustainable Industry 4.0 framework: A systematic literature review identifying the current trends and future perspectives. Process safety and environmental protection, 117, 408–425.

Kane, G. (2019). "The technology fallacy: people are the real key to digital transformation", Research Technology Management, Vol. 62 No. 6, pp. 44–49.

Kurniawati, E., et al. (2022). Digital Transformation for Micro, Small, and Medium Enterprises to Develop Sustainable Community-Based Marine Tourism. African Journal of Hospitality, Tourism and Leisure, 11(2), 1118–1127.

Luthra, S., Mangla, S. K. (2018). When strategies matter: Adoption of sustainable supply chain management practices in an emerging economy's context. Resources, Conservation Recycling 138:194–206. https://doi.org/10.1016/j.resconrec. 2018.07.005

Mathew, P. V., & Sreejesh, S. (2017). Impact of Responsible Tourism on Destination Sustainability and Quality of Life of Community in Tourism Destinations. Journal of Hospitality and Tourism Management, 31, 83-89. https://doi.org/10.1016/j.jhtm.2016.10.001.

Manavalan, E., Thanigai M.A., & Kandasamy, J. (2021). Chapter 14 - Sustainable supply chain management in manufacturing industries, In K. Gupta & K. Salonitis (Eds.), In Handbooks in Advanced Manufacturing, Sustainable Manufacturing, (pp. 367-389). Elsevier.

Madzík, P., Falát, L., Copuš, L. and Valeri, M. (2023). "Digital transformation in tourism: bibliometric literature review based on machine learning approach", European Journal of Innovation Management, Vol. 26 No. 7, pp. 177–205. https://doi.org/10.1108/EJIM-09-2022-0531

Mehmood, Y., Farhan, A., Asma, A., Ibrar, Y., Muhammad, I., Sghaier, G. (2017). https://ieeexplore.ieee.org/stamp/stamp.jsp?tp=&arnumber=8030479 Date of access: 03.06.2024.

Moghrabi, I. A. R., Bhat, S. A., Szczuko, P., AlKhaled, R. A., Dar, M. A. (2023). "Digital transformation and its influence on sustainable manufacturing and business practices", Sustainability, Vol. 15 No. 4, pp. 1–35, doi: 10.3390/su15043010.

Moldan, B., Janousková, S. ve Hak, T. (2012). How to understand and measure environmental sustainability: ındicators and targets, Ecological Indicators, 17, 4–13.

Morelli, J. (2011). Environmental sustainability: a definition for environmental professionals, Journal of Environmental Sustainability, 1(1), 1–10.

Nadkarni, S. and Prügl, R. (2021). "Digital transformation: a review, synthesis and opportunities for future research", Management Review Quarterly, Vol. 71 No. 2, pp. 233–341, doi: 10.1007/s11301-020-00185-7.

OECD (2022), OECD Tourism Trends and Policies 2022, OECD Publishing, Paris, https://doi.org/10.1787/a8dd3019-en.

Prados-Castillo, J. F., Guaita Martínez, J. M., Zielińska, A., Gorgues Comas, D. (2023). A Review of Blockchain Technology Adoption in the Tourism Industry from a Sustainability Perspective. J Theor Appl Electronic Commerce Res 18(2):814–830. https://doi.org/10.3390/jtaer18020042

Pratisto, E. H., Thompson, N. and Potdar, V. (2022). "Immersive technologies for tourism: a systematic review", Information Technology & Tourism, Vol. 24 No. 2, pp. 181–219, doi: 10.1007/s40558-022-00228-7.

Queiroz, M. M., Wamba, S. (2019). Blockchain adoption challenges in supply chain:An empirical investigation of the main drivers in India and the USA. Int J Inform Manag 46:70–82. https://doi.org/10.1016/j.ijinfomgt.2018.11.021

Ren, S., Zhang, Yang Liu Y., Sakao T., Huisingh D., Cecilia M. V. B. A. (2019). A comprehensive review of big data analytics throughout product lifecycle to support sustainable smart manufacturing: A framework, challenges and future research directions. Journal of Cleaner Production, Volume 210: 1343–1365, ISSN 0959-6526, https://doi.org/10.1016/j.jclepro.2018.11.025

Riaz, M., Pamucar, D., Athar Farid, H. M., Hashmi, M. R. (2020). q-Rung Orthopair Fuzzy Prioritized Aggregation Operators and Their Application Towards Green Supplier Chain Management. Symmetry 12(6):976. https://doi.org/10.3390/sym12060976

Sarfraz, M., Khawaja, K. F., Han, H. et al. (2023). Sustainable supply chain, digital transformation, and blockchain technology adoption in the tourism sector. Humanit Soc Sci Commun 10, 557 https://doi.org/10.1057/s41599-023-02051-9

Sharma G. D., Yadav A., Chopra R. (2020). Artificial intelligence and effective governance: A review, critique and research agenda, Sustain. Fut. (2),100004, https://doi.org/10.1016/j.sftr.2019.100004

Sigala, M. (2020). Tourism and COVID-19: Impacts and implications for advancing and resetting industry and research. Journal of Business Research, 117, 312–321. https://doi.org/10.1016/j.jbusres.2020.06.015

Sujata, J. Aniket , D. and Mahasingh, M. (2019), "Artificial Intelligence Tools for Enhancing Customer Experience", International Journal of Recent Technology and Engineering, Vol. 8, Issue. 2S3. DOI : 10.35940/ijrte.B1130.0782S319.

Verhoef, P. C., et al. (2021). "Digital transformation: A multidisciplinary reflection and research agenda", Journal of Business Research, 122, 889–901, https://doi.org/10.1016/j.jbusres.2019.09.022.

Wen, J., Kozak, M., Yang, S. and Liu, F. (2021). "COVID-19: potential effects on Chinese citizens' lifestyle and travel", Tourism Review, Vol. 76 No. 1, pp. 74–87. https://doi.org/10.1108/TR-03-2020-0110

Wynn, Martin, and Peter Jones. (2022). "IT Strategy in the Hotel Industry in the Digital Era" *Sustainability* 14, no. 17: 10705. https://doi.org/10.3390/su141710705

Xiang, Z., Du, Q., Ma, Y., & Fan, W. (2017). A comparative analysis of major online review platforms: Implications for social media analytics in hospitality and tourism. Tourism Management, 58; 51–65. https://doi.org/10.1016/j.tourman.2016.10.001

Yıldız, Ö. (2019). Nesnelerin İnterneti, Yapay Zeka ve Turizm. (Ed. H. Kurgun, O. A. Kurgun) Turizm 4.0 Kavramlar ve Uygulamalar. 93–112 Detay Yayıncılık: Ankara.

Accor.com (2021). https://group.accor.com/en/Actualites/2021/04/world-earth-day, Date of Access: 25 May 2024

Tuigroup.com (2024). https://www.tuigroup.com/en-en/media/press-releases/2024/2024-04-10-tui-green-it-award-innovations-partnerships-and-social-responsibility Date of Access: 25 May 2024

Berna Mazlum[*] and Selda Guven[†]

Chapter 32 Service Orientation and Tourism

Introduction

Front-line employees in service businesses play a crucial role in customer interactions, engaging directly with customers (Liao & Chuang, 2004). While the service industry encompasses diverse sectors, each with its own characteristics and cognitive functions (Coelho & Augusto, 2010), customers often evaluate the quality of a service based on the behaviour of service employees due to the industry's intangible and interactive nature (Hennig-Thurau, 2004). At this juncture, personality traits significantly shape individuals' attitudes towards services and their subsequent behaviour (Chien-Wen, Hsiu-Li, Jo-Ping, Mei-Shiue, & Chen-Hui, 2013; Kim & Lee, 2016), which are influenced by these traits (Tata, Prashar, & Parsad, 2021). In this context, service orientation is reflected in an employee's inclination to meet customer demands while on duty (Brown, Mowen, Donavan, & Licata, 2002). Frimpong (2014) explains service orientation as the characteristics, attitudes, behaviors, or service cultures that facilitate proficient service performance. Indeed, it is generally believed that individuals are more likely to succeed in careers that align with their fundamental personality traits (Mechinda & Patterson, 2011).

The concept of service orientation is significant to both organizations and individuals (Popli & Rizvi, 2015). From an organizational perspective, it is considered an aspect of the organizational climate and is defined by the service activities, practices, and procedures within a workplace that ensure, promote, and reward high-quality service delivery (Lytle, Hom, & Mokwa, 1998; Mathew, Jose, G, & Chacko, 2020). These practices, policies, and procedures signal a systematic effort to reduce employee turnover (Kasekende, Byarugaba, & Nakate, 2013). Gonzalez and Garazo (2006) stress that it is crucial for effective management to ensure employees exhibit attitudes and behaviors that lead to qualified service. The extent of service orientation in relation to operational strategy plays

* Recep Tayyip Erdoğan University, Ardeşen Tourism Faculty, Department of Tourism Management, berna.sad@erdogan.edu.tr

† Çanakkale Onsekiz Mart University, Ezine Vocational School of Higher Education, Tourism and Travel Services Programme, seldaegilmezgil@comu.edu.tr

a vital role in shaping marketing strategy, serving as a key foundation for differentiating services, recruiting high-caliber employees, and evaluating their performance (Wu, Liang, Tung, & Chang, 2008). Within an organizational context, the ability to provide excellent services depends on employees being present at critical stages of service production and possessing the willingness and skills to meet customer demands (Lynn, Lytle, & Bobek, 2000). Service orientation caters to both customers' social demands and their individual economic needs (Lee, Che-Ha, & Alwi, 2021).

Effective recruitment and selection practices are crucial in service industries to maintain high service quality and ensure customer satisfaction. Recruitment criteria for front-line employees that focus solely on service qualifications are often insufficient due to the demanding interpersonal and emotional nature of their work. It is essential that companies also assess the service orientation of their employees, or their inclination to serve customers (Sousa & Coelho, 2013). Service orientation, viewed as an individual characteristic, highlights an employee's tendency to be supportive, considerate, and cooperative, reflecting how personal traits translate into workplace attitudes and behaviors (Lytle, Hom, & Mokwa, 1998). This orientation is argued to be more critical in service organizations than in other types of companies, underscoring the fundamental role of employee behavior in implementing service strategies (Saura, Contri, Taulet, & Velazquez, 2005). Adopting service orientation as a personal strength can mitigate the adverse effects of work demands and enhance customer service, particularly when coupled with organizational support. Service-oriented employees exhibit high levels of empathy, positivity, and a lower propensity for irritation or frustration in customer interactions. This enables them to provide service with a smile, minimizing emotional strain and improving the overall customer experience (Yavas & Babakus, 2010).

In today's business landscape, the customer-oriented attitudes of employees have become a key concern among various companies due to the increasing importance of the marketing system in maintaining competitive advantage. In the service industry, the behavior of employees in providing customer service is crucial for success, making this concept even more significant than in other sectors of the economy (Nouri, 2021). Moreover, service jobs, a key area in the workforce, are known to be highly stressful (Kiffin-Petersen & Soutar, 2020). The tourism industry, in particular, is centered on providing services and communicating with clients to meet their needs for goods or services, heavily relying on labor and the workforce as vital assets for businesses (Chen, Wu, & Chen, 2010). Tourism generates numerous job opportunities in various sectors, including lodging and transportation. Consequently, the presence of skilled and educated personnel is a key factor in the success of tourism development plans and programs. However, employment in the tourism sector is characterized by instability, low job status, long hours, and low pay. The industry faces significant challenges due to the difficulty in finding qualified employees and the resulting high turnover rates, which can be detrimental to its success (Aynalem, Birhanu, & Tesefay, 2016). Additionally, being employed in a field that does not align with one's personality can diminish self-assurance and lead to emotional exhaustion, fatigue, stress, and detachment from work (Köşker, Unur, & Gursoy, 2019). Therefore, this section focuses on examining the concept of service orientation in relation to tourism, exploring how service-oriented employees can mitigate the challenges faced by the industry and contribute to its overall success.

The Concept and Features of Service Orientation

The customer-focused values of a business are reflected by service industry employees, as customer satisfaction is predominantly influenced by the human aspect of service (Liu & Chen, 2006). Services are unique in that production and consumption occur simultaneously, meaning employee behavior can significantly influence customer perceptions of service quality and satisfaction. Employees who prioritize customer satisfaction make concerted efforts to address customer needs, thereby enhancing satisfaction and fostering relationship development (Sousa & Coelho, 2013). According to He, Wang, Zhu, & Harris (2015), effectively managing service employees and motivating them to deliver exceptional customer service is crucial for service management. Employees inspired by their organization's service orientation are more likely to engage in behaviors that exceed their job duties to support the institution and their co-workers (Pan, 2021). The service-oriented nature of a corporate structure can improve employee performance, and the success of service delivery is closely tied to the service orientation embedded within the corporate culture, influencing the longevity of high-quality service (Lexutt, 2020). In tourism establishments, while the commitment of staff to offering service is foundational, it is equally important to have an institutional structure that supports service orientation. Kim & Qu (2020) suggest that customer-centric employees are inclined to exhibit positive service behaviors towards customers. Therefore, it is essential to cultivate a corporate culture that values customers and promotes customer service, facilitating employees in finding satisfaction in serving customers and delivering excellent service. Reward programs and consistent training are crucial in upholding this culture and adequately preparing employees to meet customer expectations.

Customer service is shaped by two conflicting perspectives. One perspective argues that employees offer services based on their inherent personalities, while the other suggests that customer service results from proper training and a conducive service environment. In this context, individuals predisposed to serving exhibit characteristics of being attentive, pleasant, and courteous, as well as being sensitive to customers' needs, amiable, and popular. These traits contribute positively to the morale and harmony of work groups (Dienhart, Gregoire, & Downey, 1990). Mediano & Ruiz-Alba (2019) emphasized that service-oriented employees have the capability to identify, comprehend, and address customer needs, and they act accordingly. Personality traits such as adaptability, emotional stability, responsibility, activity need, and learning goal orientation are connected to service orientation (Liao & Subramony, 2008). According to Jeng (2018), employees with a strong service orientation are skilled at anticipating and addressing customer needs due to their strong belief in the significance of customer satisfaction. This belief acts as an individual's internal tool for identifying the necessary actions to handle potential customer conflicts efficiently. Yoo & Arnold (2014) suggest that this strong inclination can be used as an internal tool to deal with uncertainty, role overload, and conflicts with customers, supervisors, or colleagues. Zeglat, Aljaber, & Alrawabdeh (2014) point out that employees who possess friendly and courteous traits are likely to demonstrate a high service orientation.

Organizational factors can either enhance or diminish an individual's natural inclination towards service orientation in the workplace, despite it being a personality trait. According to Popli & Rizvi (2015), service orientation is viewed as an organizational philosophy and is considered a component of the corporate culture, organizational climate, and strategy. A company's service-focused culture encompasses its standards, convictions, ethics, and conduct, all of which influence employee productivity. Jung and Yoon (2013) argue that an organization's service orientation impacts employees' job satisfaction, organizational commitment, and intention to leave. Kim & Qu (2020) suggest that organizations can boost this inclination by concentrating on enhancing social skills such as emotional intelligence and critical thinking. This focus allows employees to better interact with customers by understanding and appreciating their emotions, resulting in more personalized and meaningful exchanges. Virgiawan, Riyanto, & Endri (2021) state that when organizations show support for their employees, those employees exhibit a positive work attitude, such as being more service-oriented. In this context, service companies should prioritize hiring individuals with a high inclination towards customer service during the personnel selection and recruitment processes. Additionally, these companies should cultivate an organizational culture and develop strategies that support and nurture service orientation. By doing so, organizations can foster a work environment that aligns with their service-oriented goals, thereby enhancing overall service quality and customer satisfaction.

Service Orientation and Tourism

Customer satisfaction perception and service quality in service marketing heavily rely on the behavior of employees, especially those who engage in face-to-face interactions with customers during service provision. The focus of service employees on providing excellent service directly impacts a service firm's performance by fostering enduring relationships between customers and employees (Hennig-Thurau & Thurau, 2003). Service orientation empowers service employees with a strong organizational identification, aligning their personal interests with the common goal of improving customer service. However, employees lacking service orientation, despite having a stronger corporate identity, struggle to offer high-quality service to customers due to a lack of intrinsic motivation (He, Wang, Zhu, & Harris, 2015). Employees who value customer satisfaction tend to have better interactions with customers, playing a vital role in delivering services. This, in turn, shapes how customers perceive the quality of service rendered by an employee and their overall satisfaction with the service (Gazzoli, Hançer, & Kim, 2013). Monitoring employees' attitudes and behaviors is crucial for service businesses to maintain quality service delivery, as customer perceptions of service encounters are largely determined by their interactions with service providers (Lee, Nam, Park, & Lee, 2006).

It is common for customer-facing staff at the front line of the tourism industry to be paid meager salaries. These employees frequently encounter difficult interactions with customers who tend to be aggressive and unkind (He, Wang, Zhu, & Harris, 2015). Additionally, the tourism industry is primarily concerned with high levels of informality, a lack of infrastructure, seasonal fluctuations, and a poorly educated workforce. These concerns directly impact the day-to-day activities of the tourism sector, posing challenges for employees in consistently meeting customer requirements (Domi, Capelleras, & Musabelliu, 2020). Despite these challenges, the role of human resources in ensuring the sustainability of the tourism industry is crucial. The attitudes and behaviors of employees help customers form an image of the service organization (Lombardi, Sassetti, & Cavaliere, 2019). Moreover, individuals who do not find a good fit between their personality and their job may not find satisfaction in the industry (Köşker, Unur, & Gursoy, 2019). In fact, an employee's likelihood of leaving their current position is determined by the compatibility between their personality and the organization's conditions (Jung & Yoon, 2013).

A crucial aspect in the tourism industry is the employees' orientation towards providing exceptional service. Tajeddini (2010) emphasizes that catering to customer needs is of utmost importance in the tourism industry compared to other sectors, given the distinctive features of tourism services. Marques, Correia, & Costa (2018) suggest that an enhanced service orientation within tourism companies at the organizational level results in improved employee emotions, leading to greater customer involvement. Peña, Jamilena, & Molina (2013) point out that tourism employees are considered invaluable assets to their businesses and are viewed as the cornerstone for achieving desired work outcomes. According to Hen & Myagmarsuren (2013), when tourism organizations implement a service-oriented approach, it enhances employees' service delivery. In the view of Domi & Domi (2021), tourism is a specialized field distinguished by its emphasis on service, demanding that employees engage closely with customers to provide exceptional service. Sustainability and development hinge on this essential prerequisite. A service-oriented mindset is linked to stability in tourism jobs, increased productivity, reduced turnover intention, and pursuing a career in the tourism industry. These characteristics can offer significant advantages to organizations, resulting in positive outcomes. Therefore, fostering a service-oriented culture is not only beneficial for employee satisfaction and retention but also critical for the long-term success and sustainability of the tourism industry.

Results

In summary, the tourism industry faces various challenges and evolving dynamics that impact its performance and competitive advantage. To succeed in this environment, tourism organizations must prioritize factors such as productivity drivers, customer satisfaction, and employee service orientation. Today's tourism landscape is characterized by increased competition, changing customer expectations, regulatory updates, global connectivity, and technological advancements. In response, tourism companies must focus on establishing lasting connections with clients by offering personalized products and experiences tailored to their specific needs and desires. The key to delivering such experiences lies in having qualified and service-oriented employees who excel at understanding and meeting customer expectations. To cultivate a service-oriented workforce, tourism companies should prioritize the recruitment and hiring of individuals with a natural inclination towards excellent service. Additionally, ongoing training programs should be implemented to further develop and reinforce service orientation among employees. Integrating service orientation into the organizational culture ensures that providing exceptional customer service becomes a core value within the company. In an industry where interpersonal skills are paramount for maintaining competitiveness, employees should be viewed as valuable assets that contribute to the success of the company, its products, and the overall customer experience. By prioritizing service orientation and investing in employee development, tourism organizations can enhance their performance, profitability, and sustainability in today's dynamic market

References

Aynalem, S., Birhanu, K., & Tesefay, S. (2016). Employment opportunities and challenges in tourism and hospitality. *Journal of Tourism and Hospitality, 5*(6), 1–5.

Brown, T. J., Mowen, J. C., Donavan, D. T., & Licata, J. W. (2002). The customer orientation of service workers: Personality trait effect on self and supervisor performance rating. *Journal of Marketing Research, 39*(1), 110–119.

Chen, C.-F., & Myagmarsuren, O. (2013). Exploring the moderating effects of value offerings between market orientation and performance in tourism industry. *International Journal of Tourism Research, 15*, 595–610.

Chen, S.-C., Wu, M.-C., & Chen, C.-H. (2010). Employee's personality traits, work motivation and innovative behavior in marine tourism industry. *Journal of Service Science and Management, 3*, 198–205.

Chien-Wen, T., Hsiu-Li, H., Jo-Ping, L., Mei-Shiue, L., & Chen-Hui, W. (2013). The relationships among employee personality traits, service attitude, and service behavior in the leisure farm. *Journal of Tourism and Hospitality Management, 1*(2), 75–88.

Coelho, F., & Augusto, M. (2010). Job characteristics and the creativity of frontline service employees. *Journal of Service Research, 13*(4), 426–438.

Dlenhart, J. R., Gregolre, M. B., & Downey, R. G. (1990). Service orientation of restaurant employees. *Hospitality Research Journal, 14*(2), 421–429.

Domi, S., & Domi, F. (2021). The interplay effects of skill-enhancing human resources practices, customer orientation and tourism SMEs performance. *European Journal of Training and Development, 45*(8/9), 737–761.

Domi, S., Capelleras, J.-L., & Musabelliu, B. (2020). Customer orientation and SME performance in Albania: A case study of the mediating role of innovativeness and innovation behavior. *Journal of Vacation Marketing, 26*(1), 130–146.

Frimpong, K. (2014). Service orientation in delivery: Perspectives from employees, customers, and managers. *Services Marketing Quarterly, 35*(1), 54–67.

Gazzoli, G., Hançer, M., & Kim, B. P. (2013). Explaining why employee-customer orientation influences customers' perceptions of the service encounter. Journal of Service Management, 382–400." Şeklinde düzenlenecektir.

Gonzalez, J. V., & Garazo, T. G. (2006). Structural relationships between organizational service orientation, contact employee job satisfaction and citizenship behavior. *International Journal of Service Industry Management, 17*(1), 23–50.

He, H., Wang, W., Zhu, W., & Harris, L. (2015). Service workers' job performance: The roles of personality traits, organizational identification, and customer orientation. *European Journal of Marketing, 49*(11/12), 1751–1776.

Hennig-Thurau, T. (2004). Customer orientation of service employees: Its impact on customer satisfaction. *International Journal of Service Industry Management, 15*(5), 460–478.

Hennig-Thurau, T., & Thurau, C. (2003). Customer orientation of service employees: Toward a conceptual framework of a key relationship marketing construct. *Journal of Relationship Marketing, 2*(1–2), 23–41.

Jeng, S. P. (2018). Enhancing the creativity of frontline employees the effects of job complexity and customer orientation. *The International Journal of Logistics Management, 29*(1), 387–408.

Jung, H. S., & Yoon, H. H. (2013). The effects of organizational service orientation on person–organization fit and turnover intent. *The Service Industries Journal, 33*(1), 7–29.

Kasekende, F., Byarugaba, K. J., & Nakate, M. (2013). Employee satisfaction: Mediator of organizational service orientation and employee retention. *Journal of Business and Management, 19*(3), 41–61.

Kiffin-Petersen, S. A., & Soutar, G. N. (2020). Service employees' personality, customer orientation and customer incivility. *International Journal of Quality and Service Sciences, 12*(3), 281–296.

Kim, B., & Lee, J. (2016). Relationships between personal traits, emotional intelligence, internal marketing, service management, and customer orientation in Korean outpatient department nurses. *Asian Nursing Research, 10*(1), 18–24.

Kim, H., & Qu, H. (2020). Effects of employees' social exchange and the mediating role of customer orientation in the restaurant industry. *International Journal of Hospitality Management, 89*, 1–8.

Köşker, H., Unur, K., & Gursoy, D. (2019). The effect of basic personality traits on service orientation and tendency to work in the hospitality and tourism industry. *Journal of Teaching in Travel & Tourism, 19*(2), 140–162.

Lee, C. M., Che-Ha, N., & Alwi, S. F. (2021). Service customer orientation and social sustainability: The case of small medium enterprises. *Journal of Business Research, 122*, 751–760.

Lee, Y.-K., Nam, J.-H., Park, D.-H., & Lee, K. A. (2006). What factors influence customer-oriented prosocial behavior of customer-contact employees? *Journal of Services Marketing, 20*(4), 251–264.

Lin, M.-Y., Chiang, C.-F., & Wu, K.-P. (2018). How hospitality and tourism students choose careers: Influences of employer branding and applicants' customer orientation.Journal of Hospitality & Tourism Education, 30(4), 229–240." Lütfen referans kısmında çıkarabilir misiniz? Metinde atıflanmamıştır.

Lexutt, E. (2020). Different roads to servitization success – A configurational analysis of financial and non-financial service performance. *Industrial Marketing Management, 84*, 105–125.

Liao, H., & Chuang, A. (2004). A multilevel investigation of factors influencing employee servce performance and customer outcomes. *Academy of Management Journal, 47*(1), 41–58.

Liao, H., & Subramony, M. (2008). Employee customer orientation in manufacturing organizations: Joint influences of customer proximity and the senior leadership team. *Journal of Applied Psychology, 93*(2), 317–328.

Lin, M.-Y., Chiang, C.-F., & Wu, K.-P. (2018). How hospitality and tourism students choose careers: Influences of employer branding and applicants' customer orientation. *Journal of Hospitality & Tourism Education, 30*(4), 229–240.

Liu, C.-M., & Chen, K.-J. (2006). Personality traits as antecedents of employee customer orientation: A case study in the hospitality industry. *International Journal of Management, 23*(3), 478–485.

Lombardi, S., Sassetti, S., & Cavaliere, V. (2019). Linking employees' affective commitment and knowledge sharing for an increased customer orientation. *International Journal of Contemporary Hospitality Management, 31*(11), 4293–4312.

Lynn, M. L., Lytle, R. S., & Bobek, S. (2000). Service orientation in transitional markets: Does it matter? *European Journal of Marketing, 34*(3/4), 279–298.

Lytle, R. S., Hom, P. W., & Mokwa, M. P. (1998). Serv*or: A managerial measure of organizational service-orientation. *Journal of Retailing, 74*(4), 455–489.

Marques, G. S., Correia, A., & Costa, C. M. (2018). The influence of customer orientation on emotional labour and work outcomes: a study in the tourism industry. *European Journal of Tourism Research, 20*, 59–77.

Mathew, S., Jose, A., Goplakrishnan, R., & Chacko, D. P. (2020). Examining the relationship between e-service recovery quality and e-service recovery satisfaction moderated by perceived justice in the banking context. Benchmarking: An International Journal, 27(6), 1961–1980." Şeklinde düzenlenecektir.

Mechinda, P., & Patterson, P. G. (2011). The impact of service climate and service provider personality on employees' customer-oriented behavior in a high-contact setting. *Journal of Services Marketing, 25*(2), 101–113.

Mediano, J. M., & Ruiz-Alba, J. L. (2019). New perspective on customer orientation of service employees: A conceptual framework. *The Service Industries Journal, 39*(13–14), 966–982.

Nouri, B. A. (2021). The impact of service climate and personality traits on customer oriented behaviour of employees. *Transnational Marketing Journal, 9*(2), 425–444.

Pan, Z. (2021). Paradoxical leadership and organizational citizenship behaviour: The serial mediating effect of a paradoxical mindset and personal service orientation. *Leadership & Organization Development Journal, 42*(6), 869–881.

Peña, A. I., Jamilena, D. M., & Molina, M. Á. (2013). Impact of customer orientation and ICT use on the perceived performance of rural tourism enterprises. *Journal of Travel & Tourism Marketing, 30*(3), 272–289.

Popli, S., & Rizvi, I. A. (2015). Exploring the relationship between service orientation, employee engagement and perceived leadership style: A study of managers in the private service sector organizations in India. *Journal of Services Marketing, 29*(1), 59–70.

Saura, I. G., Contri, G. B., Taulet, A. C., & Velazquez, B. M. (2005). Relationships among customer orientation, service orientation and job satisfaction in financial services. *International Journal of Service Industry Management, 16*(5), 497–525.

Sousa, C. M., & Coelho, F. (2013). Exploring the relationship between individual values and the customer orientation of front-line employees. *Journal of Marketing and Management, 29*(15–16), 1653–1679.

Tajeddini, K. (2010). Effect of customer orientation and entrepreneurial orientation on innovativeness: Evidence from the hotel industry in Switzerland. *Tourism Management, 31*(2), 221–231.

Tata, S. V., Prashar, S., & Parsad, C. (2021). Intention to write reviews: influence of personality traits, attitude and motivational factors. *Journal of Systems and Information Technology, 23*(2), 218–242.

Virgiawan, A. R., Riyanto, S., & Endri, E. (2021). Organizational culture as a mediator motivation and transformational leadership on employee performance. *Academic Journal of Interdisciplinary Studies, 10*(3), 67–79.

Wu, C. H.-J., Liang, R.-D., Tung, W., & Chang, C.-S. (2008). Structural relationships among organisation service orientation, employee service performance, and consumer identification. *The Service Industries Journal, 28*(9), 1247–1263.

Yavas, U., & Babakus, E. (2010). Relationships between organizational support, customer orientation, and work outcomes: A study of frontline bank employees. *International Journal of Bank Marketing, 28*(3), 222–238.

Yoo, J. J., & Arnold, T. J. (2014). Customer orientation, engagement, and developing positive emotional labor. *The Service Industries Journal, 34*(16), 1272–1288.

Zeglat, D., Aljaber, M., & Alrawabdeh, W. (2014). Understating the impact of employee empowerment on customer oriented behavior. *Journal of Business Studies Quarterly, 6*(1), 55–67.

Handan Özçelik Bozkurt*, Emin Arslan† and Hakan Kendir‡

Chapter 33 Examination of Historical Mosques in Tokat Province Within the Scope of Cultural Heritage Tourism

Introduction

Heritage tourism, as an activity that guides tourism movements, pertains to the utilization of natural and cultural heritage assets by tourist destinations as a product to attract visitors. Natural and cultural heritage areas may nevertheless deteriorate as a result of the high volume of tourism. Consequently, there must be some agreement between the notions of sustainable development and cultural heritage tourism. (Fyall & Garrod, 1998: 213; MacDonald & Jolliffe, 2003).

Supply sources such as important architectural structures, ancient cities, museums, and archaeological sites are frequently visited by tourists participating in cultural heritage tourism (Kendir et al., 2019). The traditional Turkish architecture and belief systems exhibited in mosques captivate a considerable number of both domestic and international visitors. The initial objective of this study is to examine Tokat's historical mosques in the context of cultural heritage tourism. According to an alternative perspective, the purpose of this research is to help preserve every cultural heritage asset in Tokat.

In light of the information presented above, it can be said that the historical mosques in Tokat have pioneering architectural value in terms of Anatolian Turkish culture. From this vantage point, the study aims to highlight the significant historical mosques in Tokat province for cultural heritage tourism. In addition, the mosques in Tokat were evaluated and interpreted in detail regarding cultural heritage tourism.

* PhD., Sinop University, Tourism Faculty, Department of Gastronomy and Culinary Arts, hozcelikbozkurt@sinop.edu.tr

† PhD., Tokat Gaziosmanpaşa University, Zile Dinçerler School of Tourism and Hotel Management, Department of Gastronomy and Culinary Arts, emin.arslan@gop.edu.tr

‡ PhD., Tokat Gaziosmanpaşa University, Zile Dinçerler School of Tourism and Hotel Management, Department of Gastronomy and Culinary Arts, hakan.kendir@gop.edu.tr

Cultural Heritage Tourism

One of the most important components of tourism is cultural heritage. Especially in developed countries, cultural heritage elements are a great economic resource for the tourism sector. In addition, tourists visit cultural heritage sites in order to have cultural experiences in the countries they travel to and create economic dynamism by spending in those places (Garrod & Fyall, 2000). This mobility has led to the emergence of the concept of "cultural heritage tourism." Cultural heritage tourism is the travel that tourists undertake because they want to see historical ruins and cultural assets on site and demand activities such as local culinary experiences, local events, and handicrafts. In recent years, the relevant literature has frequently examined the tourist attractions created by heritage elements, which are considered within the scope of cultural tourism (Chhabra et al., 2003; Nyaupane et al., 2015; Kendir et al., 2017; Dönmez et al., 2017; Arslan & Küçük, 2017; Arslan et al., 2019; Huang & Chu, 2020; Kendir, 2020; Butler et al., 2022).

Cultural heritage sites are widely used in countries with nation-state characteristics to reinforce national-spiritual feelings towards the local people and to establish a bond of patriotism in the society. Therefore, the national and spiritual feelings of the local people also play an important and decisive role in terms of cultural heritage tourism. Monuments of national heroes, mausoleums, battlefields, and other items of national and spiritual importance are important in terms of cultural heritage tourism (Timothy & Boyd, 2006: 3). On the other hand, many academic studies (Zhuang et al., 2014; Liang & Hui, 2016; Feng et al., 2017; Butler et al., 2022) show that as a result of the increasing interest of tourists in cultural heritage items, the society re-evaluates its own culture and cultural heritage It shows that awareness about protection is increasing. Therefore, it is noteworthy that the preservation of heritage elements and their transfer to future generations is important (Bec et al., 2019). Preserving archaeological remains, ensuring the sustainability of cultural assets, highlighting local cultural identity, and creating new social opportunities are of strategic importance for the development of cultural heritage (Elfadaly et al., 2018).

UNESCO (United Nations Educational, Scientific, and Cultural Organization) registers tangible and intangible cultural heritage items and announces them with updated and published lists every year. Tangible heritage assets, especially those built or created by people, have long been one of the most basic sources of cultural tourism (Richards, 2018). Tourists frequently visit places of faith with historical and cultural value, such as mosques, churches, and synagogues, which are among the tangible heritage assets. It is known that tourism activities are effective in the visits of many people to religious places that are not affiliated with their faith, as well as pilgrims who visit the holy places of their own faith (Nyaupane et al., 2015; Huang & Chu, 2020). Tourists of different faiths frequently visit mosques, places of faith with historical and cultural value, as part of cultural heritage tourism. For instance, the Hagia Sophia Mosque in Istanbul welcomed approximately 14 million local and foreign visitors in 2022 (Anadolu Agency, 2023). Approximately 5.8 million people visited the Sheikh Zayed Grand Mosque, located in Abu Dhabi, the capital of Dubai, in 2017 (Huang & Chu, 2020). Within the framework of this information, the mosques in Tokat province, which is located in the Central Black Sea Region of Turkey and has attracted attention in recent years with its historical and cultural values, were examined.

Historical Mosques in Tokat Province

Tokat province, located in Turkey's Black Sea Region, has Samsun and Ordu in the north, Amasya in the west, and Yozgat and Sivas in the south and east. Tokat, which has 11 districts (Map 1), has a surface area of 10,072 km^2 and a total population of 606,934 as of 2023 (TUIK, 2024).

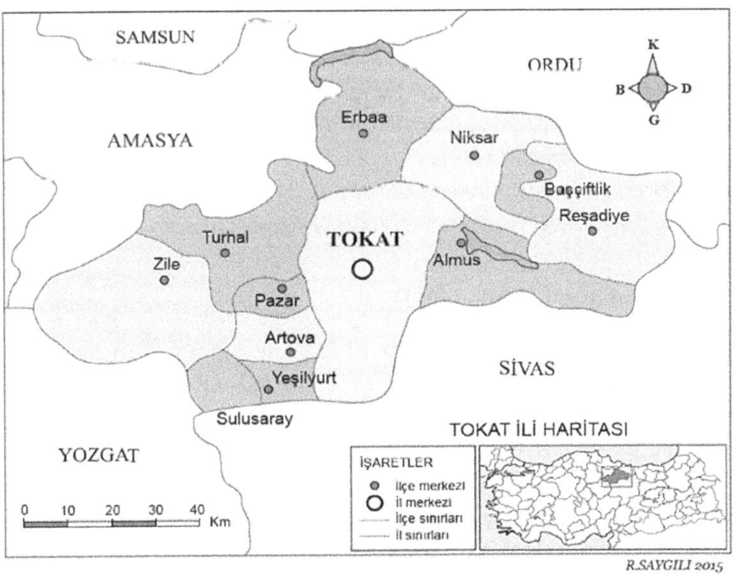

Map 1. Study Area Location
(Source: Geography Map, 2024)

Tokat, which has hosted many different civilizations throughout history, respectively, came under the rule of Hatti, Hittite, Phrygian, Roman, Byzantine, Danishmend, Ilkhanid, Seljuk, and Ottoman. On the other hand, during these periods, the province of Tokat was known by various names such as Comano Pontika, Komana, Evdoksia, Dokia, Dokat, Kah-Cun, Sobaru, Darün-Nusret, and Darün-Nasr. Additionally, Tokat was one of the first regions settled by the Turks, who conquered Anatolia in the Battle of Manzikert in 1071. The Danishmend State and the Seljuk State played a significant role in concentrating Turks in the region, shaping the architectural structure, and building many of the first works of Turkish-Islamic architecture during these periods. A substantial portion of them have undergone restoration. Conversely, the Ottoman period saw the construction of architectural works in Tokat. Tokat gained provincial status after the Republic's declaration in 1923 (Tokat Provincial Directorate of Culture and Tourism, 2024).

Tokat's cultural and tourist attractions include mosques, ancient cities, baths, caravanserais, mansions, and many other historical monuments. It is known that the use of wooden materials in architecture is widespread, especially in Tokat, which attracts attention with its historical mosques. Notably, some mosques in Tokat center and districts feature wooden pillars and ceilings. On the other hand, the geographical structure of the city explains the use of wood as both a construction and decoration material in traditional houses (Atak, 2016). The presented information suggests that the architectural structures of the historical mosques in Tokat are diverse. From this vantage point, the study aims to highlight the significant historical mosques in Tokat province for cultural heritage tourism. Furthermore, we thoroughly evaluated and interpreted the historical mosques in Tokat in the context of cultural heritage tourism.

Garipler Mosque (1080–1090)

Built between 1080 and 1090, it is one of the oldest mosques in Anatolia. It is located in the Pazarcık neighborhood of Tokat. It leans on the hillside where Tokat Castle is located along the western wall (Encyclopedia of Islam, 1996). The original mosque structure is important in terms of early Turkish-Islamic architecture. According to some sources (AA, 2019; Açkel, 2023), the Turks built the Garipler Mosque as their oldest mosque after settling in Anatolia. The Karakhanids, the first Turkish-Islamic state in Central Asia, built the Degaron Mosque in 1040, whose architectural structure is similar to the mosque's (Tokat Provincial Directorate of Culture and Tourism, 2024).

Picture 1. Garipler Mosque
(Source: Tokat Provincial Directorate of Culture and Tourism, 2024).

From this point on, it is certain that the Garipler Mosque, seen in Picture 1, is one of the oldest buildings built by the Turks in Anatolia. Therefore, it holds a significant place in Anatolian Turkish history. This makes the mosque a highly valuable asset for Tokat's cultural heritage tourism.

Tokat Ulu Mosque (13th Century)

It was first constructed in the 13th century during the Danish Principality Period but underwent a complete renovation in 1679 under the Ottoman Empire.

Picture 2. Grand Mosque
(Source: Kumcağız, M. – Anatolian Agency, 2020).

The mosque's design features a rectangular plan. The mosque employs columns and column headings. The mosque typically has a wooden ceiling. Red and green colors dominate the ceiling. We used the kundekari technique, one of the traditional Turkish arts, and decorated the ceiling core with gold gilding and floral figures. Figures with Rumi motifs decorate the columns and arches carrying the wooden ceiling, adhering to the decoration features of the 13th century (Culture Portal, 2024). Ulu Mosque is one of Tokat's important buildings in terms of cultural heritage tourism, with its interior architecture featuring very artistic and ornate motifs, as shown in Picture 2.

Malum Seyyid Mosque (13th Century)

It is estimated that the mosque, located in Dutluca village in Tokat's Sulusaray district, was built in the 13th century. The mosque's most notable feature is its complete construction from wood and cut stone, without any use of nails. The geometric shapes on the wooden sections of the mosque are quite striking. With these features, the mosque reflects the Turkish-Islamic architecture of the Seljuk period. Recently, the Malum Seyyid Mosque underwent restoration to its original state and welcomed visitors. (AA, 2016).

Picture 3. A Wooden Section of the Malum Seyyid Mosque
(Source: Türkoğlu, E. – Anatolian Agency, 2016).

The UNESCO World Heritage List plans to include the Malum Seyyid Mosque due to its unique features. Figure 3 depicts the Malum Seyyid Mosque, which holds significant potential for cultural heritage tourism because of its exceptional status as the most remarkable mosque in Tokat.

Meydan (Hatuniye) Mosque (1485)

It was constructed by Gülbahar Hatun, the mother of Ottoman Sultan Bayezid II, in 1485 during the Ottoman Empire. The most magnificent and striking part of the mosque is the entrance gate. The mosque's door boasts intricate and magnificent architecture. The Meydan Mosque is a building with a large main dome supported by side domes. With these features, it reflects the Ottoman architecture of the late 15th century (Encyclopedia of Islam, 1996).

Picture 4. Meydan (Hatuniye) Mosque
(Source: DHA, 2019).

The Meydan Mosque in Picture 4, which has a large fountain in its garden, is an important work bearing traces of the Ottoman Empire's rise. In terms of cultural heritage tourism, Meydan Mosque is also one of the structures that attracts the most visitors in Tokat.

Ali Paşa Mosque (1572–1573)

The mosque, built by Ali Pasha in 1572–1573, during the reign of Ottoman Sultan Suleyman, is located in the liveliest location of Tokat city center. Built with cut stone, the Ali Paşa Mosque is a large and magnificent structure. A square plan guided the mosque's construction. Carving niche rooms into the very thick stone walls expanded the space. There is a large dome on top of octagonal supports. Ali Paşa Mosque represents 16th-century Ottoman architecture very well (Encyclopedia of Islam, 1989; Afyoncu, 2001). There is a large fountain in the courtyard of the Ali Paşa Mosque, seen in Picture 5, which was built later than the mosque. Additionally, Ali Paşa's tomb is in the northeast corner of the mosque courtyard (Encyclopedia of Islam, 1989).

Picture 5. Ali Paşa Mosque
(Source: Directorate General of Foundations, 2019).

Ali Paşa Mosque is one of the most visited places of worship in Tokat. The Ali Paşa Mosque stands out among Tokat's tourism attractions in terms of cultural heritage tourism and faith tourism.

Elbaşoğlu Mosque (1795)

Elbaşoğlu Mosque, located in the Zile district of Tokat, was built in 1795 by Seyit Ahmet Ağa, the prominent feudal lord of Zile. A tiled roof covers the mosque, built with cut and rubble stones. There is a small courtyard on the north side and a minaret with a single balcony in the northwest corner.

Picture 6. Elbaşoğlu Mosque
(Source: Zile Municipality, 2021)

Recent restoration works have revealed rich hand-carved works and examples of wooden decoration within the mosque's interior. These values in interior architecture make the Elbaşoğlu Mosque unique (Aktemur, 2013; Zile Municipality, 2021). Elbaşoğlu Mosque (Picture 6) hosts one of the richest examples of traditional Turkish hand-carved decoration and wood art. As a result, Elbaşoğlu Mosque is a very important attraction element in terms of cultural heritage tourism in Zile. The table below displays other historical mosques in Tokat province, each possessing unique cultural and historical characteristics.

Table 1. Other Historical Mosques in Tokat Province

Mosque Name	District of location	Date of construction
Takyeciler Mosque	Centre	15th century
Behzad-ı Veli Mosque	Centre	1536
Zile Nasuh Paşa Mosque	Zile	1267
Niksar Ulu Mosque	Niksar	1145
Kesikbaş Mosque	Turhal	1759

Table 1. Continued

Mosque Name	District of location	Date of construction
Silahtar Ömer Paşa Mosque	Erbaa	1707
Çöreğibüyük Mosque	Niksar	13th century
Mahmutpaşa Mosque	Centre	17th century
Alaca Mescit Mosque	Centre	14th century
Boyacı Hasan Ağa Mosque	Zile	1497

Kaynak: Türkiye Culture Portal, 2022

Table 1 shows other mosques in Tokat Province that have historical and cultural heritage value. Notably, the construction of the historical mosques in Tokat coincided with the initial dates of Turkish settlement in Anatolia. As a result, Tokat contains significant traces of all periods of Anatolian Turkish history. Tokat is a treasure trove for those looking to learn about Anatolian Turkish history. This situation reveals that Tokat has a very strong potential to attract both local and foreign tourists in terms of cultural heritage tourism.

Conclusion and Recommendations

In terms of cultural heritage tourism, Tokat province boasts numerous mosques of historical significance. Between the 12th century and the 19th century, the Turkish-Islamic architectural tradition guided the construction of almost all of these mosques. Studies in the field of culture and faith tourism (Kartal et al., 2015; Bilim & Ay, 2022) have determined that mosques with historical and architectural value particularly attract local and foreign tourists. Therefore, effective promotion and marketing of the historical mosques in Tokat can lead to intense tourist attractions.

As a result, historical mosques in Tokat province are an important source of supply for cultural heritage tourism. Effectively using these values in terms of tourism will support Tokat's economy and create new employment. Therefore, it would be beneficial to conduct studies on this issue at academic and sectoral levels. This study aims to increase awareness about mosques' role in cultural heritage tourism in Tokat. We plan to develop a groundbreaking study that will serve as a guide for future academic research in this field. On the other hand, all stakeholders, especially academics, local decision-makers, and sector representatives, can benefit from this study.

References

AA, Anatolian Agency. (2016). *Çivi kullanılmadan yapılan cami 8 asırdır ayakta.* https://www.aa.com.tr/tr/kultur-sanat/civi-kullanilmadan-yapilan-cami-8-asirdir-ayakta/661231. Erişim Tarihi: 24.05.2024.

AA, Anatolian Agency. (2019). *Tarihi cami bin yıldır Müslümanların hizmetinde.* https://www.aa.com.tr/tr/kultur-sanat/tarihi-cami-bin-yildir-muslumanla rin-hizmetinde/1494807. Erişim Tarihi: 22.05.2024.

AA, Anatolian Agency. (2023). *Ayasofya-i Kebir Cami-i Şerifi'ni 2022'de 13 milyon 635 bin 229 kişi ziyaret etti.* https://www.aa.com.tr/tr/kultur/ayasofya-i-kebir-cami-i-serifini-2022de-13-milyon-635-bin-229-kisi-ziyaret-etti/2777 987. Erişim Tarihi: 15.05.2024.

Açıkel, A. (2023). Anadolu'nun en eski İslam mabedi: Tokat Garipler Camisi ve vakıfları. *Vakıflar Dergisi,* 59, 9–19.

Afyoncu, E. (2001). XVI. yüzyılda Osmanlı Beylerbeyleri I. Temerrüd Ali Paşa. *Belleten, 65* (244), 1007–1034.

Aktemur, A. M. (2013). Zile Elbaşoğlu Camii'nin sıvalar altında kalan gizemi. *Journal of Turkish Studies, 8*(8), 1621–1642.

Arslan, E., Kendir, H. & Özkoç, A. G. (2019). *Niksar'ın Kültürel Miras Turizmi Potansiyelinin Değerlendirilmesi,* VIII. National IV. International Eastern Mediterranean Tourism Symposium, Mersin Üniversitesi 19–20 Nisan 2019. Mersin/Anamur. ss: 262–271.

Arslan, E. & Küçük, M. A. (2017). *Tokat-Zile'deki Asırlık "40 Hatim" Geleneğinin Somut Olmayan Kültürel Miras ve Kültür Turizmi Açısından İncelenmesi,* DOKAP Bölgesi Uluslararası Turizm Sempozyumu Özet Bildiri Kitabı, Karadeniz Teknik Üniversitesi 23–24 Ekim 2017. Trabzon. s: 36.

Atak, E. (2016). Tokat yöresindeki ahşap minareli camiler. *Atatürk Üniversitesi Güzel Sanatlar Enstitüsü Dergisi,* 37, 240–247.

Bec, A., Moyle, B., Timms, K., Schaffer, V., Skavronskaya, L. & Little, C. (2019). Management of immersive heritage tourism experiencs: A conceptual model. *Tourism Management, 72,* 117–120.

Bilim, Y. & Ay, S. (2022). Young Turkish tourists' religious tourism motivations, perceptions and intentions about a religious destination: The case of Konya, Turkey. *International Journal of Religious Tourism and Pilgrimage, 10*(2), 13–27.

Butler, G., Szili, G., & Huang, H. (2022). Cultural heritage tourism development in Panyu District, Guangzhou: community perspectives on pride and preservation, and concerns for the future. *Journal of Heritage Tourism,* 17(1), 56–73.

Chhabra, D., Healy, R. & Sills, E. (2003). Staged authenticity and heritage tourism. *Annals of Tourism Research, 30*(3): 702–719.

Coğrafya Harita. (2024). *Tokat İli Haritası.* http://cografyaharita.com/haritalarim/4l_tokat_ili_haritasi.png. Access Date: 20.05.2024.

Cultural Portal, (2024). Ulu Cami – Tokat https://www.kulturportali.gov.tr/turkiye/tokat/gezilecekyer/ulu-cam, Access Date: 20.05.2024..

DHA (2019). *Hatuniye Camisi zamana meydan okuyor.* https://www.youtube.com/watch?v=n5DQ9hXf650&t=17s. Erişim Tarihi: 21.05.2024.

Directorate General of Foundations (2019). Ali Paşa Camii ve Türbesi. https://www.vgm.gov.tr/haberler/tokat-ali-pasa-camii-ve-turbesi-yeniden-ibadete-acildi. Erişim Tarihi: 22.05.2024.

Dönmez, Y., Kendir, H., Türkmen, F. & Çabuk, S. (2017). *Tarihi Evlerin Kültür Turizmine Etkilerinin Araştırılması: Tokat-Zile Örneği.*, Uluslararası Taşköprü Pompeipolis Bilim Kültür Sanat Araştırmaları Sempozyumu, Kastamonu Üniversitesi 10–12 Nisan 2017. Kastamonu. ss: 1114–1127.

Elfadaly, A., Attia, W., Qelichi, M. M., Murgante, B., & Lasaponara, R. (2018). Management of cultural heritage sites using remote sensing indices and spatial analysis techniques. *Surveys in Geophysics*, 39, 1347–1377.

Encyclopedia of Islam. (1989). *Ali Paşa Camii.* https://islamansiklopedisi.org.tr/ali-pasa-camii--tokat. Erişim Tarihi: 21.05.2024.

Encyclopedia of Islam. (1996). *Garipler Camii.* https://islamansiklopedisi.org.tr/garipler-camii. Erişim Tarihi: 21.05.2024.

Encyclopedia of Islam. (2004). *Meydan Camii.* https://islamansiklopedisi.org.tr/meydan-camii. Erişim Tarihi: 23.05.2024.

Feng, J., Li, Y., & Wu, P. (2017). Conflicting images of the Great Wall in cultural heritage tourism. *Critical Arts, 31*(6), 109–127.

Fyall, A. & Garrod, B. (1998). Heritage tourism: At what price? *Managing Leisure, 3*(4), 213–228.

Garrod, B. & Fyall, A. (2000). Managing heritage tourism. *Annals of Tourism Research, 27*(3), 682–708.

Huang, J. & Chu, J. (2020). Tourist experience at religious sites: A case study of the Chinese visiting the Sheikh Zayed Grand Mosque, *Journal of China Tourism Research, 16*(3), 447–471.

Kartal, B., Tepeci, M. & Atlı, H. (2015). Examining the religious tourism potential of Manisa. *Turkey with a marketing perspective. Tourism Review, 70*(3), 214–231.

Kendir, H., Dönmez, Y., Türkmen, F. & Çabuk, S. (2017). *Korunan Tarihi Kentlerde Turizmin Sürdürülebilirliği: Safranbolu Kenti,* AL-FARABI 1st International Congress on Social Sciences, 11–14 Mayıs 2017. Gaziantep. ss: 1205–1218.

Kendir, H., Arslan, E., & Türkmen, F. (2019). Evaluation of Navruz (Yengi-Kün) celebrations in Turkish world in terms of festival tourism: Shymkent-Van comparison. *Journal of Tourism & Gastronomy Studies, 7*(4), 2668–2684.

Kendir, H. (2020). Turizm eğitimi alan lisans öğrencilerinin kültürel miras algılarının belirlenmesi. *Türk Turizm Araştırmaları Dergisi, 4*(3), 2296–2310.

Kendir, H., Arslan, E. & Bozkurt, H. Ö. (2023). Examining the accommodation experience in historical buildings with content analysis: Amasya mansions example. *Journal of New Tourism Trends, 4*(2), 28–40.

Kumcağız, M. – Anatolian Agency (2020). *Tokat Ulu cami, turistlerin ilgisini çekiyor.* https://www.aa.com.tr/tr/pg/foto-galeri/tokat-ulu-cami-turistlerin-ilgisini-cekiyor-#. Erişim Tarihi: 23.05.2024.

Liang, Z. & Hui, T.-K. (2016). Residents' quality of life and attitudes toward tourism development in China. *Tourism Management, 57,* 56–67.

MacDonald, R. & Jolliffe, L. (2003). Cultural rural tourism evidence from Canada. *Annals of Tourism Research, 30*(2), 307–322.

Nyaupane, G. P., Timothy, D. J. & Poudel, S. (2015). Understanding tourists in religious destinations: A social distance perspective. *Tourism Management, 48,* 343–353.

Richards, G. (2018). Cultural tourism: A review of recent research and trends. *Journal of Hospitality and Tourism Management, 36,* 12–21.

Timothy, D. J. & Boyd, S. W. (2006). Heritage tourism in the 21st century: Valued traditions and new perspectives. *Journal of Heritage Tourism, 1*(1), 1–16.

TÜİK (Türkiye İstatistik Kurumu). (2024). *Adrese Dayalı Nüfus Kayıt Sistemi Verileri.* https://biruni.tuik.gov.tr/medas/?kn=95&locale=tr. Access Date: 20.05.2024.

Türkiye Culture Portal (2022). *Genel Bilgiler–Tokat.* https://www.kulturportali.gov.tr/turkiye/tokat/genelbilgiler. Erişim Tarihi: 25.05.2024

Türkiye Culture Portal (2024). *Gezilecek Yerler–Tokat, Ulu Cami.* https://www.kulturportali.gov.tr/turkiye/tokat/gezilecekyer. Erişim Tarihi: 23.05.2024

Tokat Provincial Directorate of Culture and Tourism (2024). *Tokat ili hakkında genel bilgiler.* https://tokat.ktb.gov.tr/TR-60574/genel-bilgiler.html. Access Date: 20.05.2024.

Zhuang, X., Hong, Z. & Deng, S. (2014). Institutional ethical analysis of resident perceptions of tourism in two Chinese villages. *Tourism Geographies, 16*(5), 785–798.

Zile Municipality, *Zile Elbaşoğlu Cami* (2021). https://site.zile.bel.tr/index.php/2021/02/01/zile-elbasoglu-cami/, Access Date: 22.05.2024.

Eren Erkiliç*

Chapter 34 A Research on Route Perceptions and Preferences of Local Motorcycle Users: The Case of Rize

Introduction

The relationship between motorcycles and tourism extends beyond leisure and recreation to encompass broader economic impacts, and motorbike use has been found to have a positive impact on the growth of economically developing nations because they provide an alternative mode of transportation and cut down on travel time (Bastos et al., 2020). Multiple research studies have taken aim at the cultural and social aspects of motorcycling. It has been highlighted how notable motorcycles are to leisure activities. On the other side, because riding a motorcycle is a part of both recreational and commuting activities, it becomes essential in a variety of social contexts (Jderu, 2020).

Driven by enthusiasts from all over the world, motorcycle tourism has grown to become a substantial niche market within the mainstream tourism sector (Sykes & Kelly, 2015). Inside of the bigger picture of tourism business sector, this type of tourism is becoming more and more commonplace drawing enthusiasts looking for distinctive two-wheeled travel experiences. Research underscores how crucial it is to comprehend the patterns and demand drivers of motorcycle leisure in order to create experiences tailored to this particular subculture (Sykes & Kelly, 2014). Motorbike tourism is prominent given that it is an immersive travel option that gives visitors a feeling of freedom and adventure while they discover diverse environments (Rymar, 2023).

As a matter of fact, motorcycle tourism refers to the practice of riding a motorbike to observe tourist attractions and places an emphasis on actively engaging with the travel experience (Zrobek-Sokolnik et al., 2021).

In order to support the growth of motorcycle tourism, Mahrani et al. (2020) research demonstrates that regional tourism organizations that specialize in the motorcycle industry should be organized in rural areas and benefit the local population. Wang et al. (2023) also highlight the relevance of actively promoting

* Assoc. Prof. Dr., Recep Tayyip Erdoğan University, Ardeşen Tourism Faculty, Department of Tourism Management, eren.erkilic@erdogan.edu.tr

rural tourism in economically disadvantaged regions in order to establish systems for allocating tourism benefits and minimizing poverty in rural areas. For instance, European Union funds have been utilized by locations like Southern Scotland and Northern Ireland for establishing motorcycle tourism initiatives (Cater, 2017).

It makes sense to have motorcycle routes in order to promote destinations along with enhancing visitor experiences. According to Dorocki (2021), motorcycle enthusiasts in particular have displayed an increasing need for autonomous and distinctive travel experiences, which these routes could cater to. Understanding the demands of motorcycle tourists and adding value to their experiences is crucial for tourism offices and companies to foster brand loyalty (Sykes and Kelly, 2014). Yet, according to Luo (2023), the improvements of the destination's image through motorcycle route development can boost competitiveness and attract a broader range of riders (Luo, 2023).

When creating motorcycle routes in a destination, it's imperative to take into account an assortment of variables, including tourism models, prospects for rural tourism, and the incorporation of smart technologies (Cater, 2017). Furthermore, signage and promotional investments can be implemented by policy makers and tourism associations to enhance traffic routes connecting destinations for tourists (Shih, 2006). Apart from destination development, the amalgamation of intelligent tourism technology and remarkable experiences might substantially impact tourist contentment and destination allegiance (Azis et al., 2020). Additionally, by accentuating the value of unique experiences, brands can enhance visitors' pleasure and contentment by comprehending the needs of motorbike tourists and developing brand experiences (Sykes and Kelly, 2014). Meeting the demands of motorcycle tourists requires the involvement of stakeholders such as destination marketing organizations, the National Park Service, and tourism authorities (Quadri-Felitti et al., 2019).

Scientifically constructed tourism routes may reinforce the distinctive qualities of destinations for long-term growth and maximize visitor satisfaction (Xie, 2022).

Consequently, motorbike tourism holds the capacity to have far-reaching effects that expand beyond personal freedom and the pursuit of adventure. These effects might encompass local cultural exchange, economic growth, and industrial diversification. The popularity of motorcycle tourism enriches tourism destinations both domestically and abroad and contributes to the economic resurgence of rural regions and the general well-being of the populace. One approach that will immediately benefit local economies and raise tourist contentment in the current scenario is the strategic planning and development of motorcycling routes. Tourism companies, destination marketers, and policy-makers may encourage the expansion of this distinctive kind of tourism and therefore contribute to sustainable development by comprehending and satisfying the demands and expectations of motorcyclists.

Methodology

The primary objective of this book section is to recommend motorcycle routes based on the feedback of motorcycle riders in the Turkish province of Rize. Motorcycle riders responded to open-ended questions derived from Gökhan and Duran's (2022) study. Accordingly, recommendations have been submitted to identify potential motorcycle routes in Rize and construct different routes. Content analysis was performed on the data that was collected using the snowball sampling approach. In addition, a few criteria were established prior to the questions being posed. For instance, the motorcyclist must be a regular rider, possess a motorcycle, and be familiar with a variety of routes of Rize Province. Three motorcycle riders were interviewed in the first phase, and eighteen more riders were contacted through the snowball sampling technique, resulting in a total of twenty motorcycle riders from whom data was collected. The length of the interviews was not limited, and they generally lasted between 25 and 45 minutes. In keeping with the goal of the study, users were also addressed the following questions in addition to demographic ones:

- Could you evaluate the motorcycle routes in Rize? Could you share the routes you've experienced, along with their starting and ending points?
- What activities did you experience during motorcycle tours in Rize and what activities would you like a motorcyclist to visit?
- What are the advantages and disadvantages of Rize in terms of motorcycle tourism?
- Can you evaluate the motorcycle routes in Rize in terms of accessibility and touristic attractions? When evaluating, you can also add your needs such as accommodation, food and beverage.
- What can be done to develop motorcycle routes in Rize and which needs do you think are most consequential?

Likewise, the website www.wikiloc.com was utilized, wherein multiple respondents with motorcycle travel experiences in Rize shared the routes they visited. The routes that were accessible were incorporated into the research study.

Findings

According to the data obtained, the research's sample of motorcycle riders is made up of people who are between the ages of 31 and 52, have acquired at least three and no more than 27 years of motorcycle experience, and hold a high school diploma or a doctorate degree.

The participants were requested to grade the motorbike routes in Rize and submit any routes they had personally applied, along with the areas where they started and ended. and their responses are presented below:

- Because of the physical features of Rize, it was found that the majority of motorcycle riders (14 participants) had no recommendations other the main roads they traveled throughout the province. It has been proposed that in order to experience the routes inside the province borders, the right sort of motorcycle—a Motocross or Enduro motorcycle—should be considered.
- Six participants who possess enduro and motocross motorcycles indicated they may go on motorcycle excursions anywhere in the province of Rize that has a road, whether it is made of asphalt or dirt, and that motocross riders can even design their own routes in the absence of any roads. Some remarks by the motocross type motorcycle users indicated that "*I live in Rize province. Our roads trace valleys perpendicular to the sea. The valleys come together in the passes in the hills, and we pass through another valley and return to the coast. In this way, I experienced all the routes that are passable in the region.*"

> *"Since I live in Rize, plateau tours are more prominent. The route I started from the Çat valley and ended in the Galler plain is very great."*

In addition to that, riders (14 participants) with various motorbike kinds (sport touring, touring, racing, and naked) additionally discussed routes that departed from Rize and included accommodations on the same day in various provinces:

> *"Some of the attractive routes you can expereince without leaving the main roads are: Rize center – Hopa - Artvin center – Yusufeli – İspir – İkizdere - Rize center. Rize center – Araklı - Bayburt – İspir – İkizdere - Rize or, Rize- Araklı – Bayburt – Gümüşhane – Trabzon - Rize. These are fine routes that you can take during the day, without leaving the main road."*
> *"There are especailly trips to the plateaus in the place I live. Because of the fact that I do not own a motorcycle for plateau trips, I favor the Rize-Artvin, Rize-İspir, Rize-Bayburt-Gümüşhane, Rize-Ayder plateau, Rize-Çaykara-Uzungöl routes as a day trip."*
> *"I elect longer routes. Rize – Samsun – Nevşehir (Cappadocia) – Sivas – Erzincan – Trabzon – Rize. Rize – Kars – Iğdır – Van – Erzurum – Rize. These are the routes I have experienced. I have have motorcycle rides to some provinces plenty of times."*

As the responses to the question of "*What activities did you experience during motorcycle tours in Rize and what activities would you like a motorcyclist to visit?*" are investigated, the following responses were obtained.

- It has been determined that motorcycle riders hardly ever engage in motorcycling activities in Rize, and those who do are usually unaware of these problems. On the other hand, they maintained that on notable days like October 29, Republic Day, they took part in the motorcycle cortege that was coordinated by a motorcycle club in the heart of Rize.
- Owners of enduro and motocross motorcycles reported that they packed their own food and drinks for group or solo trips to the plateaus, and that they shared meals and drinks at areas for relaxation. They asserted that they considered this social setting to be an activity in and of itself.
- Motorcyclists who rode differing types of motorcycles (sport touring, touring, racing, and naked) revealed that they attended motorcycling festivals in nearby provinces or took part in activities like meals or rides arranged by motorcycling clubs in other regions:

  "Festivals - food events - meet-and-greet events. Let them visit motorcycle groups in the provinces. They will both meet people from the region and have information about the region."
  "I take part in the Kafkasor plateau motofestival whenever it is organized."
  "I participated in festivals held in Ordu province."

  Motorcyclists took into account that riding their motorcycles for pleasure is an activity in and of itself, according to the data that was analyzed. The following two sample responses are offered as examples:

  "Since the motorcycle is already an event in itself, get it on and go ride it!"
  "Every event is fantastic if it happens on two wheels."

As we go over the question of "*What are the advantages and disadvantages of Rize in terms of motorcycle tourism?*", some of the answers by the participants are given below:

- In terms of advantages, motorcyclists characterized Rize Province as having a distinctive character overall. With an abundance of routes, riders can enjoy this city provided they choose the appropriate motorcycle. The city boasts winding roads that make for enjoyable riding, as well as both asphalt and off-road options. The coexistence of cultural assets and the mountain-sea was lastly cited as a benefit. The climatic conditions were highlighted by all motorcycle users and the fact that it is a rainy city. These are the two most noticeable drawbacks. Here are some user reviews addressing benefits and drawbacks:

> *"The plateaus are incredibly advantageous for those with a Motocross - Enduro motorcycle. Otherwise, I can't do much riding due to the climatic conditions."*
> *"Rize is not a climate-friendly region. If you come here with the idea of "Let's take a black sea tour by motorcycle and see two plateaus", it may turn into a torture. If you like to explore, tackle challenges and do them on a motorcycle, the region is just for you."*
> *"Its advantages are the natural beauties, the abundance of winding roads that motorcyclists love, and the availability of both off-road and asphalt road options. The disadvantage is the high rainfall rate due to climatic conditions."*

The participants were asked if "*they could evaluate the motorcycle routes in Rize in terms of accessibility and touristic attractions*" and "*when evaluating, they were invited to add their needs such as accommodation, food and beverage as well.*" Some answers are provided below.

- They have acknowledged that there is an abundance of potential routes in the city, yet riders need to have the right kind of motorcycle. In contrast to other cities, Rize lacks adequate food, drink, and accommodation due to the fact that it is in a remote location throughout the country in terms of transportation. On the other hand, it was also disclosed that the camp's amenities were insufficient and that the costs of food, drink, and accommodation were exorbitant. Some answers given by users are as follows:
- *"On almost every one of the routes that surround us, accommodation and food and beverage facilities are obtainable. Though these opportunities do occasionally decline in extremely high plateaus, there is also a good chance of finding local support. There are attractions for tourists along nearly every route. These are places where everyone desires to return."*

> *"Motorcycle transportation is not too troublesome to navigate, though expected restaurants and drinking establishments may raise their prices as a result of materialism. That's why we commonly drive to places we know."*
> *"All routes are extremely entertaining, but you'll need a motorcycle that is suitable fort his occasion."*
> *"I am convinced there are not sufficient camping areas in this region for motorcycle routes."*

As we look into the responses by the participants to the question of "*What can be done to develop motorcycle routes in Rize and which needs do you think are most eminent?*", it has been achieved that:

- Motorcyclists offered comparable responses along with a range of perspectives. They contend that it is critical to keep in mind that motorcycles are vehicles, that drivers should behave appropriately, that roads should be maintained and asphalted more often, that appropriate signage should be erected and executed, that barriers should be made biker-friendly, that social media should be used to raise awareness, that camping areas and safe environments should be established, and that motorcycle festivals should be planned. Here are a few responses:

"Signs and markers should be increased. Roads suitable for motorcycles should be built."
"There should be biker-friendly barriers on the routes, and there should be a secure garage and parking area for accommodation."
"Mapping and route works need to be done according to motorcycle types."
"New routes should be determined and new routes should be taken. Festivals are very useful."

However, in keeping with the investigation done for the subject's literature review, a webpage known as www.wikiloc.com was found. Individual routes taken by multiple motorbike riders in various regions have been included on the website. The appropriate webpage has been utilized in gathering the motorbike routes where several riders discussed their experiences with the routes they had taken and encountered in the province of Rize. It has been noted that the routes on the website and in the information about the city of Rize are primarily intended for motocross and enduro riders. In addition to automobile roads, these routes have walkways that are frequently used by riders of motocross and enduro motorbike types.

- İkizdere – Ortayayla – Sıraköy – Kale – Çat valley – Zil Castle – Şenyuva – Çamlıhemşin
- Çayeli - Zargistal plateau - Gito plateau – Çamlıhemşin – Ardeşen
- Badara plateau - Hazindağ plateau – Sal plateau – Pokut plateau
- Çat valley – Elevit plateau – Trovit plateau – Palovit plateau – Amlakit plateau –Samistal plateau – Kavrun plateau
- Gito plateau – Zargistal plateau – Tahbur – Verçenik plateau – Çat valley
- Çamlıhemşin - Pokut plateau - Amlakit plateau - Palovit plateau - Trovit plateau - Elevit plateau – Çat valley
- İkizdere – Çağrankaya plateau – Eğnaçor plateau – Tahpur plateau – Verçenik plateau – Çiçekli plateau - Çat valley
- Şenyuva - Palovit waterfall – Amlakit plateau – Palovit plateau
- Çamlıhemşin - Şenyuva – Sal plateau – Pokut plateau
- Zilkale - Baldaş Pass - Cimil plateau
- Ayder plateau – Huser plateau – Lordeçur - Lower Kaçkar Plateaus
- Çamlıhemşin - Ayder plateau - Galler Plain - Yukarı Kavrun plateau
- Ardeşen – Plateau of Sırt

Conclusion

The research findings demonstrate that the geographical features and natural elements of Rize province make it a desirable destination for motorbike riders. In general, the province of Rize delivers motocross and enduro motorbike riders a number of benefits as well as the possibility to check out a variety of routes. Still, the routes available to riders of other kinds of motorcycles are frequently restricted to major thoroughfares.

Consequently, the routes that motorcyclists have designed by utilizing the geography and scenic qualities of the province of Rize can be classified as either main road routes or plateau trail routes. The principal roads typically lead to the districts of Rize and other provinces; these itineraries can be accomplished in a single day or with accommodation. Nevertheless, motocross and enduro riders prefer more challenging, exploration-focused rides on plateaus and trails that put them in close proximity to the natural world. In fact, this arrangement makes Rize's potential for motorbike tourism obvious and makes it possible for riders to experience the surrounding region.

Within the scope of the research, participants in the study presented a range of recommendations for the creation of motorcycling routes in Rize:

- Road quality should be improved and frequent maintenance of roads should be performed,
- The number of road signs on secondary roads should be augmented, taking the motorcycle users into account,
- Rider-friendly barriers should be employed,
- The number of camping areas should be increased and secure accommodation places should be supplied,
- Social awareness campaigns for motorcyclists should be launched,
- Motorcycle festivals should be hosted in the province.

Besides, this study proposes vital implications for policy makers.

- Rize's potential for motorcycle tourism should be reinforced through infrastructure enhancements,
- Routes should be created for motorcycle users based on the appropriate sorts of motorcycles,
- Promotions regarding the established routes should be undertaken as needed,
- Motorcycling-related events should to be scheduled,
- It should be made sure that this kind of tourism benefits the community economically and experientially,

- Training programs for motorcycle users should be organized and safety measures should be improved in the province.

There are few restrictions concerning this research. Participants must fall within a specific age range and have varying levels of experience. The number of participants is restricted. Thus, the conclusions' ability to be applied generally is constrained by these limitations. Subsequent studies ought to encompass a more extensive participant pool and seek input from motorbike riders with varying demographic profiles. At the same time, exploring motorcycle routes under various seasonal circumstances and evaluating these experiences will contribute to the motorcycle tourism potential of Rize province.

References

Azis, N., Amin, M., Syafruddin, S., & Aprilia, C. (2020). How smart tourism technologies affect tourist destination loyalty. *Journal of Hospitality and Tourism Technology*, 11(4), 603–625.

Bastos, S., Gama, F., Assis, T., & Milosz, M. (2020). Is there a relationship between the use of motorcycles and the level of development of countries? *Bulletin of Geography Socio-Economic Series*, 50(50), 43–53.

Cater, C. (2017). Tourism on two wheels: patterns of motorcycle leisure in Wales. *Tourism Management*, 61, 180–189.

Dorocki, S. (2021). Tourism on a motorcycle: an attempt to describe motorcycle tourism in Poland based on the results of an internet survey conducted before the COVID-19 pandemic. *Studies of the Industrial Geography Commission of the Polish Geographical Society*, 35(3), 320–332.

Gökhan, S., & Duran, E. (2022). İzmir motosiklet rotalarının değerlendirilmesine yönelik nitel bir araştırma. *22. Ulusal Turizm Kongresi*, 27–29 Ekim, Burdur.

Jderu, G. (2020). Motorcycling in a repair society: A maintenance and repair-centered approach to automobility. *East European Politics and Societies and Cultures*, 36(2), 604–625.

Luo, Y. (2023). Study on the destination image improvement of Guangxi rural tourism. *Scientific and Social Research*, 5(2), 19–24.

Mahrani, M., Ruliana, P., & Ritonga, R. (2020). Building the image of Semarang city through motocross world championship. *Jurnal Komunikasi Ikatan Sarjana Komunikasi Indonesia*, 5(2), 243–253.

Quadri-Felitti, D., Sykes, D., & Chen, F. (2019). Is motorcycle tourism ready to rev up in Pennsylvania? An exploratory study of suppliers' business attitudes of motorcycle tourism. *International Journal of Culture Tourism and Hospitality Research*, 13(1), 1–15.

Rymar, Y. (2023). Perspective directions of motorcycle tourism development in the western territories of Ukraine during the war and post-war period. *Економіка І Регіон Науковий Вісник*, 3(90), 31–37.

Shih, H. (2006). Network characteristics of drive tourism destinations: An application of network analysis in tourism. *Tourism Management*, 27(5), 1029–1039.

Sykes, D., & Kelly, K. (2014). Motorcycle tourism demand generators and dynamic interaction leisure. *International Journal of Culture Tourism and Hospitality Research*, 8(1), 92–102.

Sykes, D., & Kelly, K. (2015). Motorcycle drive tourism leading to rural tourism opportunities. *Tourism Economics*, 22(3), 543–557.

Wang, F., Du, L., Tian, M., Liu, Y., & Zhang, Y. (2023). Sustainability of rural tourism in poverty reduction: Evidence from panel data of 15 underdeveloped counties in Anhui Province, China. *PLOS ONE*, 18(3), e0283048.

Xie, N. (2022). Personalized route planning of rural ecotourism using mobile computing and random forest algorithm. *Wireless Communications and Mobile Computing*, 2022, 1–8.

Zrobek-Sokolnik, A., Dynowski, P., & Zrobek, S. (2021). Preservation and restoration of roadside tree alleys in line with sustainable development principles-mission (im)possible? *Sustainability*, 13(17), 9635.

www.ingramcontent.com/pod-product-compliance
Ingram Content Group UK Ltd.
Pitfield, Milton Keynes, MK11 3LW, UK
UKHW021847100926
8556IPUK00020B/1106

9 783631 925102